Sławomir Wiak and Ewa Napieralska-Juszczak (Eds.)

Computational Methods for the Innovative Design of Electrical Devices

Studies in Computational Intelligence, Volume 327

Editor-in-Chief
Prof. Janusz Kacprzyk
Systems Research Institute
Polish Academy of Sciences
ul. Newelska 6
01-447 Warsaw
Poland
E-mail: kacprzyk@ibspan.waw.pl

Further volumes of this series can be found on our homepage: springer.com

Vol. 303. Joachim Diederich, Cengiz Gunay, and James M. Hogan
Recruitment Learning, 2010
ISBN 978-3-642-14027-3

Vol. 304. Anthony Finn and Lakhmi C. Jain (Eds.)
Innovations in Defence Support Systems, 2010
ISBN 978-3-642-14083-9

Vol. 305. Stefania Montani and Lakhmi C. Jain (Eds.)
Successful Case-Based Reasoning Applications-1, 2010
ISBN 978-3-642-14077-8

Vol. 306. Tru Hoang Cao
Conceptual Graphs and Fuzzy Logic, 2010
ISBN 978-3-642-14086-0

Vol. 307. Anupam Shukla, Ritu Tiwari, and Rahul Kala
Towards Hybrid and Adaptive Computing, 2010
ISBN 978-3-642-14343-4

Vol. 308. Roger Nkambou, Jacqueline Bourdeau, and Riichiro Mizoguchi (Eds.)
Advances in Intelligent Tutoring Systems, 2010
ISBN 978-3-642-14362-5

Vol. 309. Isabelle Bichindaritz, Lakhmi C. Jain, Sachin Vaidya, and Ashlesha Jain (Eds.)
Computational Intelligence in Healthcare 4, 2010
ISBN 978-3-642-14463-9

Vol. 310. Dipti Srinivasan and Lakhmi C. Jain (Eds.)
Innovations in Multi-Agent Systems and Applications – 1, 2010
ISBN 978-3-642-14434-9

Vol. 311. Juan D. Velásquez and Lakhmi C. Jain (Eds.)
Advanced Techniques in Web Intelligence, 2010
ISBN 978-3-642-14460-8

Vol. 312. Patricia Melin, Janusz Kacprzyk, and Witold Pedrycz (Eds.)
Soft Computing for Recognition based on Biometrics, 2010
ISBN 978-3-642-15110-1

Vol. 313. Imre J. Rudas, János Fodor, and Janusz Kacprzyk (Eds.)
Computational Intelligence in Engineering, 2010
ISBN 978-3-642-15219-1

Vol. 314. Lorenzo Magnani, Walter Carnielli, and Claudio Pizzi (Eds.)
Model-Based Reasoning in Science and Technology, 2010
ISBN 978-3-642-15222-1

Vol. 315. Mohammad Essaaidi, Michele Malgeri, and Costin Badica (Eds.)
Intelligent Distributed Computing IV, 2010
ISBN 978-3-642-15210-8

Vol. 316. Philipp Wolfrum
Information Routing, Correspondence Finding, and Object Recognition in the Brain, 2010
ISBN 978-3-642-15253-5

Vol. 317. Roger Lee (Ed.)
Computer and Information Science 2010
ISBN 978-3-642-15404-1

Vol. 318. Oscar Castillo, Janusz Kacprzyk, and Witold Pedrycz (Eds.)
Soft Computing for Intelligent Control and Mobile Robotics, 2010
ISBN 978-3-642-15533-8

Vol. 319. Takayuki Ito, Minjie Zhang, Valentin Robu, Shaheen Fatima, Tokuro Matsuo, and Hirofumi Yamaki (Eds.)
Innovations in Agent-Based Complex Automated Negotiations, 2010
ISBN 978-3-642-15611-3

Vol. 320. xxx

Vol. 321. Dimitri Plemenos and Georgios Miaoulis (Eds.)
Intelligent Computer Graphics 2010
ISBN 978-3-642-15689-2

Vol. 322. Bruno Baruque and Emilio Corchado (Eds.)
Fusion Methods for Unsupervised Learning Ensembles, 2010
ISBN 978-3-642-16204-6

Vol. 323. xxx

Vol. 324. Alessandro Soro, Vargiu Eloisa, Giuliano Armano, and Gavino Paddeu (Eds.)
Information Retrieval and Mining in Distributed Environments, 2010
ISBN 978-3-642-16088-2

Vol. 325. Quan Bai and Naoki Fukuta (Eds.)
Advances in Practical Multi-Agent Systems, 2010
ISBN 978-3-642-16097-4

Vol. 326. Sheryl Brahnam and Lakhmi C. Jain (Eds.)
Advanced Computational Intelligence Paradigms in Healthcare 5, 2010
ISBN 978-3-642-16094-3

Vol. 327. Sławomir Wiak and Ewa Napieralska-Juszczak (Eds.)
Computational Methods for the Innovative Design of Electrical Devices, 2010
ISBN 978-3-642-16224-4

Sławomir Wiak and Ewa Napieralska-Juszczak (Eds.)

Computational Methods for the Innovative Design of Electrical Devices

Prof. Dr. Sławomir Wiak
Technical University of Lodz
Institute of Mechatronics and Information
ul. Stefanowskiego 18/22
90-924 Lodz
Poland
E-mail: swiak@wp.pl

Ewa Napieralska-Juszczak
Université d'Artois
Technoparc Futura
Laboratoire Systèmes Electrotechniques
et Environnement (LSEE)
62400 Bethune
France
E-mail: ewa.napieralskajuszczak@univ-artois.fr

ISBN 978-3-642-16224-4   e-ISBN 978-3-642-16225-1

DOI 10.1007/978-3-642-16225-1

Studies in Computational Intelligence   ISSN 1860-949X

Library of Congress Control Number: 2010937107

© 2010 Springer-Verlag Berlin Heidelberg

This work is subject to copyright. All rights are reserved, whether the whole or part of the material is concerned, specifically the rights of translation, reprinting, reuse of illustrations, recitation, broadcasting, reproduction on microfilm or in any other way, and storage in data banks. Duplication of this publication or parts thereof is permitted only under the provisions of the German Copyright Law of September 9, 1965, in its current version, and permission for use must always be obtained from Springer. Violations are liable to prosecution under the German Copyright Law.

The use of general descriptive names, registered names, trademarks, etc. in this publication does not imply, even in the absence of a specific statement, that such names are exempt from the relevant protective laws and regulations and therefore free for general use.

Typeset & Cover Design: Scientific Publishing Services Pvt. Ltd., Chennai, India.

Printed on acid-free paper

9 8 7 6 5 4 3 2 1

springer.com

Preface

This volume in the Studies in Computational Intelligence book series published by Springer includes the extended version of a number of selected papers presented at the International Symposium on Electromagnetic Fields in Electrical Engineering ISEF'09. The Symposium was jointly organized by the LSEE (Laboratory of Electrical Systems and Environment), University of Artois, France, and the Institute of Mechatronics and Information Systems, Technical University of Lodz, Poland. The venue was Arras, a beautiful historical town in the north of France.

The aim of ISEF symposia is to discuss recent developments in modelling and simulation, control systems, testing, measurements, monitoring, diagnostics and advanced software methodology and their applications in electrical and electronic devices and mechatronic systems. ISEF is a forum for electronic and electrical engineers, applied mathematicians, computer and software engineers, to exchange ideas and experiences ranging from fundamental developments of theory to industrial applications. The conference has become a popular event among academics, researchers and practising engineers. Due to discussions during the conference, it has been decided to prepare a book in the domain of innovative methods for the electrical machine design.

Over the past thirty-five years, ISEF has gained a prominent position in electromagnetic community. Since the first meeting held in Uniejow Palace near Lodz in 1974 – at that time organised as a National Symposium on "Electrodynamics of Transformers and Electrical Machines" – ISEF has travelled around Europe visiting, in addition to various venues in Poland, several interesting places such as Pavia (twice), Southampton, Thessaloniki, Maribor, Baiona, Prague, and finally Arras in 2009.

For the meeting in Arras, more than 300 papers had been submitted as digests, and after the reviewing process 276 papers were accepted for presentation at the Conference. Those versions were considered by session chairs for possible inclusion in the post-conference special issue. The programme of the conference included three invited papers, five oral and eight dialogue sessions. All well established conference topics were covered, and supplemented by two new areas:

- artificial and computational intelligence in electrical engineering,
- noise and vibration in electrical machines.

Another novelty was a special session with presentations by PhD students working in the field of electromagnetism.

The small but very active and prominent group of 'electromagneticians' regularly attending ISEF will hopefully continue to support future meetings, so providing a particular flavour and focus. However, it is also very pleasant to see other areas strongly emerging as new conference topics, in particular computer engineering, software methodology, CAD techniques, artificial intelligence and material sciences.

This special issue of Studies in Computational Intelligence incorporates 17 chapters selected by the Guest Editors as a result of a two-stage evaluation process: first, recommendations of the chairpersons of the sessions, and next, reviews by two independent referees. Computational and modelling aspects are the main ones, although design, measurement and performance issues are considered as well.

As the Editors of this special issue, we would like to express our thanks to Springer for giving us the opportunity to share the ISEF symposium with a wider community; thanks are due also to our colleagues, in particular Dr Stéphane Duchesne and Dr Jean-Philippe Lecointe for their help, efficiency and valuable contribution to the reviewing and editing process.

At the end of these remarks, let us thank our colleagues who have contributed to the book by peer-reviewing the papers at the conference as well as in the publishing process. We also convey our thanks to Springer for their effective collaboration in shaping this editorial enterprise. As ISEF symposia are organised biannually, we hope to keep our fruitful links with Springer in the future.

Sławomir Wiak
Chairman of the ISEF Symposium

Ewa Napieralska-Juszczak
Chairwoman of the ISEF2009
Organizing Committee

Contents

1 Integrated Computer Models of 3-D Comb Drive Electrostatic MEMS Structures... 1
Sławomir Wiak, Krzysztof Smółka
 1.1 Introduction 1
 1.2 Solid Modeling of 3D MEMS Structures 2
  1.2.1 Case Study – Accelerometer 4
  1.2.2 Case Study – Actuator for Micromirror Driving 5
 1.3 Complex Strategy 7
  1.3.1 Case Study – Accelerometer 9
  1.3.2 Case Study – Actuator for Micromirror Driving 14
 1.4 Conclusions 17
 References 17

2 Multi-objective Design Optimization of Slotless PM Motors Using Genetic Algorithms Based on Analytical Field Calculation............ 19
L. Belguerras, L. Hadjout
 2.1 Introduction 19
 2.2 Electromagnetic Modeling of the Study Slotless PMSM 20
  2.2.1 Motor Topologies 20
  2.2.2 Analytical Modeling 20
  2.2.3 Electromagnetic Torque 22
  2.2.4 Model Validating 24
 2.3 Multi-objective Optimization 25
  2.3.1 Dominance 26
  2.3.2 Pareto Optimality 26
  2.3.3 Evolutionary Multi-objective Optimization Algorithms ...27
  2.3.4 Elitist Non-dominated Sorting Genetic Algorithm (NSGAII) 28
   2.3.4.1 NSGAII Algorithm 30
   2.3.4.2 Crowding Distance 31
 2.4 Design Optimization 31
  2.4.1 Objective Functions 32
  2.4.2 Design Variables 33
  2.4.3 Inequality Constraints 33
 2.5 Optimization Results 34
 2.6 Conclusion 37
 References 37

3 The FEM Parallel Simulation with Look Up Tables Applied to the Brushless DC Motor Optimization............ 39
Jakub Bernat, Jakub Kołota, Sławomir Stępień
- 3.1 Introduction 39
- 3.2 The Parallel Simulation Technique............40
- 3.3 Look-Up Tables with Integral Coefficients 42
- 3.4 The Parallel System Realization 43
- 3.5 Brushless DC Motor Model............44
- 3.6 The Optimization Problem............46
- 3.7 The Simulated Annealing Algorithm............47
- 3.8 The Rosenbrock Valley Problem 48
- 3.9 The Rastragin's Problem 50
- 3.10 Simulation Experiment Results 52
- 3.11 The Brushless DC Motor Torque Analysis............55
- 3.12 Conclusions 57
- References 57

4 Fast Algorithms for the Design of Complex-Shape Devices in Electromechanics............ 59
Eugenio Costamagna, Paolo Di Barba, Maria Evelina Mognaschi, Antonio Savin
- 4.1 Introduction 59
- 4.2 Reducing Three Dimensions to Two 59
- 4.3 The Need for Fast Solvers: Schwarz-Christoffel Mapping as a Possible Answer............60
- 4.4 Modern Numerical Methods for SC Mapping 61
- 4.5 Towards a Coordinated Approach for Field Analysis, Using Both FE and SC............65
- 4.6 Analysis of a Magnetic Levitator with Superconductors............66
  - 4.6.1 A Principle of Equivalence 66
  - 4.6.2 Application in 2D Magnetostatics............68
  - 4.6.3 Outer-Field Problem: Results............72
  - 4.6.4 Test Problem: Diamagnetic Model of a Superconductor 77
  - 4.6.5 Case Study: Trapped-Flux Model of a Magnetic Bearing............78
  - 4.6.6 An Application in Automated Optimal Design 83
- 4.7 Conclusion 84
- References 84

5 Optimization of Wound Rotor Synchronous Generators Based on Genetic Algorithms............ 87
Xavier Jannot, Philippe Dessante, Pierre Vidal, Jean-Claude Vannier
- 5.1 Introduction 87
- 5.2 Wound Rotor Synchronous Generator Model 88
  - 5.2.1 Electromagnetic Modeling of the WRSG 88
    - 5.2.1.1 Air Gap Magnetic Flux Density Computation at No Load 89

Contents

        5.2.1.2 Computation of Magnetic Flux Density in the Iron Paths 91
        5.2.1.3 Magnetic Field Circulation in the WRSG 91
        5.2.1.4 Computation of Rotor Magnetomotive Force 93
    5.2.2 Losses and Thermal Model 95
        5.2.2.1 Losses' Estimation 95
        5.2.2.2 Thermal Modeling 97
        5.2.2.3 Thermal Loop 98
5.3 Optimization Problem 99
    5.3.1 Interest of Designing with Optimization Procedure 99
    5.3.2 Choice of the Optimization Technique 100
5.4 Optimization with Genetic Algorithm 100
    5.4.1 Objectives of Optimization: Single and Multi-objective Optimization 100
    5.4.2 Principle of Genetic Algorithm 101
        5.4.2.1 Initialization 103
        5.4.2.2 Evaluation 103
        5.4.2.3 Ranking 103
        5.4.2.4 Selection 103
        5.4.2.5 Crossover 104
        5.4.2.6 Mutation 104
        5.4.2.7 Discrete Variable Handling 104
5.5 Mono-objective Optimization of a WRSG Range 104
    5.5.1 Independent Optimization of Three Different Machines 105
    5.5.2 Classical Design Approach Covering a WRSG Range 107
    5.5.3 Simultaneous Optimization of a Series of Machines 108
5.6 Multi-objective Optimization of a WRSG Range 110
5.7 Conclusion 112
References 112

6 Simple and Fast Algorithms for the Optimal Design of Complex Electrical Machines 115
Kazumi Kurihara, Tomotsugu Kubota, Yuki Imaizumi
6.1 Introduction 115
6.2 Experimental Motor and Circuit 116
6.3 Method for Analysis 117
    6.3.1 Finite Element Analysis (FEA) 117
    6.3.2 Response Surface Methodology (RSM) 120
6.4 Steady-State Performance Characteristics 120
    6.4.1 EMF Due to PMs 121
    6.4.2 No-Load Performance Characteristics 121
    6.4.3 Load Performance Characteristics 123
    6.4.4 Torque Ripple and Efficiency 125
6.5 Starting Performance Characteristics 127
6.6 Conclusions 127
Acknowledgments 128
References 128

7 The Flock of Starlings Optimization: Influence of Topological Rules on the Collective Behavior of Swarm Intelligence 129
Francesco Riganti Fulginei, Alessandro Salvini
- 7.1 Introduction 129
- 7.2 Standard PSO Overview 130
- 7.3 Influence of Topological Rules: From Particle Swarm to Flock of Starlings Optimization 132
- 7.4 FSO Validation and Comparison with the PSO Results 135
- 7.5 Remarks 142
- 7.6 Conclusions 144
- References 145

8 Multilevel Data Classification and Function Approximation Using Hierarchical Neural Networks 147
M. Alper Selver, Cüneyt Güzeliş
- 8.1 Introduction 147
- 8.2 Theoretical Framework 149
  - 8.2.1 Function Approximation Using HNN 150
  - 8.2.2 Classification Using HNN 152
- 8.3 Applications 153
  - 8.3.1 Transfer Function Initialization for Three Dimensional Medical Volume Visualization 154
  - 8.3.2 Quality Classification of Marble Slabs 158
  - 8.3.3 Classification of Radar Data 160
- 8.4 Discussions and Conclusions 162
- References 164

9 Parametric Identification of a Three-Phase Machine with Genetic Algorithms 167
L. Simón, J.M. Monzón
- 9.1 Introduction 167
- 9.2 Continuous Model FEM and Differential Formulation of Electromagnetic Field 169
  - 9.2.1 Magnetostatic 169
  - 9.2.2 Magnetodynamic 170
- 9.3 Discrete Model FEM 171
- 9.4 Boundary Conditions of the Continuous Model 172
- 9.5 Lumped Parametric Model Identification 174
  - 9.5.1 Parameter Estimation to All the Geometries 175
- 9.6 FEM Analysis 176
- 9.7 Parametric Adjustment by Genetic Algorithms 177
  - 9.7.1 Optimality 177
  - 9.7.2 Fitness Procedure 178
- 9.8 Results 179
- 9.9 Conclusions 183
- References 183

10 Ridge Polynomial Neural Network for Non-destructive Eddy Current Evaluation 185
Tarik Hacib, Yann Le Bihan, Mohammed Rachid Mekideche, Nassira Ferkha
- 10.1 Introduction 185
- 10.2 Higher Order Neural Network 187
  - 10.2.1 Pi-Sigma Neural Network 187
  - 10.2.2 Ridge Polynomial Neural Network 189
- 10.3 Forward Problem Definition 189
- 10.4 Implementation of RPNN for Solving the Inverse Problem 191
- 10.5 Results 192
  - 10.5.1 Training Without Noise 194
  - 10.5.2 Training with Noise 196
- 10.6 Conclusion 198
- References 199

11 Structural-Systematic Approach in Magnetic Separators Design 201
Vasiliy F. Shinkarenko, Mikhaylo V. Zagirnyak, Irina A. Shvedchikova
- 11.1 Introduction 201
- 11.2 Search Methodology Substantiation 202
- 11.3 Determination of Electromagnetic Separators Species Diversity 205
- 11.4 Systematics Rank Structure as the Basis for a New Concept of Design Procedures Dataware 208
- 11.5 Interspecific Synthesis of New Magnetic Separator Structures Using the Law of Electromechanical Systems Homologous Series 210
- 11.6 Conclusion 216
- References 217

12 Weight Reduction of Electromagnet in Magnetic Levitation System for Contactless Delivery Application 219
Do-Kwan Honga, Byung-Chul Woo, Dae-Hyun Koo, Ki-Chang Lee
- 12.1 Introduction 219
- 12.2 Passive Guidance Control and Optimization of Electromagnet 221
- 12.3 Optimum Design 222
- 12.4 Response Surface Methodology 224
- References 226

13 Genetic Algorithm Applied in Optimal Design of PM Disc Motor Using Specific Power as Objective 229
Goga Cvetkovski, Lidija Petkovska, Sinclair Gair
- 13.1 Introduction 229
- 13.2 Permanent Magnet Disc Motor Description 230
- 13.3 GA Optimization Method Description 230
  - 13.3.1 Reproduction 231
  - 13.3.2 Crossover 232
  - 13.3.3 Mutation 232
  - 13.3.4 Fitness Scaling 232

		13.3.4.1 Linear Scaling 233
		13.3.4.2 Power Law Scaling 234
		13.3.4.3 Boltzmann Selection 234
13.4	GA Optimal Design of Permanent Magnet Disc Motor 236
13.5	GA Optimal Design Results of PM Disc Motor 240
13.6	PM Disc Motor FEM Modeling and Magnetic Field Analysis 240
13.7	Conclusion 245
References 246

14 Magnetically Nonlinear Iron Core Characteristics of Transformers Determined by Differential Evolution. 247
Gorazd Štumberger, Damir Žarko, Amir Tokic, Drago Dolinar
- 14.1 Introduction 247
- 14.2 Dynamic Model of a Single-Phase Transformer 248
- 14.3 Approximation Functions 248
  - 14.3.1 Simple Analytic Saturation Curve 249
  - 14.3.2 Analytic Saturation Curve with Bend Adjustment 250
  - 14.3.3 Approximation with Exponential Functions 251
- 14.4 Determining Approximation Function Parameters by Differential Evolution 251
  - 14.4.1 Description of Differential Evolution Algorithm 251
  - 14.4.2 Objective Function 253
- 14.5 Results 254
- 14.6 Conclusion 258
- Acknowledgments 258
- References 258

15 Different Methods for Computational Electromagnetics: Their Characteristics and Typical Pratical Applications 261
Arnulf Kost
- 15.1 Introduction 261
- 15.2 The Finite Element Method (FEM) 262
- 15.3 The Boundary Element Method 263
- 15.4 The Moments Method (MoM) 264
- 15.5 The Finite Difference Time Domain Method (FDTD) 264
- 15.6 The Transmission Line Matrix Method (TLM) 265
- 15.7 Other Single Method and Need for Hybrid and Even More Effective 265
- 15.8 Hybrid FEM/BEM – Method for Quasistationary Fields 266
- 15.9 Hybrid FEM/BEM Method for Waves 267
- 15.10 Hybrid MoM/GTD Method 267
- 15.11 Hybrid BEM/IBC and FEM/IBC Method 269
- 15.12 Non Linear FEM and FEM/IBC with Complex Effective Reluctivity 270
- 15.13 Laser Trimming of IC Resistors Industrial Production Using 271

15.14	Equation Solvers	271
15.15	Improvements in Adaptative Mesh Generation	274
15.16	Conclusions	274
References		275

16 Methods of Homogenization of Laminated Structures...........261
Nabil Hihat, Piotr Napieralski, Jean-Philippe Lecointe,
Ewa Napieralska-Juszczak

16.1	Introduction	277
16.2	Optimization Methods	279
	16.2.1 Solving Optimization Problem	280
	16.2.2 Linear Programming	280
	16.2.3 Nonlinear Programming	283
	16.2.4 Unconstrained Optimization Problems	286
	16.2.5 Direct Search Method: Hook-Jeeves	290
16.3	Homogenization Method Principle	293
	16.3.1 Magnetic Characteristics $vi(Bi,\alpha i)$	296
	16.3.2 Homogenization Method Application	298
16.4	Conclusion	301
References		301

17 Applications Examples..........303

17.1	Comparative Finite-Elements and Permeance-Network Analysis for Design Optimization of Switched Reluctance Motors	304

Dan Ilea, Frédéric Gillon, Pascal Brochet,
Mircea M. Radulescu

	17.1.1 Introduction	304
	17.1.2 Permeance Network Analysis	304
	17.1.2.1 Airgap Permeance	304
	17.1.2.2 Poles and Back-Iron Permeances	305
	17.1.2.3 Complete Permeance Network	306
	17.1.3 Finite-Element Analysis	306
	17.1.4 Motor Specifications and Simulation Results Analysis	307
	17.1.5 Conclusions	310
17.2	Lumped Parametric Model of an Electrostatic Induction Micromotor	311

Francisco Jorge Santana-Matin, José Miguel Monzon-Verona,
Santiago Garcia-Alonso, Juan Antonio Montiel-Nelson

	17.2.1 Introduction	311
	17.2.2 Lumped Parametric Equivalent Circuit Model	312
	17.2.3 Fitting of Lumped Parameters Using GA in Steady State	313
	17.2.3.1 Objective Function	313
	17.2.3.2 GA Parameters	316
	17.2.3.3 Lumped Parameters in Stationary State	317

		17.2.4	Time Domain Analysis	318
			17.2.4.1 Transient State of the Physical Model	318
		17.2.5	Conclusions	321
	17.3	Contribution to Optimization and Modeling by Reluctances Network		322

Abdelghani Kimouche, Mohamed Rachid Mekideche,
Ammar Boulassel, Tarik Hacib, Hicham Allag

		17.3.1	Introduction	322
		17.3.2	Linear Stepping Motor Study	322
			17.3.2.1 Finite Element Method Analysis	322
			17.3.2.2 Magnetic Equivalent Circuit Modeling	324
		17.3.3	Magnetic Force Computation	327
		17.3.4	Optimization	328
		17.3.5	Conclusion	331
	17.4	Optimization and Computer Aided Design of Special Transformers		332

Marija Cundeva-Blajer, Snezana Cundeva, Ljupco Arsov

		17.4.1	Introduction	332
		17.4.2	Optimization Design Procedure of CCVIT	332
			17.4.2.1 CCVIT Mathematical Model: FEM-3D Coupled with Least Squares Method	333
			17.4.2.2 CCVIT GA Optimization	335
			17.4.2.3 Results CCVIT	336
		17.4.3	Optimization Design Procedure of RWT	337
			17.4.3.1 Initial Study of RWT	337
			17.4.3.2 RWT GA Optimization	338
			17.4.3.3 Results CAD RWT	339
		17.4.4	Conclusions	341
	17.5	Effects of Geometric Parameters on Performance of a SRM by Numerical-Analytical Approach		342

A. Bentounsi, R. Rebbah, F. Rebahi, H. Djeghloud, H. Benalla,
S. Belakehal, B. Batoun

		17.5.1	Introduction	342
		17.5.2	Preliminary Design Process	342
		17.5.3	Results by FEM Simulation	343
		17.5.4	Results by EMC Simulation	347
		17.5.5	Conclusion	349
	17.6	Multiphysics Method for Determination of the Stator Winding Temperature in an Electrical Machine		350

Zlatko Kolondzovski

		17.6.1	Introduction	350
		17.6.2	Method	351
			17.6.2.1 2D Axi-symmetric Numerical Multiphysics Model	351
			17.6.2.2 3D Numerical Heat-Transfer Model	353
		17.6.3	Results and Discussion	354

		17.6.3.1	Results for the Temperatures in a Steady State Performance 354
		17.6.3.2	Experimental Validation of the Multiphysics Method 356
	17.6.4	Conclusion 358
17.7	Particle Swarm Optimization for Reconstitution of Two-Dimentional Groove Profiles in Non Destructive Evaluation 359
	Ammar Hamel, Hassane Mohellebi, Mouloud Feliachi, Farid Hocini
	17.7.1	Introduction 359
	17.7.2	Particle Swarm Optimization 360
		17.7.2.1 Validation with Test Functions 361
	17.7.3	Problem Description 364
	17.7.4	Results 365
	17.7.5	Conclusion 366
References 366

Chapter 1
Integrated Computer Models of 3-D Comb Drive Electrostatic MEMS Structures

Sławomir Wiak and Krzysztof Smółka

Institute of Mechatronics and Information Systems, Technical University of Lodz; ul. Stefanowskiego 18/22, 90-924 Lodz, Poland
Tel.: +48(+42) 631-25-71, 631-25-81; Fax: +48(+42) 636-23-09
`wiakslaw@p.lodz.pl, ksmolka@p.lodz.pl`

Abstract. This paper deals with the numerical modeling of 3D structure of surface micromachined MEMS different comb drive geometries. The paper is devoted to the complex strategy of modeling, proposed by authors, of micro-actuators of comb structure. This methodology is based on vector field 3-D structural models and the object oriented models and lumped parameters as well, applied to the different structure of MEMS.

1.1 Introduction

We could stress that one of the most rapidly growing group of devices is Micro-Electro-Mechanical Systems (MEMS). MEMS structures are fabricated using semiconductor manufacturing technologies. These devices have tiny parts micrometers (or even nanometers), and are frequently combined with integrated circuits on a single chip providing intelligence programming and signal processing (Solutions 2009).

A huge number of MEMS structures have been successfully commercialized in recent years, namely: inkjet printer nozzles, silicon pressure sensor, crash bag accelerometers and micromachined gyroscopes, etc.

Microelectromechanical systems and other microsystems may be characterized as multi-domain systems: mechanical, electrical, thermal, fluidic, or optical phenomena have to be considered very detailed on the component level as well as coupled together from the systems point of view. These requirements result in a very complicated design demand (Fedder and Mukherjee 1996).

Actually, the task of the designer is highly complicated by the evidence that different energy domains, in general being mutually coupled, are involved when modeling micro-electro-mechanical devices (MEMD), in the frame of more general micro-electro-mechanical systems (MEMS). Moreover, the designers have to take into account the constraints imposed by the process technology, that limit the feasibility of innovative devices. Some specific codes based on Finite Element

Method (FEM) (Di Barba et al. 2008) could be successfully applied to MEMS designing while micro-domain physics are also taken into account. Such a sophisticated software could allow creating 3D structure of the device, full model analysis (structural mechanics, electrostatics, vibration, even coupled phenomena, etc.).

This paper deals with the methods of computer analysis and simulations of electrostatic accelerometers (comb drive structure). It is worth to point out that accelerometers are important devices in the range of variety applications such as air bag actuation (by Analog Device, Berkeley Sensor and Actuator Center, and Sandia), microrobots, etc. The accelerometers available on the market are capable of measuring high values of accelerations.

Micromachined accelerometers can be classified by either the method of the position detection of the seismic mass or the micromachined fabrication process of the sensing elements. Micromachined sensing elements for accelerometers can be fabricated either by bulk micromachining, surface micromachining or the LIGA process (Solutions 2009).

1.2 Solid Modeling of 3D MEMS Structures

The base method of the proposed strategy lays in solid modeling technology, moreover nowadays commonly exploited in CAD techniques. It is rapidly emerging technique as a central area of research and development in such diverse applications as engineering and product design, computer-aided manufacturing, electronic typing, etc.

The solid model, which contains the external surfaces, edges and internal volume information could be used for design representation, verification, simulation, analysis for processing, manufacturability and costing, and for both rapid prototyping, interactive design, and rapid tooling. Today, nearly all integrated CAD/CAE/CAM systems are supported by the solid modeling at CAD modeled structures. All these systems represent the shapes of solid physical objects. Such representations basic operations on these systems could be provided by solid modeling.

Recently, we observe rapid growth of methodologies and their applications in creating 3D virtual object structures (Zhang 1998, Wiak et al. 2004). With the point of view of CAD modeling, the basic aim of this work is the study of complex field/circuit models for three-dimensional MEMS structures in the electrostatic micro actuators working as the sensors of acceleration, micromirror driving system to be implemented in special switching system (optical fibres), etc. Accelerometers are one of the early successful MEMS applications. Inertial sensors are a class of micromachined which rely on the movement of the suspended proof mass.

Accelerometers are important sensors in a variety of applications ranging from air bag actuation and anti-skid braking to navigation and flight control (Boser 1998). Fields of action of accelerometer are shown in Figure 1.1.

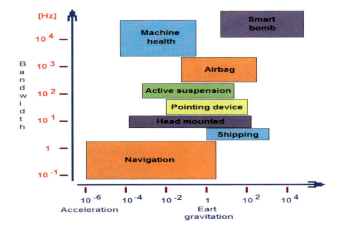

Fig. 1.1 Fields of action of accelerometer (Boser 1998)

The base of proposed strategy is solid modeling. Schematic view of the evolution of CAD modeling, from its 2-D computer-aided drawing in early 70s to recently invented technology of 3-D solid modeling-based computer-aided design, is presented in Figure 1.2.

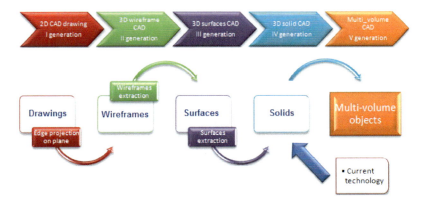

Fig. 1.2 Evolution of CAD Modeling (Hoffmann 1989)

A solid modeling techniques, which contains the external surfaces, edges and internal volume predefined could be used for design representation, verification, simulation, analysis for processing, manufacturability and costing, and for rapid prototyping and rapid design as well. Today, nearly every integrated CAD/CAE/CAM systems are supported by solid modeling at CAD modeled structures.

All these systems require representing the shapes (predefined simple parameterized volumes as library components in the CAD package) of solid physical objects, and such representations and basic Boolean operations on them could be provided by solid modeling.

Building of the geometry could be arranged by use of simple parameterized volumes like: BLOCK, CYLINDER, DISC, SPHERE, TORUS and PRISM /PYRAMID combining with Boolean Operations (like: Union With Regularization, Union Without Regularization, etc.). Boolean operations are performed on bodies, not cells. The main distinction between a cell and a body is that a cell is a volume of the model, while a body is a hierarchical assembly of cells, faces, edges and points (Wiak et al 2004, Sulima and Wiak 2008). Hence, it is necessary to pick the new cylinders created as bodies not cells. Briefly summarizing, in the next step it is possible to select from each volume the following parts: Bodies, Cells, Faces, and Edges.

Additionally we could also define transformation operations, sweeping and morphing operations. Solid modeling could be arranged as two stage process. Surfaces of volumes (cells) are initially discretized into triangles. Controls are available to define the exactness of the representation of curved surfaces. Then, using the surface mesh, each cell is meshed automatically into tetrahedral elements.

Element size can be controlled by defining a maximum element size on vertices, edges, faces or cells within a model. This allows the mesh to be concentrated in areas of interest, where high accuracy is required, or where the field is changing rapidly. One of the most sophisticated software exploiting Boolean operations and predefined parameterized volumes are reported in the literature (Zhang 1998, Wiak et al. 2004, OPERA-3D 2009).

1.2.1 Case Study – Accelerometer

The comb-drive accelerometers, shown in Figure 1.3, are very popular devices in the MEMS community and have been well characterized. The capacitive accelerometer consists of moving comb teeth, suspended by spring beams on both sides, and fixed teeth. The suspension is designed to be in the x direction of motion and to be stiff in the orthogonal direction (y), and z direction as well, to keep the comb fingers aligned. When the voltage is applied on the force unit, a net electrostatic force pulls the movable part in the desired direction. Apart of it the accelerometer could also move in other directions, not expected. Due to phenomena complexity only field models are fully acceptable in 3-D structure designing. Movable part displacement toward not desired direction "could even destroy" mathematical model, due to introducing to equivalent circuit model additional capacitances.

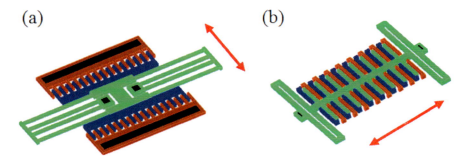

Fig. 1.3 Two types of electrostatic accelerometers MEMS (by solid modeling) (Wiak et al. 2008, Wiak and Smółka 2009)

The parallel plate capacitor is the most fundamental configuration of comb drives. The stored energy is expressed by the following formula (while C is function of design parameters):

$$W = \frac{1}{2}CU^2 \qquad (1)$$

When the plates of the capacitor move towards each other, the work done by the attractive force between them can be computed as the change in W (stored energy) versus displacement (x).

1.2.2 Case Study – Actuator for Micromirror Driving

Micromirrors with electrostatic drive find wider and wider application in different kinds of equipment, from bar-code reader to multimedia projectors. A series of drive solutions based on the electrostatic effect is presented in numerous papers and projects. Capacitive designs have the characteristic feature that they can be relatively cheaply and arbitrarily minimized down to the very limits of production technology possibilities. Their basic advantage is the fact that they have favorable F~U2 characteristic, as each electrostatic actuator. However their disadvantage is the necessity of applying quite high control voltages, which in some cases makes it impossible to use them in given situations, e.g. in medicine the safety standards do not allow to apply such voltages in direct contact with live organism.

Electrostatic micromirror comb-drive actuators are realized in many variants, so we can distinguish:

1. Monolithic actuators – built of a system of plan electrodes working in cooperation,
2. Comb-drive actuators –built of electrodes in the form of comb.

There are several design solutions of comb actuators which can be in principle divided in: horizontal, vertical, angular, and rotary. Monolithic actuators are commonly applied in optical switches and multimedia DLP projectors.

MEMS switches with micromirrors permit to switch signals directly from input to output which saves unnecessary transformation and removes the "bottle neck" effect and accelerates the whole network. An optical switch with micromirror is based on precise positioning of the mirror using an actuator, e.g. an electrostatic one. An important demand is the proper selection of the control signal so that the control unit – the mirror, rotates around the required angle with as much as possible the lowest amplitude of oscillations.

Other group of micromirror drives is the comb-drive actuators. A huge group of them are plane comb actuator structures, which of course are the simplest.

In that solution the scanning mirror is attached to a beam connected to moving combs. As a result of vibrations of the comb the mirror is moved along groves in which optical fibres are placed. Such a switch is described by (Li J. et al 2003). While R. A. Conant (Conat et al. 2000) presents an other kind of comb-drive called for short STEC. Such design allows for high mirror rotation angle up to ab. 25° with low static and dynamic mirror distortions, not over 30 nm.

P. R. Patterson (Patterson and Hah 2002), describes a similar design In his paper; it allows scanning angles even up to 40° to be reached in actuators comparable to the STEC ones. The maximum scanning angle is a function of the tooth geometry, particularly of its thickness, of the teeth overlap and shift. For the angular design this is by 1,5 higher than for a comparable STEC one.

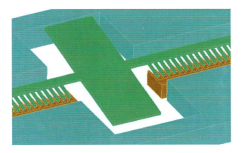

Fig. 1.4 A micromirror with a torsional comb-drive actuator (Conant et al. 1999)

The microswitch consists of two comb-electrodes in STEC configuration (Figure 1.5 – solid structural model). A single comb consists of several, up to some hundreds rectangular plates (or of other shape – Figure 1.6). When no voltage is applied, the elastic beam to which the plates are fixed, keeps them above the electrodes of the lower comb attached to the base.

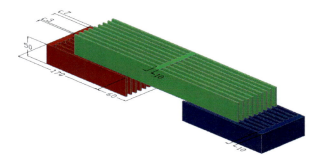

Fig. 1.5 Comb-drive microactuator (solid model - data given in μm) (Sulima and Wiak 2008)

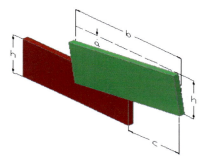

Fig. 1.6 Comb microactuator with rectangular electrodes (after angular displacement) (Sulima et al. 2008).

1.3 Complex Strategy

Authors propose (their own) a complex strategy of modeling and the optimization technique, based on the vector field model and the circuit methodology of MEMS comb structure. This strategy is fully satisfactory in order to simulate the electromechanical characteristics of actuator of different structures, and dynamic behavior of the object (Figure 1.7).

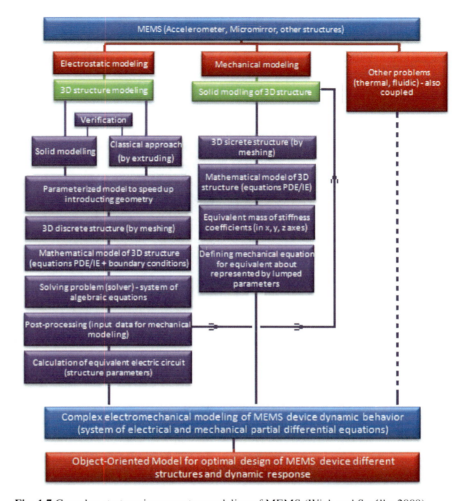

Fig. 1.7 Complex strategy in computer modeling of MEMS (Wiak and Smółka 2009).

A novel complex strategy in computer modeling of accelerometer MEMS, based on the solid modeling is proposed by the authors in this paper. This strategy is fully satisfactory in order to simulate the electromechanical characteristics of different accelerometer structures (MEMS). This methodology consists of the following general steps:

1. Solid modeling of 3-D structure of inertial devices,
2. Parameterized model to speed up geometry introducing,
3. Creation of the mathematical model of 3D structure,
4. Creation of the complex electromechanical models of MEMS dynamic behavior (system of electric and mechanical partial differential equations – based on lumped parameters models).

5. Creation of the Library of Object-Oriented MEMS components.
6. In authors' opinion, only such a new methodology makes the complex simulation of device dynamic behavior and MEMS optimal design possible as well.

Solid modeling strategy is successfully applied to the calculation of electrostatic field in microactuators with comb drive structure.

1.3.1 Case Study – Accelerometer

We also could stress that only three-dimensional structure, by means of solid modeling, of the object could give the full view of the phenomena occurring in accelerometers. Figure 1.8 presents 3-D structural model of comb drive structure (corresponding to Figure 1.3) with mesh (by use of OPERA-3D code).

A proper strategy in designing of MEMS should capture the essential static and dynamic behavior of the device using a minimal set of equations, which are in terms of the physical design parameters and material properties. Analytical models are useful to describe coupling between different directions of motion, and are derived by use of the energy methods.

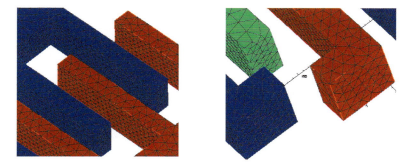

Fig. 1.8 Full 3-D structure mesh in the model of two types accelerometers from Figure 1.3 (over one million elements).

The structure of typical electrostatic accelerometers consist of elastic suspending elements (springs), shuttle mass, and electrostatic elements. Behavior of accelerometer is described by a lumped second-order equation of motion. For any generalized displacement ζ, we can write (Iyer and Mukherjee 2000, Iyer et al. 1999, Jing et al. 2002):

$$F_{e\zeta} = a_{e\zeta} m = m_{\text{eff}\,\zeta}\, \ddot{\zeta} + B_\xi\, \dot{\zeta} + k_\zeta\, \zeta \tag{2}$$

where $F_{e\zeta}$ is the external force (in the x-mode this force is compatible with required motion of fingers in comb drives), $a_{e\zeta}$ is external acceleration, $m_{\text{eff}\,\zeta}$ is the effective mass, B_ζ is damping coefficient, and k_ζ is the springs constant. The x-mode frequency for this model is given by

$$\omega_x = 2\pi f_x = \sqrt{k_x/m_x} \qquad (3)$$

The other modes are modeled similarly.

In this second-order mass-spring-damper (Figure 1.9) system shows that displacement is a function of external acceleration.

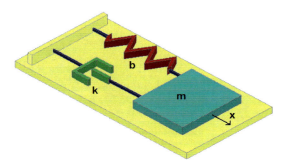

Fig. 1.9 Simplified diagram of the accelerometer structure

The above statement is basic rule in accelerometers mechanical model. In next sections we derive the mathematical formulas of other components of this model accelerometer. The effect of spring mass on resonant frequency of different modes is taken into account by an effective mass model.

Effective mass for each mode of interest is calculated by normalizing the total maximum kinetic energy of the spring by the maximum proof-mass velocity, v_{max}, where mi and L_i are the mass and the length of the i'th beam in the spring (Yong 1998, Iyer and Mukherjee 2000).

The vector field analysis (mechanical and electrical) has given the knowledge about the structures of the micro accelerometer, as well as the dynamic behavior of the analyzed object. Finite Element Method yields in good agreement with the analytical method, giving only small discrepancies.

$$m_{eff} = \sum_i^N \frac{m_i}{L_i} \int_0^{L_i} \left(\frac{v_i(\zeta)}{v_{max}} \right)^2 d\zeta \qquad (4)$$

where μ is the viscosity of air, d is the spacer gap, d is the penetration depth of airflow above the structure, Ac is the gap between comb fingers, and A_s, A_t, A_b, and A_c are bloated layout areas for the shuttle, truss beams, flexure beams, and comb-finger sidewalls, respectively (Fedder and Mukherjee 1996).

Springs designing are a very important step in the global design process of inertial sensors (Iyer and Mukherjee 2000). The suspension creates mechanical force acting against the displacement and the net displacement is obtained when the electrostatic force is equal to this mechanical force. Various designs for the spring of the suspensions are proposed in the literature (one solid model, selected for our structure, is shown in Figure 1.10).

Fig. 1.10 Exemplary suspending system (springs structures) of the accelerometer (Wiak et al. 2008).

Linear equations for the spring constants are derived using energy methods. A force (or moment) is applied to the free end(s) of the spring, in the direction of interest, and the displacement is calculated symbolically (as a function of the design variables and the applied force).

When forces (torques) are applied at the end-points of the flexure, the total energy of deformation U, is calculated as:

$$U = \sum_{beam\ i=1}^{N} \int_{0}^{L_i} \frac{M_i(\xi)^2}{2EL_i} d\xi \tag{5}$$

where, L_i is the length of the i'th beam in the flexure, M_i is the bending moment transmitted through beam i, E is the Young's modulus of the structural material and I_i is the moment of inertia of beam i, about the relevant axis (Iyer et al. 1999).

Next, the analytical expressions were derived using the Castigliano's theorem. The effective stiffness obtained analytical expressions was compared with the results from finite element method.

Authors clearly point out that only three-dimensional object structure, built by use of solid modeling methodology, could give the complex information about the physical phenomena at the accelerometers. Mechanical (stress and stiffness modeling) modeling has been carried out for some selected spring geometries (see Figure 1.10).

Several mechanical modes of the movable part of the device have been calculated, and an example of how the structure might bend due to external forces has been computed. First few mode shapes for the spring, from Figure 1.10, are shown in Figure 1.11. For all the types of suspending systems, the basic (for correct working of the inertial device) is the first mode. The field analyses (mechanical and electrical) have given the knowledge about structure of micro accelerometer simulation. System of electric and mechanical partial differential equations is also the effect of this work. Finite Element Method yields in good agreement with the analytical calculations, which gives small discrepancies.

Fig. 1.11 Examples of mode shapes of mechanical structures springs in accelerometer.

Accelerometer object-oriented modeling: Computer simulation as the method and integrated tool is a very important way to gain insight in complex systems, to make virtual experiments to get deeper understanding, and to verify new designs (Schwarz, 2004). Modeling, simulation and analysis tools are needed to support the MEMS designer involved in optimal design process. In the design of microsystems, sophisticated CAD systems based on Finite Element Method (FEM), like OPERA from Vector Fields, COSMOS and ANSYS, are exploited to perform modeling and simulation of the behavior of the system components with high accuracy. Finite element analyses are the most commonly used methods for numerical mechanical and electrostatic simulations. These methods are accurate for creating the fine meshes. However, as they are layout-based, any change of the geometry requires a new mesh, leading to inconvenient design iteration (Jing et al. 2002). We would stress that FEM simulations are time consuming.

One of the most important step in creating the complex electromechanical models of MEMS dynamic behavior (system of electric and mechanical partial differential equations – based on lumped parameters models) is defined by the Library of Object-Oriented MEMS components (Matlab/Simulink - Figure 1.12). Simulink blocks library of accelerometers components (springs and comb-drives and additional components) has been created to carry out the simulation of accelerometer dynamic behavior. The general idea is to create universal topology for accelerometer comb structure; the same for different selected structures to be investigated. It should be recommended to introduce only the data for each block, which corresponds to the investigated structure.

Therefore special block Comb-Spring-Comb have been defined, which plays role as adapter to transform each investigated specific structure to universal blocks topology. Moreover elaborated by authors algorithms and codes enables to transform the data from field model to object oriented model (block components) continuously, even during the dynamic movement.

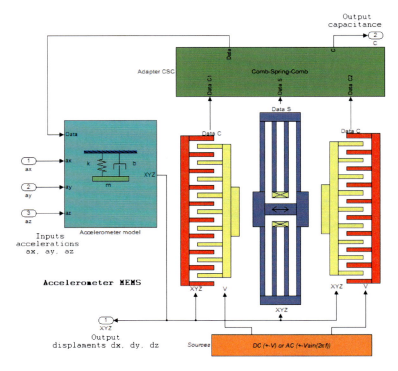

Fig. 1.12 Object-Oriented Model (OOM) by Matlab-Simulink (Wiak and Smółka 2009).

Modelling of the dynamic accelerometer behaviour by use of <u>New Object-Oriented Model</u> is based on Complex Computer 3D-structures Field Models (Electrostatic and Mechanical).

General idea of building Object-Oriented Model (OOM) consists of the following steps:

- defining the total capacitance of the object C as the function of displacements (in x, y and z axes) and geometry.
- defining the effective mass $m_{eff\zeta}$ from 3D structural mechanical analysis.

Obtained data files from 3D structural analysis are in the next step introduced as the input data to OOM.

Proposed by authors OOM could be successfully implemented to any kind of comb structure actuators, even those being manufactured by Research Centres or Companies. As the test calculations of comb element of the actuator (with the structure presented in Figure 1.12) we have performed (see Figure 1.13) for the following constraints: supply voltage in x and y axes in sine wave form, in z axis constant force. As the results (dynamic response) of complex computer simulation (joining 3D field model with object oriented models) we observe resultant electrostatic force F between electrodes and damping coefficient b_x changes along x axis.

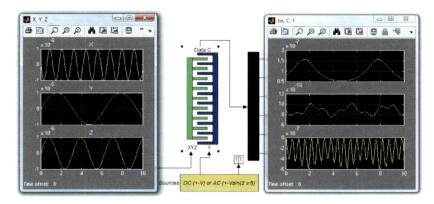

Fig. 1.13 Test calculations for Object-Oriented Model (OOM) of accelerometer

1.3.2 Case Study – Actuator for Micromirror Driving

Determination of the maximum deflection angle of the moving comb for a STEC type comb actuator (see Figure 1.6) (Hoffmann 1989) is given by:

$$\theta_{\max} = \frac{h}{(b-c)+c} \qquad (6)$$

The upper comb marked red (see Figure 1.5 and 1.6) is the moving element, in the direction of the x axis and rotation by the α angle. The system presented above is an electrostatic system thus the displacement of the electrode depends on the supply voltage. An electrostatic field is created in the air gap by applying suitable voltage, which causes the electrode supported on spring beams to be pulled in, where the force with which the electrode is pulled in depends on the difference of potential between the electrodes and on the change in the capacitance of the air capacitor formed by the system: fixed electrode-air gap–moving electrode.

The capacitance depends in turn directly on the active surface of electrodes and on the distance between them. We assume that the active electrode can move only in the direction of the y axis and rotate by the angle of α.

In order to calculate the force produced by the actuator we compute the partial derivatives of the electrostatic filed stored energy We versus variables (displacements: linear – y and angular-α), we obtain (see Figure 1.6):

$$F_y = \frac{\partial W_e}{\partial y} = \frac{\partial}{\partial y}\frac{1}{2}UC^2 = \frac{\varepsilon \cdot U^2}{2d}\frac{\partial S_{total}}{\partial y} \qquad (7)$$

or

$$M_\alpha = \frac{\partial W_e}{\partial \alpha} = \frac{\partial}{\partial \alpha}\frac{1}{2}UC^2 = \frac{\varepsilon \cdot U^2}{2d}\frac{\partial S_{total}}{\partial \alpha} \quad (8)$$

where: S_{total} – electrodes active surface, d – distance between electrodes, U - supply voltage, ε - electric permittivity.

The moving electrode is supported by an elastic beam which is exposed to torsional stresses during operation. The beam is also subjected to bending as a result of being loaded by the mass of the mechanism (Figure 1.14).

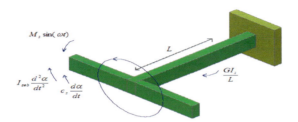

Fig. 1.14 Torques acting on suspending beam during torsional movement.

The mechanical energy W_m of such a system can be expresses by the equation:

$$W_m = \frac{1}{2}k_y y^2 + \frac{1}{2}k_\alpha \alpha^2 \quad (9)$$

where: k_y, k_α, rigidity coefficients of the beam. Detailed data related to the beam rigidity can be found in (Conat et al. 2000), and k_e is defined as follows: k_e = 2 GJ/L. The elastic beam, in our case, made of silicon, supporting the moving comb has a rectangular cross-section, where one side is many times longer than the other one. Such case we consider as a stress in a bar with a narrow cross-sectional area (a Lim bar). The distribution of stresses in a slim bar can be determined with the sufficient accuracy, assuming that the stress function depends only on the transverse variable, i.e. in our case. The rigidity index of the cross-section for torsion: $I_s = g^3h/3$. A more accurate value of this index is given by an empirical formula as (Conat et al. 2000):

$$I_s = \frac{1}{3}g^3 h \left(1 - \frac{g}{h}\frac{192}{\pi^5}\sum_{n=1,3,5...}^{\infty}\frac{1}{n^5}\tanh\frac{n\pi h}{2g}\right) \quad (10)$$

where: g –beam width, h – beam height, G –transversal elasticity modulus, k_y= $2gEh^3/L^3$, L – beam length, E –Young modulus.

The structural material of the beam is monocrystaline silicon, for which the values of the above parameters are respectively: E=129,5 GPa and G=78 GPa.

For a simplified formula of the rigidity index at torsion the total energy is as follows:

$$W = W_m - W_e = \left(E \frac{gh^3}{L^3} y^2 + \frac{G}{L} \frac{g^3 h}{3} \alpha^2 \right) - \frac{1}{2} U^2 \frac{\varepsilon_0}{d} S_{total} \qquad (11)$$

Object Oriented Model of the actuator is based on 3D structural models - electrostatic and mechanical (see Figure 1.15).

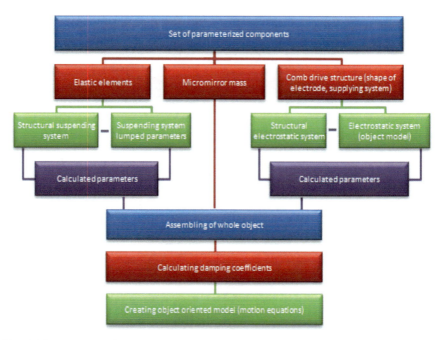

Fig. 1.15 Procedure of creating the global model of actuator (leading to object oriented model)

Defined above object oriented model allows us to design optimal structure of the actuator, leading to minimizing the amplitude and time of oscillations.

<u>Test Dynamic Analysis:</u> Carrying out a dynamic analysis of a STEC comb system, a load mass of the actuator comb is a mirror of approximately $0{,}25\mu m^2$ surface being assumed. The proposed by authors strategy of complex modeling leads to optimal design of actuator structure. One of the important results of such strategy is reducing oscillation duration to less than 0.5ns. Such parameters allow for fast and efficient switching the light signal from the waveguide net (see Figure 1.16, while the natural frequency is about 6,6kHz) (Sulima and Wiak 2008).

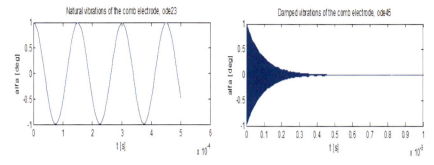

Fig. 1.16 Dynamic characteristic of the model of a comb-drive actuator – natural free vibrations of the system (Sulima and Wiak 2008).

1.4 Conclusions

The vector field analysis (mechanical and electrical) has given the knowledge about the structures of the micro- actuators, as well as the dynamic behaviour of the analyzed object.

On the basis of this research the following conclusions could be drawn:

- only the three-dimensional design with the application of the solid modelling provides the full view of the phenomena occurring in the micro- actuators,
- such complex, object oriented solution makes possible the global simulation of device dynamic states and optimization MEMS devices.

The above mentioned methodology determines the following benefits:

- time reduction between the device idea and its implementation as a final product,
- the reduction of the costly experiments for the huge number of prototypes.

The new object oriented Matlab/Simulink library blocks could be successfully used to different MEMS design, exemplary: resonators, gyroscopes, etc.

References

Boser, B.E.: Surface micromachining An IC-compatible sensor technology, Berkeley, Sensor & Actuator Center Dept. of Electrical Engineering and Computer Sciences University of California, Berkeley (1998)

Conant, R.A., Nee, J., Hart, M., Solgaard, O., Lau, K.Y., Muller, R.S.: Robustnes and Reliability of Micoromachined scanning Mirrors, Header for MEOM (1999)

Conant, R.A., Nee, J.T., Lau, K.Y., Muller, R.S.: A flat high-frequency scanning micromirror. In: Solid-State Sensor and Actuator Workshop, Hilton Head Island, SC, pp. 6–9 (June 2000)

Davies, F.R., Rodgers, M.S., Montague, S.: Design Tools and issues of silicon micromachined (MEMS) devices. In: Presented at the 2nd International Conference on Engineering Design and Automation, August 9-12, 1998, Maui, Hawaii (1998)

Di Barba, P., Savini, A., Wiak, S.: Field Models in Electricity and Magnetism. Springer, Heidelberg (2008)

Fedder, G.K., Mukherjee, T.: Physical design for surface-micromachined MEMS. In: 5th ACM/SIGDA Physical Design Workshop, Reston, VA, USA, April 15-17, 1996, pp. 53–60 (1996)

Hoffmann, C.M.: Geometric and solid modeling. Morgan Kaufmann, San Mateo (1989)

Iyer, S.V., Mukherjee, T.: Numerical spring models for behavioral simulation of MEMS Inertial Sensors. In: Proc. SPIE, vol. 4019, pp. 55–62 (2000)

Iyer, S.V., Tamal, M., Fedder Gary, K.: Multi-mode sensitive layout synthesis of microresonator. In: International Conference on Modeling and Simulation of Microsystems, Semiconductors, Sensors and Actuators (April 1998)

Iyer, S.V., Zhou, Y., Mukherjee, T.: Analytical modeling of cross-axis coupling in micromechanical springs. In: MSM 1999, San Juan, Puerto Rico, April 19-21 (1999)

Jing, Q., Mukherjee, T., Fedder, G.K.: Schematic-based lumped parameterized behavioral modeling for suspended MEMS. In: Tech. Dig. Int. Conf. Computer-Aided Design, San Jose, CA, November 10-14, pp. 367–373 (2002)

Li, J., Zhang, Q.X., Liu, A.Q.: Advanced fiber optical swiches using deep RIE (DRIE) fabrication. Sensors and Actuators A 102, 286–295 (2003)

OPERA-3D, User Guide, COBHAM/Vector Fields Limited, Oxford, England, ver. 13.0 (2009)

Patterson, P.R., Hah, D.: A scanning micromirror with angular comb drive actuation. In: XV IEEE International Conference on Micro Electro Mechanical Systems (MEMS 2002), Las Vegas, USA, January 20-24, pp. 544–547 (2002)

Solutions – Software applications for engineering simulation and processes, vol. 1(1). Spring 1999 (2009), http://www.ansys.com

Sulima, R., Wiak, S.: Modelling of vertical electrostatic comb-drive for scanning micromirrors. Compel. 27(4), 780–787 (2008)

Sulima, R., Wiak, S., Krawczyk, A.: A Dynamic Characteristic of the Model of an Electrostatic MEMS Drive for Micromirror Control. In: Proceedings of the 2008 International Conference on Electrical Machines, Vilamoura, Portugal (2008)

Sun, W.: Multi-volume CAD modeling for heterogeneous object design and fabrication. Journal of Computer Science and Technology 15(1), 27–36 (2000)

Wiak, S., Cader, A., Drzymała, P., Welfle, H.: Virtual modeling and optimal design of intelligent micro-accelerometers. In: Rutkowski, L., Siekmann, J., Tadeusiewicz, R., Zadeh, L.A. (eds.). Subseries LNCS, pp. 942–947. Springer, Heidelberg (2004)

Wiak, S., Smółka, K.: Numerical modelling of 3-D comb drive electrostatic accelerometers structure (method of Levitation Force Reduction). Compel 28(3), 593–602 (2009)

Wiak, S., Smółka, K., Dems, M., Komęza, K.: Numerical (solid) modeling of 3-D intelligent comb drive accelerometer structure - mechanical problems. Electrical Review (3), 593–602 (2008)

Zhou, Y., Fedder, G.K. (advisor: prof.), Mukherjee, T. (co-advisor: dr.).: Layout synthesis of accelerometers, MS Project Report, Department of Electrical and Computer Engineering, Carnegie Mellon University (August 1998)

Zhang, G.: Design and simulation of a CMOS-MEMS accelerometer, Project Report, Carnegie Mellon University (May 1998)

Chapter 2
Multi-objective Design Optimization of Slotless PM Motors Using Genetic Algorithms Based on Analytical Field Calculation

L. Belguerras and L. Hadjout

Laboratoire des Systèmes Electriques et Industriels/USTHB,
BP 32 El Alia Bab-Ezzouar 16111 Alger, Algerie
`lamia_belguerras@yahoo.fr, lhadjout@usthb.dz`

Abstract. In this work, a Non-dominated Sorting Genetic Algorithm (NSGAII) is used to solve a Multi-objective Optimization problem which consists of the maximization of the average torque and the reduction of the slotless PM motors mass. Firstly a magnetic analytical model of the motor is developed to define an optimization problem and a set of design constraints. Then, the (NSGAII) algorithm is used to solve this optimization problem.

2.1 Introduction

The vibrations are undesirable effects which should be reduced [1]. For that, it is necessary to identify the means making it possible to reduce these phenomena. In this perspective, we are chosen the slotless PM Motor [2, 1]. On the other hand, considering the height cost of the magnets, copper and the other materials raised leads to optimize their size while conserving acceptable performances (flux density, torque…).

In recent years, optimization algorithms have received increasing attention by the research community as well as the industry; it looks for an equilibrium point of the different goals. Each generic admissible solution for the design problem represents one point in the multi-dimensional design space spanned by all admissible machine parameters satisfying the requirements, the technological and physical constraints. Therefore, it is necessary to use the multi-objective optimization strategy based on a rigorous mathematical formulations and automatic design aids to help the engineer's technical sensibility and experience during the process [3].

There are plethora of methods and algorithms for solving the multi-objective optimization problems, one of these methods is evolutionary multi-objective optimization (EMO) algorithms; it is one of the most active research areas in the field of evolutionary computation. A number of EMO algorithms have been proposed in the literature [4], [5]. The main advantage of Evolutionary Multi-objective algorithms over other multi-objective optimization methods is that many non-dominated solutions can be simultaneously obtained by their single run. In this perspective, we have used NSGA-2 [5] for solving the inverse problem. Indeed, this method will be detailed.

This study brings a new way for solving multi-objective evolutionary optimization applied to the optimal design problem. It is formulated as an inverse problem which is solved with a multi-objective optimization associated with an analytical model. The goal is to find a set of design parameters in the feasible design space that optimize two objective functions while keeping the given constraints.

2.2 Electromagnetic Modeling of the Study Slotless PMSM

2.2.1 Motor Topologies

The structure of the machine study is schematized on the figure 2.1. On the inside of the stator iron is an electromagnetically inactive area that is used for bearings, cooling and other auxiliaries. The stator iron consists of a slotless armature, it consists of very thin. On its inside, the stator winding region is located. On the inside of the mechanical air gap the electromagnetic shielding cylinder is located, followed by the permanent magnet array and the solid rotor iron. This study structure can only be practicable for sinusoidal current supply [6].

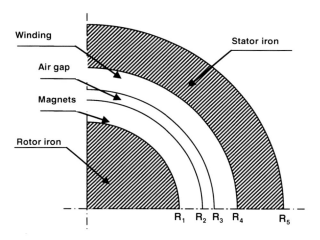

Fig. 2.1 Electromagnetic modeling of a slotless PMSM

2.2.2 Analytical Modeling

The Maxwell electromagnetic model leads to solve the following equation:

$$\overrightarrow{rot}(\frac{1}{\mu}\overrightarrow{rotA}) - \sigma\frac{\partial \overrightarrow{A}}{\partial t} = -\overrightarrow{J}_0 + \overrightarrow{rot}(\frac{\overrightarrow{M}}{\mu}) \qquad (1)$$

Where: A is the magnetic vector potential.

J_0 is the exciting current density.

M is the magnetization.

µ is the permeability

An analytical model is developed, in terms of polar coordinates, to compute the magnetic field produced by the permanents magnets of the slotless PM motors.

The study structure is divided into the air-gap region, permanent region and conductors region (fig 2.1). A magnetic vector potential formulation is used in 2D polar coordinates to describe the problem. According to the adopted assumption, the magnetic vector potential has only one component along the z-direction and only depends on the r and θ coordinates.

In polar coordinates, the Laplace's equation in both the air-gap and winding is described by:

$$\frac{\partial^2 A_a}{\partial r^2} + \frac{1}{r}\frac{\partial A_a}{\partial r} + \frac{1}{r^2}\frac{\partial^2 A_a}{\partial \theta^2} = 0 \tag{2}$$

And for the permanent magnet region, the Poisson's equation described by:

$$\frac{\partial^2 A_p}{\partial r^2} + \frac{1}{r}\frac{\partial A_p}{\partial r} + \frac{1}{r^2}\frac{\partial^2 A_p}{\partial \theta^2} = \frac{1}{r}\left[\frac{\partial M_r}{\partial \theta} - M_\theta\right] \tag{3}$$

The boundary conditions for the different regions are defined as following:

$$\left.\frac{\partial A_p}{\partial r}\right|_{r=R_1} = 0 \tag{4}$$

$$\left.\frac{\partial A_a}{\partial r}\right|_{r=R_4} = 0 \tag{5}$$

The continuity condition between the air-gap and permanent magnets leads to:

$$\begin{cases} A_p(R_2,\theta) = A_a(R_2,\theta) \\ \mu_r \left.\frac{\partial A_p}{\partial r}\right|_{r=R_2} = \left.\frac{\partial A_a}{\partial r}\right|_{r=R_2} \end{cases} \tag{6}$$

The radial and the tangential component of the flux density in each region are related to the magnetic vector potential by the well-known relations:

$$B_r = \frac{1}{r}\frac{\partial A}{\partial \theta} \qquad (7)$$

$$B_\theta = -\frac{\partial A}{\partial r} \qquad (8)$$

By using separation of variables, the general solution for the magnetic vector potential in each region has the form:
In the air-gap region

$$A_a(r,\theta) = \sum_{n=1}^{\infty}\left[a_{1n}r^{np} + a_{2n}r^{-np}\right]\sin(np\theta) \qquad (9)$$

In the permanent magnets region

$$A_p(r,\theta) = \sum_{n=1}^{\infty}\left[a_{3n}r^{np-1} + a_{4n}r^{-np-1} + \psi_n(r)\right]\sin(np\theta) \qquad (10)$$

Where: $\psi_n(r)$ is the particularly solution for equation (10).

$$\Psi_n(r) = \begin{cases} \dfrac{npM_{rn}}{n^2p^2 - 1}r & si \quad n^2p^2 \neq 1 \\ -\dfrac{1}{2}npM_{rn}r\ln r & si \quad n^2p^2 = 1 \end{cases} \qquad (11)$$

a_{1n}, a_{2n}, a_{3n} and a_{4n} are arbitrary constants calculated for each harmonic using the boundary conditions giving in (4), (5) and (6).

2.2.3 Electromagnetic Torque

For 2p-poles PM motor and sinusoidal current, the electromagnetic torque can be calculated by:

$$\Gamma(\theta) = 2pL_u \int_{R_3}^{R_4}\int_0^{\tau_p} J(\theta_s)B_r(r,\theta)r^2\partial r\partial\theta_s \qquad (12)$$

Where B_r is the radial component flux density produced by the magnets and J is the current density in the air gap winding that are expressed by the following terms:

$$B_{ear}(r,\theta) = \sum_{n=1}^{\infty} np\left[a_{1an}r^{np-1} + a_{2an}r^{-np-1}\right]\cos(np\theta) \qquad (13)$$

$$J(\theta_s) = \frac{3}{2}\sum_n J_n \cos(np\theta_s + \upsilon\omega t) \qquad (14)$$

Where: $\upsilon = \begin{cases} -1 & \text{if } n = 6m+1 \\ +1 & \text{if } n = 6m-1 \end{cases}$

$$\theta_s = \theta + \alpha_0 + \Omega t \qquad (15)$$

$$J_n = \frac{4pN_c I_m}{\pi(R_4^2 - R_3^2)} K_{bn} \qquad (16)$$

Where:

K_{bn} is the winding factor

θ_s is the stator reference frame.

θ is the rotor reference frame.

Ω is the angular speed.

Finally the electromagnetic torque and the average torque are giving respectively by the following relations:

$$\Gamma(\theta) = \frac{3}{2}\pi p \sum_{n=1}^{\infty} \Gamma_n \cos(np\theta_0 + (n+\upsilon)\omega t) \qquad (17)$$

Where:

$$\Gamma_n = nJ_n \left[\frac{a_{1n}}{np+2}(R_4^{np+2} - R_3^{np+2}) - \frac{a_{2n}}{np-2}(R_4^{-np+2} - R_3^{-np+2}) \right] \qquad (18)$$

The average torque can be calculated as following:

$$\Gamma_{av} = \frac{1}{T}\int_0^T \Gamma(t)\,dt \qquad (19)$$

Equation (20) can be writing as:

$$\Gamma_{av} = 3\frac{3}{2}L_u \pi J_1 \left[\frac{a_{11}}{5}(R_4^5 - R_3^5) - a_{21}(R_4^{-1} - R_3^{-1}) \right] \qquad (20)$$

2.2.4 Model Validating

In order to validate the proposed model, the analytical results have been compared with 2D finite element simulations.

Fig. 2.2 and Fig. 2.3 show respectively the radial flux density and electromagnetic torque developed by the slotless PMSM for parameters giving in Table 2.1. Notice that the motor is supplied by a three phase sine wave currents.

A good agreement is noticed between the FEM and analytical results.

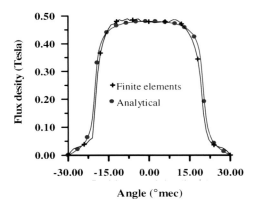

Fig. 2.2 Radial flux density distribution

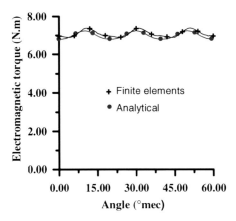

Fig. 2.3 Electromagnetic torque

Table 2.1 Fixed parameters of prototype study

Symbol	Parameter	Value
M	Magnetization of magnets	0.9 Tesla
μ_r	Relative recoil permeability	1
I_{max}	Maximum phase current	$4\sqrt{2}$ A
P	Pole number	3
α_0	(Magnet pitch/pole pitch ratio	(2/3)
N	Slot number	36
ξ	Slot opening/Tooth pitch	(1/3)
N_c	Conductors number per slot	23
L_u	machine length	40 mm
B_{fer}	Iron flux density	1.5 Tesla
f	Frequency	50 Hz
ρ_{magnet}	Magnet density	7.6 g/mm^3
ρ_{copper}	Copper density	8.93 g/mm^3
ρ_{iron}	Iron density	7.6 g/mm^3
R_5	Outer stator Radius	126.5mm
R_4	Inner stator radius	87mm
R_1	Outer rotor radius	69mm
R_0	Inner rotor radius	50mm
e_e	Air gap thickness	3mm
e_a	Permanent magnet thickness	10mm

2.3 Multi-objective Optimization

In a multi-objective (or multi-criteria) optimization problem, there are more than one objective function to be taken into consideration. Besides the existence of multiple objective functions, there may also be some constraints, which put restrictions on the search space.

It can be formulated as follows:

$$\begin{cases} optimize \ f_i(x) & for \ all \ i = 1,2,...N_{obj} \\ with \quad g_j(x) & for \ all \ j = 1,2,...J \\ and \quad h_k(x) & for \ all \ k = 1,2,...K \end{cases} \quad (21)$$

In this formalization, x represents the solution vector (a point in the search space) that holds the decision variables, the function set f represents the objective functions subject to optimization, the function set g contains the inequality constraints and finally the function set h contains the equality constraints.

In the single-criterion case, generally there exists a global maximum, and the aim of a search algorithm is to reach to that peak point. However, a multi-objective problems deal with simultaneous optimization of objectives. The outcome of such an optimization problem is a set of compromised solutions of different objectives

2.3.1 Dominance

To compare candidate solutions in multi objective optimization problems, the concepts of Pareto dominance is used. A decision vector *x* is said to dominate another *y* when it is as good as *y* regarding each objective, and there is at least one objective with respect to which *x* is better than *y* [6]. In this case, the solution *x* dominates *y*; *x* is called the non-dominated solution (Fig.2.4).

For a minimization problem, A solution vector x is said to dominate the solution vector y when:

$$\forall i \in \{1,2,...,I\}\ f_i(x) \leq f_i(y) \text{ and } \exists i \in \{1,2,...,I\} f_i(x) < f_i(y) \qquad (22)$$

2.3.2 Pareto Optimality

If there is no feasible solution y that dominates x, x is said to be a Pareto-optimal solution of the multi-objective optimization problem in [7]. The set of all Pareto-optimal solutions is the Pareto-optimal solution set. However, the solutions that are non-dominated within the entire search space are denoted as Pareto-optimal and constitute the Pareto-optimal set or Pareto-optimal front [7, 4].

A sample pareto-optimal set of a problem with two objective functions subject to minimization is illustrated in (Fig.2.4).

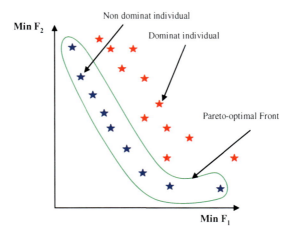

Fig. 2.4 Example Pareto-Optimal Fronts for a 2-Objectives Problem

Fig. 2.5 illustrates possible pareto-optimal fronts for problems with two objectives, with respect to their optimization configurations regarding minimization or maximization.

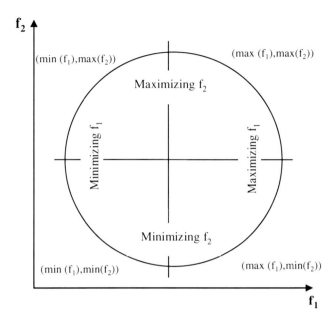

Fig. 2.5 Example Pareto-Optimal Fronts for a minimization or maximization 2-Objective Problem [7]

2.3.3 Evolutionary Multi-objective Optimization Algorithms

Evolutionary multi-objective optimization (EMO) algorithms are similar to traditional single objective evolutionary algorithms in that all genetic algorithms search complex problem spaces using the process that is analogous to Darwinian natural selection. Evolutionary algorithms use a population based search in witch height quality solutions are involved using the three basic operators (selection, crossover and mutation) similar to those in natural selection and evolution [8,9].

In this section, we briefly explain some basic concepts in genetic algorithms (GA):

1. **Selection:** according to its fitness value, an individual may reproduce, survive or die.
2. **Fitness:** the fitness of each function is determined by how well it satisfied objectives and constraints.
3. **Crossover:** occurs by combining the decision variables of height quality solution (parent) to create (child) solutions. It is needed to increase diversity among the population.

4. **Mutation:** this operator is needed because important genetic information may occasionally be lost as a result of selection and crossover.

The task of evolutionary multi-objective optimization (EMO) algorithms is to find well distributed Pareto-optimal or near Pareto-optimal solutions as many as possible. The main advantage of EMO algorithms over other multi-objective optimization methods is that many non-dominated solutions can be simultaneously obtained by their single run [4].

Generally, GAs select individuals according to the values of the fitness function. However, in a multi-objective optimization problem, some criteria being considered, the evaluation of the individuals requires that a unique fitness value, referred to as a dummy fitness be defined in some appropriate way. To achieve this, by application of the definition of non-dominance, the chromosomes are first classified by fronts. The non-dominated individuals of the entire population define front 1 ; in the subset of remaining individuals, the non-dominated ones define front 2, and so on ; the worst individuals define front n, where n is the number of fronts see (Fig.2.6) [4].

A number of EMO algorithms have been proposed in the literature [7, 9 and 10]. The NSGA-II algorithm of Deb et al. [5] is one of the most well-known and frequently-used EMO algorithms in the literature.

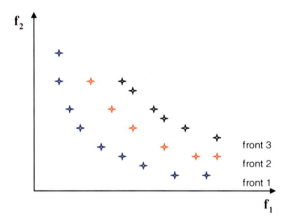

Fig. 2.6 Classification of population

2.3.4 *Elitist Non-dominated Sorting Genetic Algorithm (NSGAII)*

The method used in this paper is NSGAII, it's the second generation EMO developed by Deb and al [5] which made significant improvements to the original NSGA by

1. Using a more efficient non-domination sorting scheme a crowding measure.
2. Eliminating the sharing parameter

3. Adding an implicitly elitist selection method that greatly aids in capturing Pareto surfaces.
4. The NSGAII can handle both real and binary representations.

As in other evolutionary algorithms, first the NSGA-II algorithm generates an initial population. This is usually performed randomly. Then an offspring population is generated from the current population by selection, crossover and mutation. The next population is constructed from the current and offspring populations. The generation of an offspring population and the construction of the next population are iterated until a fixed stopping condition is satisfied.

Each solution in the current population is evaluated in the following manner. First, Rank 1 is assigned to all non-dominated solutions in the current population. All solutions with Rank 1 are tentatively removed from the current population. Next, Rank 2 is assigned to all non-dominated solutions in the reduced current population. All solutions with Rank 2 are tentatively removed from the reduced current population. This procedure is iterated until all solutions are tentatively removed from the current population (i.e., until ranks are assigned to all solutions) (Fig.2.7). As a result, a different rank is assigned to each solution. Solutions with smaller ranks are viewed as being better than those with larger ranks. Among solutions with the same rank, an additional criterion called a crowding measure is taken into account. The crowding measure for a solution calculates the distance between its adjacent solutions with the same rank in the objective space [5]. Less crowded solutions with larger values of the crowding measure are viewed as being better than more crowded solutions with smaller values of the crowding measure. A pair of parent solutions is selected from the current population by binary tournament selection based on the Pareto ranking and the crowding measure.

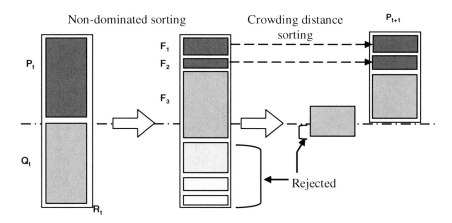

Fig. 2.7 NSGA-II procedure [5]

When the next population is to be constructed, the current and offspring populations are combined into a merged population. Each solution in the merged population is evaluated in the same manner as in the selection phase of parent solutions using the Pareto ranking and the crowding measure. The next population is constructed by choosing a fixed number (i.e., population size) of the best solutions from the merged population. Elitism is implemented in the NSGA-II algorithm in this manner.

2.3.4.1 NSGAII Algorithm

(NSGAII) uses an elite-preserving strategy as well as an explicit diversity preserving mechanism [5]. The offspring population Q_t is first created by using the parent population P_t. Then, the two populations are combined together and a non-dominated sorting is performed (Fig.2.7). To preserve diversity, a density metric called **Crowding Distance** is used. The different steps of the algorithm are described below:

- **Step 1:** A random population is initialized.
- **Step 2:** Objective functions for all objectives and constraint are evaluated.
- **Step 3:** Front ranking of the population is done based on the dominance criteria.
- **Step 4:** Crowding distance is calculated.
- **Step 5:** Selection is performed using crowded binary tournament selection operator.
- **Step 6:** Crossover and mutation operators are applied to generate an offspring population.
- **Step 7:** Parent and offspring populations are combined and a non-dominated sorting is done.
- **Step 8:** The parent population is replaced by the best members of the combined population.

In **Step 3**, each solution is assigned a non-domination rank (a smaller rank to a better non-dominated front). In **Step 4**, for each ith solution of a particular front, density of solutions in its surrounding is estimated by taking average distance of two solutions on its either side along each of the objective [5]. This average distance is called the crowding distance.

Selection is done based on the front rank of an individual and for solutions having same front rank, selection is done on the basis of their crowding distances (larger distance is the better solution).

To create new offspring, simulated binary crossover (**SBX**) operator and polynomial mutation operator are used [11, 12].

In **Step 8**, initially solutions of better fronts replace the parent population. When it is not possible to accommodate all solutions of a particular front, that front is sorted on the basis of crowding distance and as many individuals are selected on the basis of higher crowding distance, which makes the population size of the new population same as the previous population.

Inspiration of using the **SBX** operator arrives from the property of creating off springs in proportion to the distance between two parent solutions. The **SBX** operator biases solutions near each parent more favourably than solutions away from the parents [11].

2.3.4.2 Crowding Distance

This concept is first introduced by Deb et al [5] in their NSGA II algorithm that makes use of the density of solutions around a particular point on the non-dominated front. The density is estimated by calculating the so-called crowding distance of point i, which is the average distance of two point i-1 and $i+1$ on either side of this point i along each of the objectives (Fig.2.8).

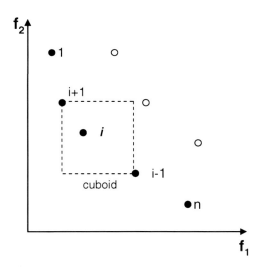

Fig. 2.8 Crowding distance [6]

2.4 Design Optimization

An automatic procedure that combines the developed analytical model with a multi objective optimization routine based on non-dominated Sorting Genetic Algorithm (NSGAII) [5] has been performed for slotless PMSM motors (Fig.2.9). The goal is to find a set of design parameters (Table 2.2), $x=(e_a, g_0, \alpha_0, \zeta, h_{cs}, h_{cr}, e_e)$ in the feasible design space (Table 2.3) that maximizes the average torque $\Gamma(x)$ and minimizes the total mass $M(x)$ while keeping the inequality constraints corresponding to the flux density and the outside radius (equation (29)).

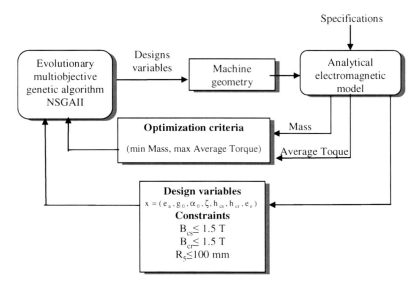

Fig. 2.9 General synoptic of the optimization

2.4.1 Objective Functions

The optimal design of the study structure has been here performed. The adopted Objective Function is a combination of the different goals to be fulfilled. A first term $\Gamma(-)$ takes into account the average torque. It is given by the following relation:

$$\Gamma_{av} = 3\frac{3}{2} L_u \pi J_1 \left[\frac{a_{11}}{5} \left(R_4^5 - R_3^5 \right) - a_{21} \left(R_4^{-1} - R_3^{-1} \right) \right] \tag{23}$$

The second term of the Objective Function takes into account the total mass *(M)* is defined by the total mass of magnets, winding, iron rotor and iron stator, it can be written as following:

$$M(x) = M_{magnets}(x) + M_{winding}(x) + M_{iron\,rotor}(x) + M_{iron\,stator}(x) \tag{24}$$

Where:

$$M_{magnets}(x) = \rho_{magnets} 2pL_u \alpha_0 ((R_1 + e_a)^2 - R_1^2) \tag{25}$$

$$M_{winding}(x) = \rho_{copper} 0.5 N_{con} L_u \zeta ((R_3 + g_0)^2 - R_3^2) \tag{26}$$

$$M_{iron\,rotor}(x) = \rho_{iron} \pi L_u ((R_0 + h_{cr})^2 - R_0^2) \tag{27}$$

$$M_{iron\,stator}(x) = \rho_{iron} \pi L_u ((R_4 + h_{cs})^2 - R_4^2) \tag{28}$$

And N_{con} is the number of conductors.

2.4.2 Design Variables

Design variables are parameters that the designer adjust in order to modify the system he is designing. For the slotless PMSM considered motors, this study is restricted to the design parameters giving in Table 2.2.

Table 2.2 Optimization parameters

e_a	magnet length
g_0	conductors thickness
α_0	magnet pitch
ξ	slot opening/tooth pitch
h_{cr}	Stator iron thickness
h_{cs}	Rotor iron thickness
e_e	Air gap thickness

2.4.3 Inequality Constraints

These are two kinds of inequality constraints witch are defined by:

- The flux density in the iron and the outside radius are less then respectively 1.5 Tesla and 100mm (equation 30).

$$\begin{cases} B_{iron} \leq 100mm \\ R_5 \leq 1.5 Tesla \end{cases} \qquad (29)$$

And

- The space parameters given in the (Table 2.3)

Table 2.3 Space variables of the study slotless PMSM

Parameters	Range	Units
e_a	3,5 … 10	(mm)
g_0	2,5…7,5	(mm)
α_0	25…66,66	(%)
ξ	50…100	(%)
h_{cr}	5…20	(mm)
h_{cs}	5…20	(mm)
e_e	0.2…0.8	(mm)

2.5 Optimization Results

In our problem, the obtained Pareto optimal front is depicted in (Fig. 2.11). From this figure we can observe that the NSGAII converge near to the true Pareto-optimal front as well as maintain the diversity of population on the Pareto-optimal front (Fig.2.10). The optimization is conducted with the multi-objective genetic algorithm (NSGA 2) with the parameters giving in the Table 2.4.

This algorithm can provides a well distributed Pareto front for real coded and the space parameters keeping the inequality constraints defined in the optimization problem.

For sine-wave current supplies, the preliminary and optimal design parameters are given in (Table 2.5). It can be seen that the slotless PMSM mass is strongly reduced and the average torque is increased for the given constraints.

Fig.2.12 and Fig.2.13 show respectively the torque waveforms and flux density corresponding to the preliminary and the final points Pre-design and Opti-design whose parameters are given in Table 2.5. Notice that the preliminary design allows obtaining the optimal average torques 7.11, 10.32 and 12.27 N.m.

Table 2.4 NSGAII Parameters

Population size	150
Number of generations	250
Crossover operator	20
Mutation operator	20

Table 2.5 Parameters of the study Motor before and after optimization

Parameters		Pre-Design	Opti-Design Lim/inf	Middle	Lim/sup
e_a	(mm)	10	7.81	9.8	10
g_0	(mm)	7.4	2.5	2.5	2.5
α_0	(%)	44	30	49.45	60
ξ	(%)	5	3.5	3.65	2.46
h_{cr}	(mm)	14	10.7	11.11	20
h_{cs}	(mm)	15	5	5.48	7.02
e_e	(mm)	0.8	0.4	0.2	0.2
Γ_{av}	(N.m)	7.20	7.11	10.32	12.27
M	(Kg)	6.78	3.17	4.16	6.42

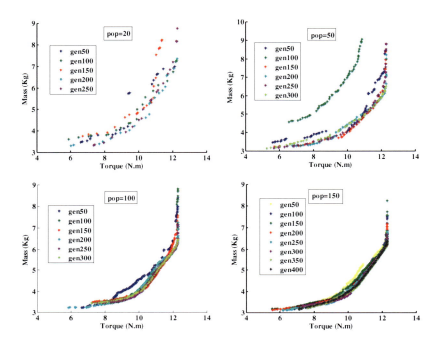

Fig. 2.10 Pareto front for different population sizes and generations

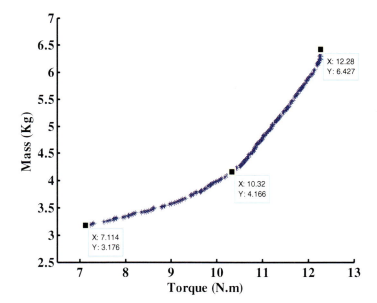

Fig. 2.11 Pareto-optimal front

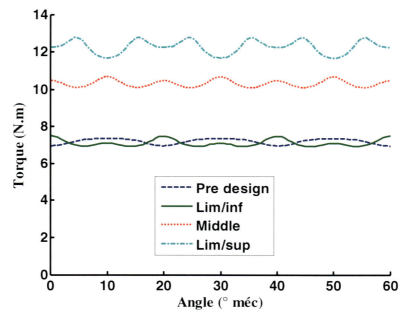

Fig. 2.12 Torque waveforms (Before and after optimization)

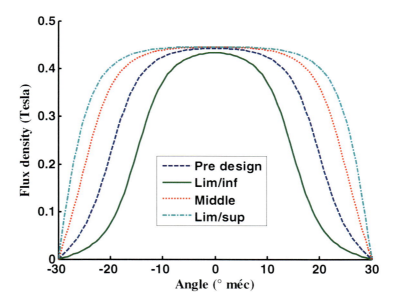

Fig. 2.13 Flux density (Before and after optimization)

2.6 Conclusion

We have developed a new technique of design study which permits to determine rapidly the optimal structure parameters able to match very closely the behavior of the slotless PM motor. The interest of the proposed method is that we can make a previous dimensioning of such a structure.

Exploiting optimization results shows that the overall mass of the machine designs can be reduced by about 40% and the average torque can be increased by about 50%. So, this process constitutes a very efficient design tool if performances and coupling analysis issues are concerned.

A method for solving a vector optimization problem is applied to a problem with two objectives and seven design parameters. The NSGA-II algorithm provides the best optimal machine designs for the given constraints. This algorithm can be extended to solve problems with more than two objectives, a large number of variables and considering other types of constraints.

References

Nogarede. B, Etude de moteurs sans encoches à aimants permanents de forte puissance à basse vitesse. Thèse de Doctorat. Institut National Polytechnique de Toulouse (1990)

Belguerras. L, Optimisation multi objectifs d'une machine à aimants montés sur la surface du rotor et à induit sans encoches. Mémoire de Magister. USTHB Alger (2008)

Regnier. J, Conception de systèmes hétérogènes en Génie Electrique par optimisation évolutionnaire multicritère. Thèse de Doctorat. INP de Toulouse, France (2003)

Coello, C., Carlos, A.: A Short Tutorial on Evolutionary Multiobjective Optimization. In: Zitzler, E., Deb, K., Thiele, L., Coello Coello, C.A., Corne, D.W. (eds.) EMO 2001. LNCS, vol. 1993, pp. 21–40. Springer, Heidelberg (2001)

Deb, K., Agrawal, S., Pratap, A., Meyarivan, T.: A fast and elitist multi-objective genetic algorithm: NSGA-II. IEEE Transactions on Evolutionary Computation 6(2), 182–197 (2002)

Holm, S.R.: Modeling and optimization of a permanent magnet machine in a flywheel. Doctorate thesis. Technical university of Delft (2003)

Coello, C., Carlos, A., Van Veldhuizen, D.A., Lamont, G.B.: Evolutionary Algorithms for Solving Multi-Objective Problems, p. 576. Kluwer Academic Publishers, New York (2002)

Holland, J.H.: Adaptation in natural and artificial systems. PhD thesis, University of Michigan Press (1975)

Goldberg, D.E.: Genetic algorithms for search, optimization, and machine learning. Addison-Wesley, Reading (1989)

Srinivas, N., Kalyanmoy, D.: Multiobjective Optimization Using Non dominated Sorting in Genetic Algorithms. Evolutionary Computation 2(3), 221–248 (1994)

Deb, K., Agrawal, R.B.: Simulated binary crossover for continuous search space. Complex Syst. 9, 115–148 (1995)

Raghuwanshi, M.M., Kakde, O.G.: Survey on multiobjective evolutionary and real coded genetic algorithms. In: Proceedings of the 8th Asia Pacific Symposium on Intelligent and Evolutionasy Systems, pp. 150–161 (2004)

Chapter 3
The FEM Parallel Simulation with Look Up Tables Applied to the Brushless DC Motor Optimization

Jakub Bernat, Jakub Kołota, and Sławomir Stępień

Poznań University of Technology, Poland
jakub.bernat@put.poznan.pl

Abstract. The Finite Element Method (FEM) is widely used to model and simulate electromechanical devices. It enables us to analyze device's properties and improve its performance. Although, its relevance for electric machine design has been proved, its drawbacks are time consuming calculations. Therefore, it is proposed a novel parallel solver to the FEM modeling technique. However, the presented simulation algorithm is limited to magnetic linear models without eddy currents. The FEM parallel simulation technique is applied to simulate brushless DC motor. The analysis of the field distribution and movement characteristics are included. Moreover, the torque ripple problem is described and its influence of motor movement is shown. The brushless DC motor optimization is performed based on the simulated annealing algorithm. The objective function variables are a dimension of the stator teeth and a width of the permanent magnets. The optimization algorithm characteristics are presented and analyzed. Additionally, the reasons for and against of the simulated annealing algorithm as multivariable optimization tool for the FEM models are examined.

3.1 Introduction

The optimization problem of electromechanical devices has recently become the area of vigorous investigation. High performance applications are in demand by the industry [9]. The optimization requires many calculations and is a time consuming task. In particular, nonlinearity of the system increases difficulty of the solving these problems. Many techniques trying to simplify the optimization process have been presented in previous work. Modeling techniques involve approximate functions to predict the shape of objective formula. Surrogate modeling is example of mentioned techniques. It is based on polynomials which describe the function neighborhood [3]. More extended model is used in Kriging modeling, where Gaussian process is introduced. Sensitivity analysis is alternative technique [8]. This approach evaluate change of objective function respectively to change of design parameters. Additionally, the optimization problem can be solved using genetic algorithms or heuristic methods. However, "no free lunch" says that the efficiency of each method is highly dependent on the problem statement. This says that each optimization problem has to be analyzed independently [3,5].

The optimization algorithms described above, perform the objective function for many values of design parameters. It forces us to use modeling techniques of the electromagnetic devices. The well known method is the Finite Element Method, which has been presented in many publications [2,4,6,7,9,13,14].

The parallel techniques has been already applied to the FEM analysis. However, in the paper [15] only the calculation of inductance has been parallized.

In this paper, new contribution is proposed to solving FEM models. To improve the transient simulation of electromechanical devices, parallel system is introduced with look-up tables. The presented techniques are used to optimize brushless DC motor geometry using simulated annealing algorithm [1].

3.2 The Parallel Simulation Technique

The first step of this work is to introduce the parallel computation system into FEM analysis. This analysis will be used to compute the objective function during the optimization process, which is described in next paragraph. The presented method is based on the general equation of electromagnetic the field [2,4,7,12,13], which is given by:

$$\nabla \times (\frac{1}{\mu} \nabla \times \mathbf{A}) = \mathbf{j} + \frac{1}{\mu_0} \nabla \times \mathbf{M} \qquad (1)$$

where $\mathbf{A}$ is the magnetic vector potential, $\mathbf{M}$ is magnetization, $\mathbf{j}$ is current density, μ is the magnetic permeability and μ_0 is the magnetic permeability of the air. The parallel computation method splits the solution of equation (1) to each current source. Nevertheless, the right hand of equation (1) includes current sources (or voltage sources including Kirchoff law) and permanent magnets. However, from the point of view of macroscopic material behavior, the magnets can be replaced by virtual current source. Hence, the equation (1) is rewritten in form:

$$\nabla \times (\frac{1}{\mu} \nabla \times \mathbf{A}) = \sum_{k=1}^{N} \mathbf{j}_k \qquad (2)$$

where k is excitation index and N is number of excitations. In equation (2) each source is represented by independent current density. Additionally, the superposition is applied to split left hand of equation (1). The superposition rule becomes valid with assumption of the magnetostatic field problem without movement:

$$\nabla \times (\frac{1}{\mu} \nabla \times (\sum_{k=1}^{N} \alpha_k \mathbf{A}_k)) = \sum_{k=1}^{N} \alpha_k \nabla \times (\frac{1}{\mu} \nabla \times \mathbf{A}_k) = \sum_{k=1}^{N} \alpha_k \mathbf{j}_k \qquad (3)$$

where $\mathbf{j}_k$ is the current density, $\mathbf{A}_k$ is the magnetic vector potential, α_k is the proportional factor, k is the excitation index and N is the number of excitations. The benefit of superposition rule is not only split of solving equation (1) into k process.

Additionally, the properties of FEM model represented by equations (1)-(3) provide the possibility to introduce pre-calculation step into the magnetic field analysis. It is based on look-up tables with magnetic vector potential **A**. Indeed, the magnetic vector potential is found for each source acting separately and the total magnetic vector potential A is computed using following formula (superposition rule):

$$\mathbf{A} = \sum_{k=1}^{N} \alpha_k \mathbf{A}_k \qquad (4)$$

The magnetic vector potential $\mathbf{A_k}$ is independent from α_k. Thus, it enables us to create look-up table of the magnetic vector potential $\mathbf{A_k}$. The solution of equation (1), which does not consider motion, can be found using equation (4) with $\mathbf{A_k}$ read from look-up table. Thus, it is a very fast solution method of equation (1).

When consider the motion into analysis, then the superposition rule presented in equation (3) is not valid. Nevertheless, the finite set of rotor positions can be defined during analysis for each position separately. Thus, the look-up table consists of different magnetic vector potentials for different position. Additionally, solving equation (1) for each position is parallelized because analysis does not depend on itself. Hence the total magnetic vector potential can be found using following rule:

$$\mathbf{A}(\varphi_i) = \sum_{k=1}^{N} \alpha_k \mathbf{A}_k(\varphi_i) \qquad (5)$$

where α_k is magnitude excitation, φ_i is rotor position, $\mathbf{A}_k(\varphi_i)$ is magnetic vector potential, k is excitation index, i is position index and N is number of excitations. The main benefits of this approach is that the equation (1) is solved only once for each position of rotor and for each current source. The calculations are executed parallel with many computer nodes.

The angular position and speed are found using motion equation.

$$J \frac{d\omega}{dt} + B\omega = T \qquad (6)$$

where J is rotor inertia, B is damping factor, T is rotor torque. The equation (6) is expressed in state space [12]:

$$\frac{d}{dt}\begin{bmatrix} \theta \\ \omega \end{bmatrix} = \begin{bmatrix} 0 & 1 \\ 0 & -\frac{B}{J} \end{bmatrix}\begin{bmatrix} \theta \\ \omega \end{bmatrix} + \begin{bmatrix} 0 \\ \frac{1}{J} \end{bmatrix} T \qquad (7)$$

The electromagnetic torque is calculated using Maxwell Stress Tensor [7, 11]. The references how to manage force calculation with the FEM technique is described in [16,17]. The calculation is based on the magnetic vector potential **A**, which is found in each iteration from equation (5) using look-up table. Employing described approach, the transient simulation is very fast.

3.3 Look-Up Tables with Integral Coefficients

The storage of many matrixes (built from solved potentials **A**) is flexible because of possibility to compute any field property. However, this approach is inefficient. The problem can be suitably solved with the motion and circuit inclusion. Then the field properties could be computed as the term of $\oint_l \mathbf{A}\mathbf{dl}$ related to mutual or self inductance and $\int_V \nabla \mathbf{P} d\mathbf{V}$ (**P** is Maxwell Stress Tensor) which describes the magnetic force. The circuit equation is given by:

$$u_k = R_k i_k + \frac{\partial}{\partial t}\oint_{l_k} \mathbf{A}(\varphi_i)\mathbf{dl} \qquad (8)$$

where u_k is source voltage, R_k is windings resistance, k is source index. The term $\oint_{l_k}\mathbf{A}(\varphi_i)\mathbf{dl}$ describes the magnetic flux in the k-th coil and it depends on the magnetic vector potential **A**, which is read from look-up tables. Hence, it is expressed as the sum of integrals using equation (5).

$$\oint_{l_k}\mathbf{A}(\varphi_i)\mathbf{dl} = \oint_{l_k}\sum_{k=1}^{N}\alpha_k \mathbf{A}_{ik}(\varphi_i)\mathbf{dl} = \sum_{k=1}^{N}\alpha_k \oint_{l_k}\mathbf{A}_{ik}(\varphi_i)\mathbf{dl} \qquad (9)$$

The value of magnetic vector potential $\mathbf{A}_k$ is known in pre-calculation step, so for transient simulation only the value of integrals $\oint_{l_k}\mathbf{A}_{ik}(\varphi_i)\mathbf{dl}$ are stored.

In the motion equation (6) the only part dependent from **A** is torque. It is computed using Maxwell Stress method given by:

$$\mathbf{T} = \int_V r \times \nabla\left(\frac{1}{\mu_0}\mathbf{B}(\mathbf{Bn}) - \frac{1}{2\mu_0}\mathbf{B}^2\mathbf{n}\right)d\mathbf{V} \qquad (10)$$

Due to express formula (10) by integrals, the equation (5) is presented by magnetic induction vector:

$$\mathbf{B}(\varphi_i) = \sum_{k=1}^{N}\alpha_k \mathbf{B}_k(\varphi_i) \qquad (11)$$

Subsisting **B** by components from look-up table, the torque defined as follows:

$$\mathbf{T}(\varphi_i) = \sum_{j=1}^{N}\sum_{k=1}^{N}\alpha_j \alpha_k \int_V r \times \nabla\left(\frac{1}{\mu_0}\mathbf{B}_j(\varphi_i)[\mathbf{B}_k(\varphi_i)\mathbf{n}] - \frac{1}{2\mu_0}[\mathbf{B}_j(\varphi_i)\mathbf{B}_k(\varphi_i)]\mathbf{n}\right)d\mathbf{V} \qquad (12)$$

The integral $\int_V r \times \nabla \left(\frac{1}{\mu_0} \mathbf{B}_j(\varphi_i)[\mathbf{B}_k(\varphi_i)\mathbf{n}] - \frac{1}{2\mu_0}[\mathbf{B}_j(\varphi_i)\mathbf{B}_k(\varphi_i)]\mathbf{n} \right) d\mathbf{V}$ is known in pre-calculation step, so only its value is stored for transient simulation. The main benefit of using equation (9) and equation (12) during simulation is lack of magnetic vector potential **A** in look-up tables.

3.4 The Parallel System Realization

The parallel system is built to create look-up tables with integrals from equation (9) and equation (12). It is implemented using popular .NET technique and remote calls are based on .Remote interfaces. There exists one master node (server), which manages computation. It creates the set of equation (1) to solve. The calculation of each equation is an independent task, so it can be executed by many nodes. The slaves (client) are connected to the master. The free node takes task, which is equation (1) to solve. As the result returns integrals from equation (9) and equation (12). The architecture of parallel system is presented in Fig. 3.1. The number of slaves N_c must be at least one. The maximum number of slaves is equal to number of sources multiplied by number of different positions.

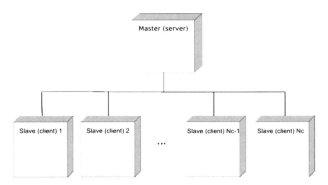

Fig. 3.1 The architecture of parallel system

The look-up tables calculation is split to the master nodes and the slave nodes. Therefore, the both node types runs with different algorithms.

The algorithm steps for master node:

1. Read the FEM model and resolve the number of current sources N_s (real or virtual from permanent magnet) and the number of moving part's positions N_p
2. Create $N_m = N_s N_p$ FEM models (for each current source and for each moving part)
3. Notify slave nodes, that models are ready to calculate
4. Wait for the integral coefficients of all models
5. Save the integral coefficients as look-up table

The algorithm steps for slave node:
1. Notify master node, that this node is ready to use
2. Wait for the FEM model to calculate
3. Calculate the FEM model from master node - find the field distribution and the integral coefficients
4. Return the results to master node
5. Ask for the next FEM model, if received go to step 3.

3.5 Brushless DC Motor Model

Brushless DC motor is modeled and simulated using the described above technique. The motor is three phased with rotor inside. There are six magnetic poles on stator and four poles on rotor. The windings are on the stator tooth. Each phase contains a two teeth set on opposite side of stator. Each tooth is equipped with wider and tighter part. Geometry is presented in Fig. 3.5 and exact dimensions in Table 3.1. The motor is connected to a three phased bridge. The inverter is switched relative to rotor angular position, which can be identified by hallotrons in real application. There exists six different switch modes for three phased BLDC motor. Applying the bridge with angular position feedback, the device has one input and it is voltage amplitude. The presented motor control schema is described in many

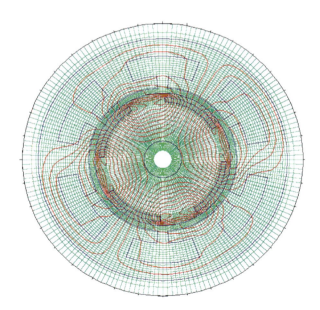

Fig. 3.2 Field distribution in BLDC motor

publications [2]. Due to check reality of calculation, the simulation results are compared with other publications. Fig. 3.3 shows the angular velocity step response. The typical oscillations caused by torque ripples are visible in steady state. The torque is presented in Fig. 3.4.

Table 3.1 Motor parameters

Motor's parameter	Value	Unit
Motor length	90	mm
Resistance/phase	15	Ω
Rotor inertia	1,098e-6	gcm^2
Damping	1e-5	Nms
Outer diameter of stator	27,5	mm
Inter diameter of stator	16,25	mm
Diameter of rotor	15	mm
Inner stator teeth width	20	deg
Outer stator teeth width	52	deg
Permanent magnet width	54	deg

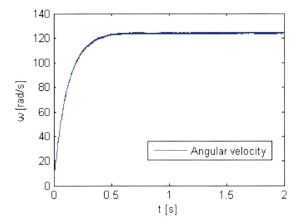

Fig. 3.3 Angular velocity step response

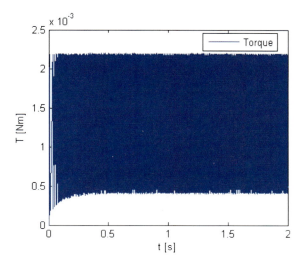

Fig. 3.4 Motor response torque

3.6 The Optimization Problem

The goal of the optimization problem is to find the maximum rotor speed at steady state. The optimization variables are: permanent magnet width ζ, outer stator teeth width λ and inner stator teeth width δ. The geometry constraints have been defined.

$$0° < \zeta \leq 90°, \quad 0° < \lambda \leq 60°, \quad 0° < \delta \leq 52°, \quad \delta \leq \lambda - 8°, \quad \zeta, \lambda, \delta \in \mathbf{Z} \qquad (13)$$

where $\mathbf{Z}$ is the set of integer numbers. The first constraint assures that permanent magnets do not overlap. The second one provide that outer tooth do not overlap. The last constraint gives the space for windings in stator teeth. The domain is discrete, because of using regular mesh in FEM. Fig. 3.5 shows the motor geometry with optimization variables.

The optimization problem is defined by expression as follows:

$$\max_{\xi,\lambda,\delta} \quad \omega_{steady}(\xi,\lambda,\delta) \qquad (14)$$

where ω_{steady} is rotor angular velocity and ζ, λ, δ is optimization variables.

In summary, the objective function has discrete domain with constraints. Moreover, the optimization problem is nonlinear, therefore the algorithm choose should be carefully analyzed.

The continuous algorithms, which are based on the line search or trust region technique, cannot be applied to problems where a model's mesh is assumed to be constant. Therefore, only the methods which are allowed for discrete domain can be applied in this case. Such approach is stochastic or genetic techniques or searching algorithms like Hooke-Jeeves.

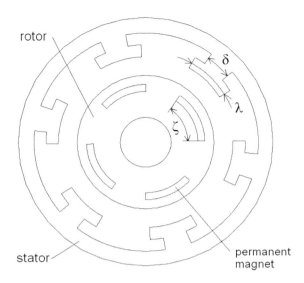

Fig. 3.5 BLDC motor geometry with optimization variables

The objective function nonlinearity must be also considered. Therefore, methods searching global optimum can be only applied in this optimization problem. The global optimization's methods like genetics or stochastic algorithms seems to be appreciate for described problems. The alternative is multiple restart of methods searching local optimum with different initial conditions.

Moreover, the no-free launch theorem says that it doesn't exist any optimization algorithm, which can be applied to any kind of optimization problem and it will give good performance. Thus, the number of steps required to reach optimal can vary for different object designs or different optimization targets.

3.7 The Simulated Annealing Algorithm

The one of the stochastic methods is Simulated Annealing algorithm. It is global optimization method, which could have discrete domain. It doesn't require such computational effort like genetic algorithms, nevertheless is very powerful optimization method. The SA approach is well known in the optimization theory. It enables us to search global optimum, for nonconvex multidimensional problems. It's origin is from annealing process, which is applied to materials to achieve better their endurance. This technique has been transferred to optimization theory, where the annealing process is the decrease of the search region. The algorithm allows to increase of the objective function from time to time, to get the better results at the end of searching. Thus, the local optimum can be omitted and the algorithm will convergence to global optimum. The algorithm steps are as follows (based on the book [10]):

1. Set up the initial temperature T and the initial search point x_0. Assign the initial search point to the current search point x_{cur}
2. Based on the current point x_{cur} find randomly new point x_{new}, which satisfy the problem constraints.
3. Calculate the difference between current point and new point $\delta = f(x_{new}) - f(x_{cur})$. If $\delta < 0$ assigns $x_{cur} = x_{new}$. Otherwise assigns the new point x_{new} to the current point x_{cur} if the random number $0 < R < 1$ is lower or equal to $e^{-\delta T}$.
4. Execute step 2 and 3 unit the limit of new points for the temperature T is reached
5. Decrease the temperature T with some predefined schedule.
6. If $T < T_{min}$ then choose the lowest current point as the optimal point

The temperature decrease (the 5th step in the algorithm) can vary along with different implementations. The popular approach is to use the geometric progression, for instance $T = \alpha T$ with $0 < \alpha < 1$. The another approach is to change T with proportional rate to $\frac{1}{\log k}$ where k is algorithm's iteration index.

The another problem is to choose properly the next point x_{new}. It must be the perturbation of the current point x_{new}, which is significant for higher temperatures T and nonsignificant for lower temperatures.

3.8 The Rosenbrock Valley Problem

The performance of the Simulated Annealing algorithm is verified based on the Rosenbrock function [18]. The purpose of such test is to check implementation and show the algorithm performance. The Rosenbrock function is nonlinear with many local optimums. Moreover, the optimization algorithms usually convergences to global optimum very slow. The Rosenbrock problem is defined as follows:

$$\min_{\mathbf{x} \in R^n} f(\mathbf{x}) \tag{15}$$

where f(**x**) is the Rosenbrock function given by:

$$f(\mathbf{x}) = \sum_{i=1}^{n-1}\left[100\left(x_{i+1} - x_i^2\right)^2 + \left(1 - x_i\right)^2 \right] \tag{16}$$

The problem is analyzed with constraints:

$$-2.048 \le x_i \le 2.048 \quad for \quad i = 1\ldots n \tag{17}$$

The global optimum is in the point, which can be found analytically:

$$x_i = 1 \quad for \quad i = 1...n \tag{18}$$

The shape of the functions is similar to banana, therefore the Rosenbrock problem is called banana problem from time to time. The 2 dimensional case, when n = 2, is presented in Fig. 3.6.

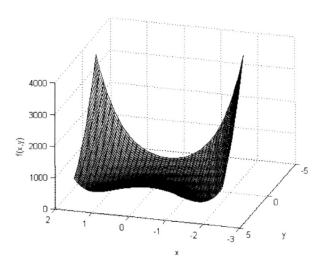

Fig. 3.6 The Rosenbrock function for n = 2

The Simulated Annealing algorithm has been applied to solve the optimization problem of the Rosenbrock problem for n = 2. The global optimum is $x_1 = 1$ and $x_2 = 1$. The algorithm has been run with the cooling function defined as:

$$T(k) = T_{start} e^{-\alpha k} \tag{19}$$

where k is iteration index, T_{start} is the initial temperature and α is the constant defined for the problem. The minimum temperature during optimization is T_{min} = 0.05 and the number of trails for each temperature is 20. The optimal point found by the algorithm is x_{1opt} = 1.003 and x_{2opt} = 1.0057, which is very close to global optimum point. The total number of trials is 1061, which gives 1060 trails for algorithm and 1 for initial point. The performance of the algorithm is presented in Fig. 3.7. The Fig. 3.8 shows the decay of the cooling temperature. It is worth to notice that the cooling temperature is at the beginning very high and very fast becomes low. It means, that at the beginning large scope of points is search and afterwards the algorithm search close to optimum. The decay ratio should be set with respect to the shape of the objective function, because it has significant effect on its performance. Unfortunately, either the shape of the optimization function or the approximate shape are not know, therefore the T_{start} and α have to be set based on the designer experience.

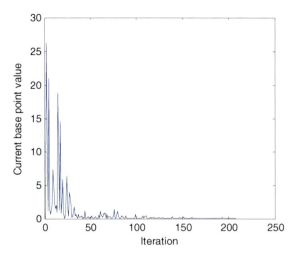

Fig. 3.7 Progress of objective function

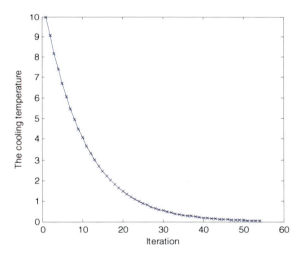

Fig. 3.8 The cooling temperature decay

3.9 The Rastragin's Problem

The Rastragin function is nonlinear with many local optimums, therefore is very difficult to find the optimum point. The Rastragin's function has been presented in [18]. The optimization problem based on that function is defined as follows:

$$\min_{\mathbf{x} \in R^n} f(\mathbf{x}) \qquad (20)$$

where f(**x**) is the Rastragin's function given by:

$$f(\mathbf{x}) = 10n + \sum_{i=1}^{n}\left[x_i^2 - 10\cos(2\pi x_i)\right] \tag{21}$$

and the objective function constraints are:

$$-5.12 < x_i < 5.12 \quad for \quad i = 1\ldots n \tag{22}$$

The Rastragin's function consists of the cosine, which produces many local optimum points. Nevertheless, there is global optimum at $x_i = 0$ for $i = 1\ldots n$. The shape of the function is presented in Fig. 3.9.

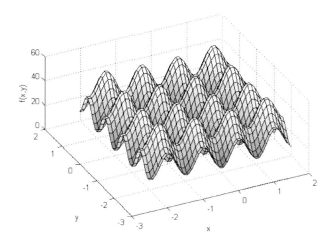

Fig. 3.9 Rastragin function for n =2

The Simulated Annealing algorithm has been applied to find an optimum point for the Rastragin's problem. The initial conditions are the same as in the previous problem. The algorithm has found the optimal point at $x_{1opt} = 0.0012$ and $x_{2opt} = 0.007$. The number of iterations in both cases is the same.

It should be noted that in the Rosenbrock's problem the objective function values decreases rapidly and after few iterations the decrease is slow. In the Rastragin's problem the function decreases slowly at the beginning and it has sharp slope in the medium. The difference is caused by the shape of both functions. The first one has wide valey, which optimization algorithm founds very easily. However, the convergence to optimal point is slow, because the valley is flat. The second one has many local optimums spread across the optimization domain, which causes that optimization points jumps from one point to another.

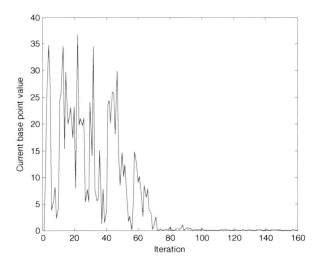

Fig. 3.10 The objective function values for the current point

3.10 Simulation Experiment Results

The Simulated Annealing algorithm has been applied to optimization of the brushless DC motor. The problem domain is 3 dimensional, which is defined as follows:

$x = \begin{bmatrix} \zeta & \lambda & \delta \end{bmatrix}^T$, where $x \in \mathbf{Z}^3$ and fulfill constraints from equation (3)

The starting point is $x = \begin{bmatrix} \zeta & \lambda & \delta \end{bmatrix}^T$ and the initial temperature is $T_0 = 20$. The optimization process is executed for 100 iterations. The same number of different geometries are analyzed as the number of iterations, because in one algorithm step one geometry is considered. The geometry change requires applying simulation of the brushless DC motor with distributed parameters. Therefore, the FEM simulation technique presented in this chapter has been applied to the optimization process. When the calculation of the objective function is required, the FEM analysis is used to simulate the brushless DC motor. The presented parallel approach to FEM analysis requires for each geometry look-up tables, which must be created whenever the motor geometry parameters have been changed. Afterwards the transient response is calculated using look-up tables. Fig. 3.11 presents the progress of the objective function. It can be noticed that the current value of the objective function can decrease from time to time. Exactly, for the initial conditions steady state speed is above $100\frac{rad}{s}$ and it decreases to around $50\frac{rad}{s}$ in 40 iteration. However, the increase of the objective function is significant after 40 iterations and the current point converge to optimal point. The best solution is for following parameters: $\zeta = 88°, \lambda = 48°, \delta = 36°$ with the rotor speed $\omega_{steady} = 470\frac{rad}{s}$ is obtained for the optimal values.

The FEM Parallel Simulation with Look Up Tables 53

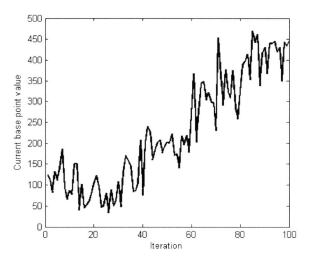

Fig. 3.11 The objective function progress

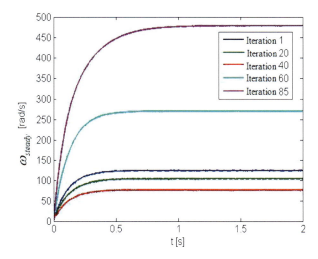

Fig. 3.12 The transient responses during optimization

The parameters variety is very significant at the beginning of the optimization. The width of teeth either inner or outer changes very rapidly and even it convergences to global solution very slowly. The permanent magnet width doesn't have so rapid changes. Hence, it is easier to notice, that the permanent magnet should have rather large size than the small. It is obvious, because the wider magnet gives more energy to motor, which acts with stator windings. The change of parameters is shown in Fig. 3.13, 3.14 and 3.15.

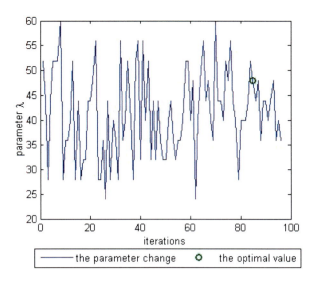

Fig. 3.13 The change of outer stator teeth width during optimization

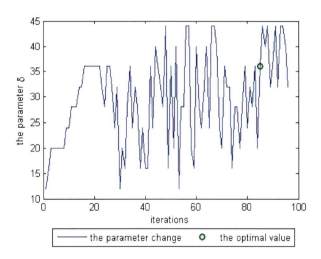

Fig. 3.14 The change of inner stator teeth width during optimization

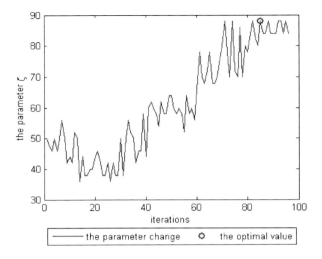

Fig. 3.15 The change of permanent magnet width during optimization

3.11 The Brushless DC Motor Torque Analysis

The optimization process changes a device's design to achieve the better performance. On the other hand the design changes influence on the other system parameters. Therefore, it is worth to ask whether the optimization process doesn't decrease the other object's properties. The optimization variables in the presented problem has influence on the motor inertia, the shape of the back electromotive force, the torque acting between a rotor and a stator, the winding's resistance and the self inductance. The analysis has shown that the electromagnetic torque has the most significant effect on the motor performance. The reason of a such behavior is straightforward related with optimization problem. The steady angular velocity is proportional to the torque between the rotor and the stator, hence the increase of the angular velocity causes the increase of the torque. Therefore, the analysis of the torque ripples and the cogging torque is required to check whether this undesired effect doesn't raise.

The torque ripple and the cogging torque causes that the brushless DC motor torque it doesn't become constant and it contains the higher harmonic components. The described effect can be measured by pulsation factor, which according to [20, 21] can be defined as follows:

$$\varepsilon = \frac{T_{max} - T_{min}}{T_{avg}} 100\% \qquad (23)$$

The electromagnetic torque in the steady state has been shown for the each iteration of the optimization process in Fig. 3.16. Additionally, the pulsation factor is presented in Fig. 3.17. Unfortunately, the increase of the electromagnetic torque causes the increase of the pulsation factor, which is undesirable effect because it produces the higher harmonics in the rotor angular velocity.

The presented optimization problem has as the objective function the rotor angular velocity at the steady state. Therefore, the optimization process doesn't consider another parameters. However, if the multiobjective optimization was analyzed, the compromise between the pulsation factor and the angular velocity at the steady state have to be defined. The range of variables where this problem occurs is called as Pareto set [3].

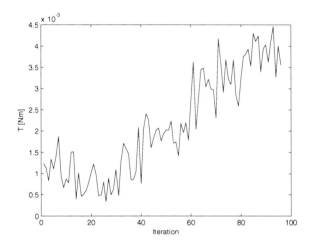

Fig. 3.16 The electromagnetic torque acting on the rotor during optimization process

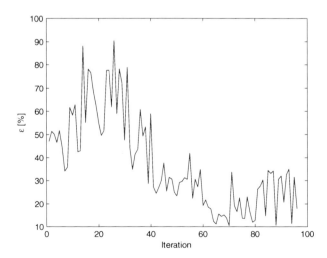

Fig. 3.17 The pulsation factor during optimization process

3.12 Conclusions

In this chapter, a novel approach is presented to electromagnetic devices simulation. It is based on parallel computations, which are lead out from superposition rule. Due to decreasing computers cost, proposed method enables us for solving FEM problem efficiently. The benefit of the presented simulation technique is very efficient calculation of device transient response. Therefore it can be used to solve optimization problems or to develop control systems. The method has been applied to optimization of BLDC motor. It is noticeable that the teeth and permanent magnet dimension has significant influence on angular motor speed. In future the parallel technique can be extended for different types of nonlinearity. The eddy current problem or B-H curve might be considered in these problem for better accuracy.

Moreover, the Simulated Annealing algorithm has been presented and applied to solve the nonlinear optimization problem. The Rastragin and Rosenbrock function has been used as the objective function.

References

[1] Kirkpatrick, S., Gelatt, C.D., Vecchi, M.P.: Optimization by simulated annealing. Science 220, 671–680 (1983)
[2] Jang, G., Chang, J., Hong, D., Kim, K.: Finite element analysis of an electromechanical field of a BLDC motor considering speed control and mechanical flexibility. IEEE Trans. Mag. 38, 945–948 (2002)
[3] Sykulski, J.K.: New trends in optimization in electromagnetics. In: The IET 7th International Conference on Computation in Electromagnetics, CEM 2008, pp. 50–51 (2008)
[4] Hawe, G.I., Sykulski, J.K.: A hybrid one-then-two stage algorithm for computationally expensive electromagnetic design optimization. In: COMPEL, vol. 26, pp. 240–250 (2007)
[5] Hawe, G.I., Sykulski, J.K.: Probality of improvement methods for constrained multi-objective optimization. In: The IET 7th International Conference on Computation in Electromagnetics, CEM 2008, pp. 44–49 (2008)
[6] Di Barba, P., Dughiero, F., Savini, A.: Multiobjective Shape Design of an Inductor for Transverse-Flux Heating of Metal Strips. IEEE Trans. Mag. 39, 1519–1522 (2003)
[7] Tärnhuvud, T., Reichert, K.: Accuracy problems of force and torque calculation in FE-systems. IEEE Trans. Mag. 24, 443–447 (1988)
[8] Dong-Hun, K., Lowther, D.A., Sykulski, J.K.: Efficient Global and Local Force Calculations Based on Continuum Sensitivity Analysis 43, 1177–1180 (2007)
[9] Ponick, B.: Miniaturisation of Electrical Machines. International Symposium on Theoretical Electrical Engineering (2009)
[10] Spall, J.C.: Introduction to Stochastic Search and Optimization. Wiley Interscience, Hoboken (2003)
[11] Sadowski, N., Lefevre, Y., Lajoie-Mazenc, M., Cros, J.: Finite Element torque calculation in electrical machines while considering the movement. IEEE Trans. on Magn. 28, 1410–1413 (1992)

[12] Stepien, S., Patecki, A.: Modeling and position control of voltage forced electromechanical actuator. COMPEL - International Journal for Computation and Mathematics in Electrical and Electronic Engineering 25, 412–426 (2006)
[13] Nowak, L., Mikolajewicz, J.: Field-circuit model of the dynamics of electromechanical device supplied by electronic power converters. COMPEL - International Journal for Computation and Mathematics in Electrical and Electronic Engineering 23, 1531–1534 (2004)
[14] Demenko, A., Stachowiak, D.: Orthogonal transformation of moving grid model into fixed grid model in the finite element analysis of induction machines. COMPEL - International Journal for Computation and Mathematics in Electrical and Electronic Engineering 23, 1015–1022 (2004)
[15] Kolota, J., Smykowski, J., Stepien, S., Szymanski, G.: Parallel computations of multiphase electromagnetic systems. Electrical Review 11, 29–32 (2007)
[16] Henrotte, F.: Handbook for the computation of electromagnetic forces in a continuous medium. International Compumag Society - Technical article (2004)
[17] Bossavit, A.: Forces inside a magnet, International Compumag Society – Technical article (2004)
[18] Molga, M., Smutnicki, C.: Test functions for optimization needs (2005), http://www.zsd.ict.pwr.wroc.pl/files/docs/functions.pdf
[19] Hanselman, D.C.: Minimum Torque Ripple, Maximum Efficiency Excitation of Brushless Permanent Magnet Motors. IEEE Trans. on Industrial Electronics 41, 292–301 (1994)
[20] Knypiński, Ł., Nowak, L.: Analysis and design the outer-rotor permanent magnet brushless DC motor. In: XX symposium Electromagnetic Phenomena in Nonlinear Circuits, pp. 21–22 (2008)
[21] Stachowiak, D.: Edge element analysis of brushless motors with inhomogeneously magnetized permanent magnets. COMPEL - International Journal for Computation and Mathematics in Electrical and Electronic Engineering 23(4), 1119–1128 (2004)

Chapter 4
Fast Algorithms for the Design of Complex-Shape Devices in Electromechanics

Eugenio Costamagna[1], Paolo Di Barba[2], Maria Evelina Mognaschi[2], and Antonio Savini[2]

[1] Department of Electronics, University of Pavia
via Ferrata 1, I-27100 Pavia, Italy
`eugenio.costamagna@unipv.it`
[2] Department of Electrical Engineering, University of Pavia
via Ferrata 1, I-27100 Pavia, Italy
`{paolo.dibarba,eve.mognaschi,savini}@unipv.it`

Abstract. The analysis of complex-shape electromechanical devices is considered. The use of numerical Schwarz-Christoffel (SC) mapping, coordinated with finite element (FE) analysis, is proposed for fast computation of 2D fields.

4.1 Introduction

Conformal-mapping procedures for field analysis, based on the well-known SC transformation [1], are a valuable tool for the simulation of two-dimensional Laplace's equations in homogeneous media to be compared with the FE method. In fact, complicated shapes can be handled in a straightforward way and no meshing algorithms are needed. Moreover, open boundary and field singularity problems can be easily considered, because the essence of SC mapping is to account for corners causing singularities which are numerically or physically undesirable. The interest towards the application of SC conformal mapping for analytical and numerical solution of two-dimensional electric and magnetic field has grown in the 1990s [2, 3]. More recently, numerical SC techniques have been extensively and successfully applied to investigate complex slotted domains in magnetic motors and actuators as well as in electrostatic motors [4, 5]. A new field of research is represented by the association of the SC field modeling to the design process of electric and magnetic devices [6].

4.2 Reducing Three Dimensions to Two

Real life is 3D and 3D CAD today is possible. There certainly exist practical problems which can only be solved in 3D; when modelling e.g. a rotating electrical machine, end effects are mainly due to fringing field, like e.g. end-winding field.

End effects dominate in an axially short machine, which could eventually call for 3D modelling. A rather extreme example is that of a permanent-magnet machine, in which the axial lengths of magnet pole, stator yoke, and rotor armature are unequal. However, end effects are relatively unimportant in axially-long machines, where a 2D transverse cross-section could be appropriately used for modelling.

In a matter of fact, very many analysis problems are based on 2D approximations. When applicable, this remark could determine a remarkable save of memory and runtime, which is especially welcome when procedures of automated optimal design are used. In other words, reducing three dimensions to two is tolerably viable for design purposes, and should be always recommended in an optimisation loop. For instance, a truly three-dimensional model could be approximated by means of orthogonal projections on suitable planes; in this case, 2D simulations in orthogonal sections are easily obtained. Then, simulation results can be recomposed, giving an acceptable accuracy at a definitely low cost. In practice, in an optimisation procedure with 2D-based models, a 3D simulation can be performed just at the end, to assess the final result and incorporate corrections.

The speed of processors has been increasing rapidly year by year; according to a prediction by computer suppliers, the power of computing – commonly expressed in terms of millions of instructions per second – double every year. As a matter of fact, using conventional technology like a 64-bit processor, commercial FE solvers in 2D take tens of seconds through few minutes of CPU time, even in the extreme case of non-linear or transient analyses with mesh adaptation, which perfectly meets optimisation procedure requirements.

A particular case is that of axisymmetric geometries, which can be often reduced to 2D domains (see later on).

4.3 The Need for Fast Solvers: Schwarz-Christoffel Mapping as a Possible Answer

Regretfully, the aforementioned remarks cannot be applied to all classes of problems in computational electromagnetism. In fact, even though solution times are smaller and smaller, they rise rapidly for certain classes of problems, like e.g. repeated field analyses, coupled field problems, multiobjective optimisation procedures. In the latter case, the most general solution is given by the set of nondominated solutions, i.e. those for which the decrease of a design objective is not possible without the simultaneous increase of at least one of the other objectives. Solution times can then grow to many hours or even days, and computational methods fail to be used as regular tools for analysis and design.

When dealing with vector – or multiobjective – optimisation problems in electromagnetic design, the evaluation of each objective requires at least a field simulation, in general based on FE method. Field simulation, in turn, might have an inherent complexity due to various reasons: complicated shape of the device implying two- or three-dimensional models, coupled-field analysis, non-linear material properties, transient conditions calling for time-stepping analysis. Even using powerful computing facilities, typical runtimes required by the direct problem limit severely e.g. the use of evolutionary algorithms of design optimisation.

To clarify this point, let us imagine a case in which the main output of nf objectives implies the FE simulation of the torque-angle curve in a rotating electrical machine. In the case of e.g. a twelve-pole machine, taking a rotation step of 1 deg and considering half-pitch symmetry, $\frac{1}{2} \times \frac{360}{12} = 15$ FE analyses are necessary to find out the torque-angle curve, for a given geometry of the device. If, additionally, the non-linear characteristic of ferromagnetic material is taken into account and 10 Newton iterations are involved, the number of FE analyses grows up to 150 per geometry. If the optimisation of the geometry relies on a standard evolutionary algorithm with a population of 15 individuals, and 20 generations are required to converge, the total number of field analyses becomes 45,000. In the favourable case of 10 s per FE analysis, the resulting runtime is 450,000 s (more than 5 days) for finding 15 non-dominated solutions (one per individual), which is unpractical for industrial design.

In general, the computational cost c [s] of a basic evolutionary algorithm can be estimated as $c \approx c_0 \cdot n_i \cdot n_p \cdot n_f$, where c_0 [s] is the processor-dependent time necessary to run a single FE analysis, n_i is the number of convergence iterations for a prescribed search accuracy, n_p is the number of individuals in the evolving population, and n_f is the number of objectives. The very point is that most evolutionary methods have been developed for solving problems in which the computational cost due to objective function computation is low or moderate.

Reducing 3D to 2D sometimes is not enough to speed up numerical procedures of field analysis in an acceptable way; so, designing fast solvers is necessary, in order to find new solutions to challenging problems, like those amenable to vector optimisation problems.

A possible answer might relies on modern algorithms devised for the numerical inversion of the Schwarz-Christoffel formula, a subject often overlooked by most of those working in the broad area of electricity and magnetism.

4.4 Modern Numerical Methods for SC Mapping

For a long time the SC conformal transformation [1, 7, 8, 9] has been a very accurate tool used to perform the analysis of real-life devices exhibiting complicated geometries, and to obtain reliable reference data to compare calculations carried out by means of other methods. Complex scalar or vector field problems were easily described thanks to the capability of working out difficulties arising from singularities and from open boundaries. Moreover, SC transformations offer a very general tool, because, quoting the introduction of [1], "virtually all conformal transformations whose analytic forms are known are SC maps, albeit sometimes distinguished by an additional change of variables". So, SC mapping functions keep the promise of the Riemann theorem, that (quoting again [1]) "any simply connected region in the complex plane can be conformally mapped into any other, provided that neither is the entire plane": see also [9], and paragraph 5.12 of [10].

So, many engineering problems were solved, whenever analytical solution for the related SC mapping problem were found, often in terms of elliptic integrals.

The most used SC formula (see, for instance, [1], page 10) maps the real axis of the complex plane (that we call SC plane) into a polygon, and the upper half-plane onto the interior of it. The images of the polygon vertices in the SC plane are called prevertices. Similar formulae provides maps from a disc, or maps of the region exterior to a bounded polygon, thus providing a flexible approach to conformal mapping problems.

When the prevertices are known, only some accuracy in computing the integrals in the SC formula is needed, and normally computation starting from the SC plane considered geometries where the ratios between consecutive sides (i.e., distances between prevertices) were not so large as to cause problems. The most important cause of integration inaccuracy were the singularities at the prevertices, ends of the sides in the SC plane, and Gaussian integration provided a satisfactory solution: probably, its first use can be traced back to the early seventies of last century, with [11] (Gauss-Jacobi quadrature) and [12] (Gauss-Chebysheff rules).

Starting from the polygon geometry, no general analytical rule is known to solve the so-called "parameter problem": the positions of the prevertices on the real axis have to be found by solving non-linear equations, normally via optimisation procedures. When digital computers provided reasonable computing speeds, procedures have been implemented to perform the numerical inversion of the SC formula (SCNI), starting from the polygonal geometry, and the "crowding" problem emerged: some distances between prevertices collapse in unpredictable way, preventing accurate calculations. In particular, a singularity at the far end of a very short contiguous side prevents the accurate integration over a relatively long side along the real axis of the SC plane.

Discussions and solutions to this problem are reported in [9] and [1]. The most effective solution was [9] the implementation of "compound" Gauss-Jacobi quadrature rules: the integration interval was subdivided in accordance to a "one-half rule": no singularity (at the far end of a contiguous side) shall lie closer to the interval of integration than half the length of the interval itself. Something similar was independently proposed in [13], with a different approach to perform the subdivision of the integration interval, in [14] and in [15]; in [15] and in [13] attention was extended to unbounded domains too. In fact, compound rules allowed us to tackle the problem of elongated regions, and to consider SC formulas and SCNI effective numerical tools, with wide spread application, as effective optimisation tools were discussed in the above papers to solve the parameter problem.

It is worth of mention an analytic procedure presented in [16], allowing us to put an end to the subdivision of the interval in a particular but very important case. When, at the end of a long integration interval, a singular factor of the type $x^{-1/2}$ appears in the SC mapping function, and a singular factor of the same type is occurring because an analogous prevertex is found at the opposite end of the contiguous side, we can account for both of them by means of an analytical formula [17, formula (195.1)], provided that other singularities in the SC mapping function are lying sufficiently far from the integration interval, and the related factors are nearly constants. So, this leads to relatively large lengths of the ending subdivision of the interval, yet allowing us to handle length ratios between consecutive sides in the SC plane exceeding 10^{20} or even more, using double precision programming. This is important, as the quoted singularities represent vertices similar to the vertices of a rectangle, and they can appear in close proximity when elongated regions are considered. An even more general formula was considered [17, (371)], to handle different exponents in the singular factors, but this leads to compute differences between similar terms and to some loss of accuracy.

When crowding effects are moderate, any numerical quadrature rule providing high density of sampling points near the end of the integration interval can be successful, in particular, the Takahasi-Mori [18] and the so called tanh transformation [19, 20]. Comparisons of results have been presented in [16, 21], with reference to elongated L-shaped regions. A phenomenon was observed in [21], which often mitigates the consequences of integration errors. It is customary to map elongated regions into the SC half-plane via SCNI, and then into rectangles via direct calculation of the SC formula. Only the polygon angles at the prevertices are changed, the sides between them in the SC plane do not change, and the errors due to crowding are similar in the inverse and direct calculation, despite imperfect location of the prevertices.

Meanwhile, SC procedures were extended to doubly connected regions, see [22] and paragraph 4.9 of [1]. This was the SC panorama when a new progress was made. A comprehensive presentation of the methods and applications related to conformal mapping and to SC formulas at that time can be found in [1], including curved boundaries, in [23] and, with an application in transmission line and capacitance calculation, in [16], references, and [21].

The progress is related to the introduction of the cross-ratio parameters as well as Delaunay triangulation in the numerical procedures, causing new capabilities both in transforming elongated regions and in observing the local behaviour of the maps. We refer for this to [24] and again to [1]: the above procedures have been included in the SC toolbox for MATLAB, a very powerful conformal mapping tool both from the analytical point of view, and from its capabilities of friendly use.

Coming back to the standard SC formulas, a powerful methods to circumvent severe crowding problems was presented in [25], where a formula is derived to map elongated regions to strips and then to elongated rectangles. The same idea of matching the features of the prevertices plane to the elongated regions was resumed in [26] following a different approach, by means of a modified mapping formula using the perimeter of a rectangle in the same way the standard one uses the real axis of the SC plane. A right-angle corner singularity is completely removed, if the corner is mapped into a rectangle vertex. Accurate calculations of elliptic functions are needed to perform ancillary computations, but this is not a problem, and good results were supplied for the L-shaped region, in comparison with other integration procedures.

It can be useful to extend similar comparisons assuming, as a reference, cross-ratio computations performed by the *crrectmap* procedure of the SC Toolbox [27].

In fact, even if in this particular case it is possible that this procedure is not providing the best results, differences from these are unimportant, and it is likely that only *crrectmap* will allow us to obtain similar accuracy in general conditions, in particular with regions elongated in various directions. We map the L-shaped region into a rectangular domain, and compare the moduli of the obtained rectangles. The L-shaped region is an importante example, because it represent the quarter of a square coaxial line cross section. Data are collected from the *rectmap*, *hplmap*, and *diskmap* procedures of the same SC Toolbox, from two-step procedures (referred as *Jac+Jac*, *HP+Jac*, *Jac+DCEL1*, *HP+DCEL1*, respectively), from the procedure in [26] using - when useful - the Takahasi-Mori transformation and the tanh transformation procedure with 48 nodes (*SCRETT-TM* and *SCRETT-th*, respectively). We consider also data collected from the analytical procedure of Bowman in [28], from the numerical procedure presented by Papamichael and Kokkinos in [29], and from a doubly-connected region obtained after the symmetrical connection of four L-shaped regions, and analysed following Hu [22]. Results are shown as percent errors in Table 4.1 and Table 4.2.

The aforementioned *Jac* and *HP* procedures use 11 segments compound Gauss-Jacobi integration with 48 nodes, being in the last the ending segment computed by the analytical formula introduced in [16] when "half pole" vertices are found both at the ends of a long side and at the far end of a contiguous short side: this is precisely the condition at the ends of the L-shaped region.

Observing the tables, we note, first of all, the good agreement of the SC *crrectmap* reference with the Bowman and Papamichael and Kokkinos values, then the similar or better agreement with the reference of the SCNI procedures based on the traditional formula when compound integration is performed. *Jac* routines are useful for a very notable range, until *HP* routines can be used. In turn, these are useful when the length 1+a, measuring in fact the ratio between the longest and the shortest sides of the L shaped region, is so large to secure a suitable distance of any other vertices from the end subinterval of the side. It is likely that the best values for very large 1+a values are supplied by *HP* routines, because we observe a very linear behaviour of the moduli in the last part of Table 4.1, where the assumed reference leads to more erratic values. Values for the tandem procedures *Jac+DCEL1* and *HP+DCEL1*, compared to those from the *Jac+Jac* and *HP+DCEL1* ones, show some masking effect for the *Jac+Jac*, being the computation via DCEL1 virtually exact, suitable to detect the errors caused by the *Jac* routines.

Finally, some capabilities in axisymmetric 3D problems are worth of attention. They have been explored in [23], chapter 6, and used in connection to simple finite difference procedures to avoid singularities of the analyzed field, making possible accurate solutions of the suitable Laplace equations (see paragraph 5.3 of [9] and paragraph 5.11 of [30]) using coarse meshes. Some examples can be found in [31, 32], and in references therein.

Fast Algorithms for the Design of Complex-Shape Devices in Electromechanics 65

Table 4.1 Percent errors of values obtained by means of various compound techniques with respect to those obtained by means of *crrectmap*.

1+a	Rectmap	Hplmap	Diskmap	Jac+Jac	HP+HP	HP+Jac	Jac+DCEL1	HP+DCEL1
1.25	$4 \cdot 10^{-7}$	$7 \cdot 10^{-7}$	$8 \cdot 10^{-7}$	$6 \cdot 10^{-7}$			$6 \cdot 10^{-7}$	
2	$3 \cdot 10^{-7}$	$7 \cdot 10^{-7}$	$7 \cdot 10^{-7}$	$7 \cdot 10^{-7}$			$7 \cdot 10^{-7}$	
5	$-1 \cdot 10^{-6}$	$-1 \cdot 10^{-6}$	$8 \cdot 10^{-3}$	$-1 \cdot 10^{-6}$			$-1 \cdot 10^{-6}$	
7	$-2 \cdot 10^{-7}$	$-1 \cdot 10^{-7}$		$-2 \cdot 10^{-7}$			$-2 \cdot 10^{-7}$	
8	$4 \cdot 10^{-8}$	$3 \cdot 10^{-6}$		$3 \cdot 10^{-7}$			$3 \cdot 10^{-7}$	
9	$-2 \cdot 10^{-6}$	$-3 \cdot 10^{-5}$		$3 \cdot 10^{-7}$			$3 \cdot 10^{-7}$	
11	$-5 \cdot 10^{-4}$			$1 \cdot 10^{-7}$	$1 \cdot 10^{-7}$	$-1 \cdot 10^{-5}$	$1 \cdot 10^{-5}$	$1 \cdot 10^{-7}$
13	-0.4			$-3 \cdot 10^{-6}$	$-3 \cdot 10^{-6}$	$-6 \cdot 10^{-3}$	$6 \cdot 10^{-3}$	$-3 \cdot 10^{-6}$
15				$-4 \cdot 10^{-6}$	$-3 \cdot 10^{-6}$	-0.24	0.25	$-4 \cdot 10^{-6}$
21					$-1 \cdot 10^{-6}$			$-1 \cdot 10^{-6}$
41					$-5 \cdot 10^{-7}$			$-5 \cdot 10^{-7}$
61					$-2 \cdot 10^{-6}$			$-2 \cdot 10^{-6}$
81					$-8 \cdot 10^{-7}$			$-8 \cdot 10^{-7}$
101					$-2 \cdot 10^{-6}$			$-2 \cdot 10^{-6}$
201					$-1 \cdot 10^{-6}$			$-1 \cdot 10^{-6}$

Table 4.2 Percent errors of values obtained by means of various non-compound techniques with respect to those obtained by means of *crrectmap*.

1+a	SCRETT-TM	SCRETT-th	Hu	Bowman	Papamichael
1.25	$3 \cdot 10^{-5}$	$-2 \cdot 10^{-6}$	$4 \cdot 10^{-6}$	$6 \cdot 10^{-7}$	$6 \cdot 10^{-7}$
2	$2 \cdot 10^{-6}$	$-3 \cdot 10^{-6}$	$2 \cdot 10^{-6}$	$7 \cdot 10^{-7}$	$7 \cdot 10^{-7}$
5	$-1 \cdot 10^{-6}$	$-3 \cdot 10^{-6}$	$3 \cdot 10^{-6}$	$-1 \cdot 10^{-6}$	$-1 \cdot 10^{-6}$
7	$-2 \cdot 10^{-7}$	$-2 \cdot 10^{-6}$	$3 \cdot 10^{-6}$		
8	$3 \cdot 10^{-7}$	$-1 \cdot 10^{-6}$	$3 \cdot 10^{-6}$		
9	$3 \cdot 10^{-7}$	$-1 \cdot 10^{-6}$	$3 \cdot 10^{-6}$		
11	$1 \cdot 10^{-6}$	$-2 \cdot 10^{-6}$	$1 \cdot 10^{-6}$		
13	$-4 \cdot 10^{-6}$	$-5 \cdot 10^{-6}$	-16.3		

4.5 Towards a Coordinated Approach for Field Analysis, Using Both FE and SC

A coordinated FE-SC approach proved to work in an effective way, offering both accuracy and speed in various computations implemented. A collection of solved problems is shortly reviewed.

- The optimal shape design of a permanent-magnet motor was performed directly in the transformed plane, by controlling the prevertex locations

[4]. The design criterion was the cogging torque reduction in a small-sized motor. Among other benefits of the SC-FE approach, it is possible to note that, this way, repeated inversions of the SC formula were avoided.

- After a sequence of field analyses, both the torque-angle curve and the friction force acting on an eccentric electrostatic micromotor were simulated [33]. The conformal mapping of a doubly-connected domain made it possible to transform the whole air-gap region, and just one transformation was enough to compute the field in an arbitrarily large number of angular positions between stator and rotor, by means of rotating supply voltages instead of electrodes along the outer circumference.
- In the air-gap of an electrostatic device the boundary conditions were synthesized in terms of potential lines and field lines [5]. Using SC transformations, it was possible to identify a small subregion, arbitrarily extracted from a large field domain. In a sense, this is a field zooming technique, avoiding time consuming FE analysis in the whole domain.
- The field analysis of the aforementioned electrostatic micromotor was revisited from the viewpoint of the multiobjective design optimisation [6]. Actually the shape design of the rotor was formulated as a vector optimisation problem with three design criteria. The relevant fronts of non-dominated solutions were identified in fast way, thanks to an effective processing of SC maps.

As in every technique, also in the SC approach some drawbacks are apparent. In this respect, the following remarks can be put forward:

- computational procedures have to be developed *ad hoc* for a given class of problems;
- a certain degree of user skill is required in order to identify the most effective transformation and processing between real geometry and complex plane;
- algorithms are often non-optimised in terms of exploitation of storage, memory and processor speed.
- the low portability of the various codes implemented is an unpractical consequence to cope with.
- a severe drawback is that, in principle, the transformation technique requires to model homogeneous materials, although some attempts were done in [23] to cope with inhomogeneity.

4.6 Analysis of a Magnetic Levitator with Superconductors

4.6.1 A Principle of Equivalence

A principle of equivalence between magnetic fields comes from the transmission condition holding at the interface between two 2D regions Ω_1 and Ω_2 with

Fast Algorithms for the Design of Complex-Shape Devices in Electromechanics 67

permeabilities μ_1 and μ_2, respectively. If there is a current of density $\bar{J}_S$ (A m^{-1}) carried by boundary $\Gamma = \Omega_1 \cap \Omega_2$, it turns out to be

$$\hat{n}_1 \times (\bar{H}_2 - \bar{H}_1) = \bar{J}_S \qquad (1)$$

where $\hat{n}_1$ is the outer normal unit-vector along $\partial \Omega_1$.

If $\bar{J}_S = 0$, the tangential component of $\bar{H}$ is continuous.

Equation (1) states the transmission condition for magnetic fields. Based on it, the following *principle of equivalence* is assumed to be valid for two-dimensional fields.

Let a simply-connected domain $\Omega = \Omega_1 \cup \Omega_2$, composed of two sub-domains Ω_1 and Ω_2, be considered (Fig. 4.1a) and let the common boundary $\Omega_1 \cap \Omega_2$ be current free. Then, the field in the whole domain Ω can be obtained by superposition of two fields:

i) the field $\bar{H}_2$ in Ω_2, assuming a zero field in Ω_1 and forcing a specific current $\bar{J}'_s = \hat{n}_2 \times \bar{H}_1$ along $\Omega_1 \cap \Omega_2$ (Fig. 4.1b);

ii) the field $\bar{H}_1$ in Ω_1, assuming a zero field in Ω_2 and forcing a specific current $\bar{J}''_s = \hat{n}_1 \times \bar{H}_2$ along $\Omega_1 \cap \Omega_2$ (Fig. 4.1c).

In general, $\bar{J}_s$ can be viewed as a fictitious current density, restoring the effect of either of the sub-domains cut off. In other words, it takes into account the mutual coupling between the two field problems.

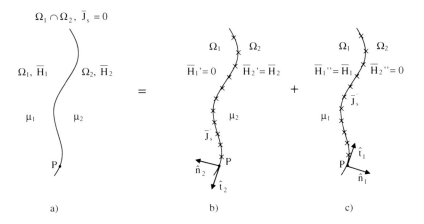

Fig. 4.1 Equivalence principle for 2D magnetostatic fields.

In particular, the equivalence principle holds when the common boundary $\Omega_1 \cap \Omega_2$ is an interior boundary of Ω_1, i.e. Ω_1 is a doubly-connected domain, (Fig. 4.2).

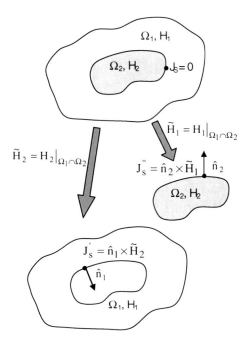

Fig. 4.2 Equivalence principle for a doubly-connected domain.

The equivalence corresponds to the substitution theorem in circuit theory. As a consequence, it is possible to model the effect of a small domain belonging to a large region by means of suitable cuts.

4.6.2 Application in 2D Magnetostatics

Let the region shown in Fig. 4.3 be considered, in which a permanent magnet is placed near a rectangular domain Ω_2.

Fast Algorithms for the Design of Complex-Shape Devices in Electromechanics 69

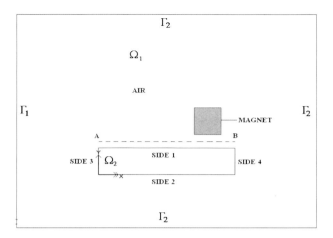

Fig. 4.3 Field domain.

The field is governed by Poisson's equation for vector potential, subject to boundary conditions.

The idea is to split the field problem defined in the whole region into two sub-problems defined on reduced domains: an inner-field problem in the homogeneous domain Ω_2, and an outer-field problem in the complementary domain Ω_1, which includes a permanent magnet. To this end, a cut is made along the border of Ω_2, restoring the boundary conditions by means of a current sheet along the border. In turn, region Ω_2 is removed in the outer-field problem, so obtaining a doubly-connected domain Ω_1, and the current sheet is forced along its inner border.

Referring to Fig. 4.3, the two-dimensional formulation of the outer-field problem in the homogeneous domain Ω_1 ($\mu_1 = \mu_0$) reads:

find vector potential $\overline{A} = (0, 0, A)$ such that $\overline{\nabla} \cdot \overline{A} = 0$ and

$$-\overline{\nabla} \cdot \mu^{-1} \overline{\nabla} A = J \chi_{\Omega_j} \text{ in } \Omega_1 \subset \mathfrak{R}^2 \tag{2}$$

with $\Omega_J \subset \Omega_1$, subject to:

$$A = 0 \text{ along } \Gamma_1 \tag{3}$$

$$\frac{\partial A}{\partial n} = 0 \text{ along } \Gamma_2 \tag{4}$$

$$\frac{\partial A}{\partial n} = J_s \text{ along } \partial \Omega_2 = \Omega_1 \cap \Omega_2 \quad \text{with } J_s = \left\| \hat{n}_1 \times \overline{H}_2 \right|_{\Omega_1 \cap \Omega_2} \right\| \tag{5}$$

The permanent magnet is modelled as a current sheet: the right-hand side of (2) takes it into account (J is the current density modelling the permanent magnet, Ω_J is the magnet domain, and χ_{Ω_J} is its support).

Equations (2) through (5) cast an *oblique-derivative problem in a doubly-connected domain* (see Fig. 4.4).

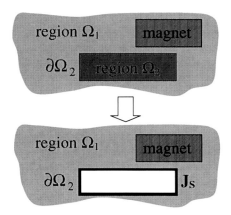

Fig. 4.4 Detail of the field problem, after cutting Ω_2 off.

As for numerical aspects, a detail of a typical finite-element mesh, composed of about 20,000 elements, is shown in Fig. 4.5.

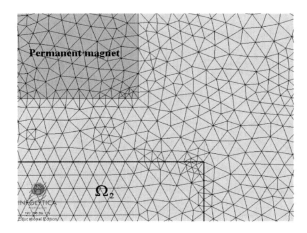

Fig. 4.5 Detail of the mesh.

For increasing the simulation accuracy, second-order triangular elements are considered.

The current density J_s, identified by means of (1) after a field analysis on the whole domain Ω ($\mu=\mu_0$), is shown in Fig. 4.6 and Fig. 4.7.

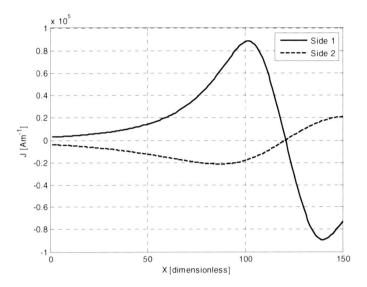

Fig. 4.6 Surface current densities set up along Ω_2 (sides 1 and 2).

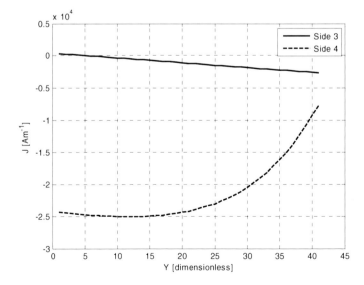

Fig. 4.7 Surface current densities set up along Ω_2 (sides 3 and 4).

For the purpose of simulation, the MagNet code by Infolytica was used [36].

4.6.3 Outer-Field Problem: Results

A comparison between the solutions obtained with the two models is shown in Fig. 4.8 and Fig. 4.9 in terms of the flux line maps.

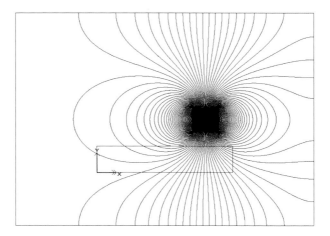

Fig. 4.8 Flux lines for the simply-connected domain Ω.

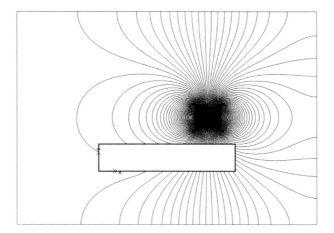

Fig. 4.9 Flux lines for the doubly-connected domain Ω_1.

Moreover, in Fig.s 4.10 through 4.13 the field components along the A-B line (dashed line in Fig. 4.3) are compared for the two cases in Fig. 4.8 and Fig. 4.9, respectively.

Fast Algorithms for the Design of Complex-Shape Devices in Electromechanics

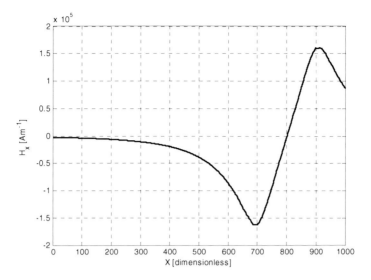

Fig. 4.10 X-directed components of the magnetic field evaluated along the line A-B (simply-connected domain Ω).

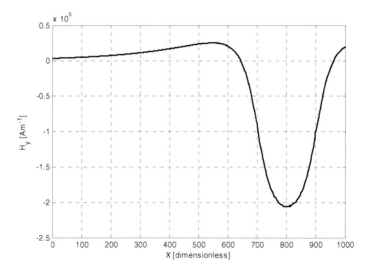

Fig. 4.11 Y-directed components of the magnetic field evaluated along the line A-B (simply-connected domain Ω).

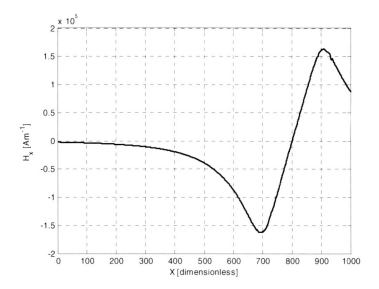

Fig. 4.12 X-directed components of the magnetic field evaluated along the line A-B (doubly-connected domain Ω_1).

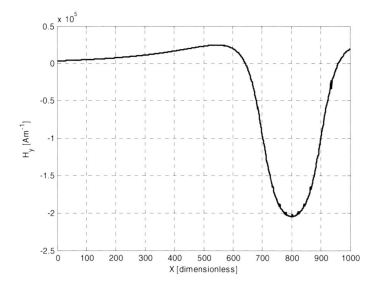

Fig. 4.13 Y-directed components of the magnetic field evaluated along the line A-B (doubly-connected domain Ω_1).

Considering Fig.s 4.8, 4.9 and Fig.s 4.10 through 4.13, it can be stated that results of the outer-field analysis agree with results of the field analysis in the whole region, in terms of both flux line distribution and field profiles as well.

A further assessment of results is proposed: the X and Y components of the magnetic field for the simply-connected domain along the A-B line are compared with the field along the same line for the doubly-connected domain Ω_1 in a two-fold case:

- the permanent magnet is removed and the boundary condition (5) with $J_s \neq 0$ is forced (problem PB1);
- the magnet is restored and the boundary condition (5) with $J_s=0$ holds (problem PB2).

The flux lines for PB1 and PB2 problems are shown in Fig. 4.14 and in Fig. 4.15, respectively.

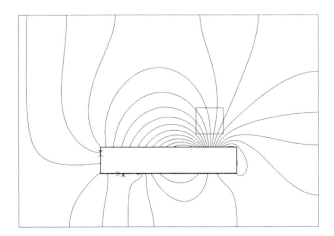

Fig. 4.14 Flux lines for the PB1 problem in Ω_1.

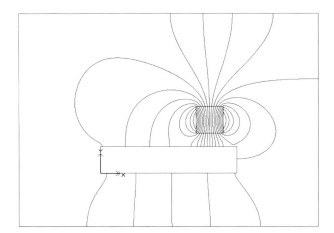

Fig. 4.15 Flux lines for PB2 problem in Ω_1.

The field along the A-B line for the two problems is shown in Fig. 4.16 and Fig. 4.17, respectively. The sum of the two contributions is also reported. These profiles should be compared with those shown in Fig.s 4.10 through 4.13.

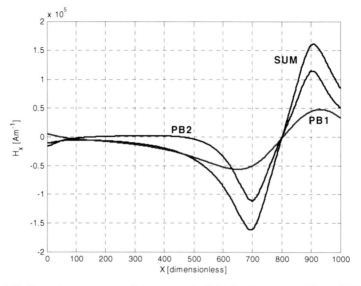

Fig. 4.16 X-directed component of the magnetic field for the two problems PB1 and PB2 and their sum.

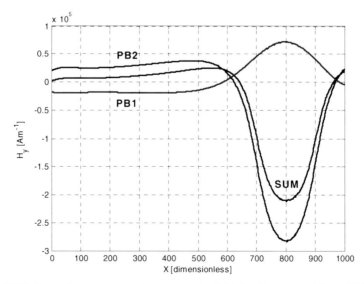

Fig. 4.17 Y-directed component of the magnetic field for the two problems PB1 and PB2 and their sum.

It can be noted that the curves match.

4.6.4 Test Problem: Diamagnetic Model of a Superconductor

Let a high-temperature superconductor (HTSC), filling domain Ω_2, be considered; the HTSC is modelled as a perfectly diamagnetic material (Meissner state). For the sake of simulation, $\mu_r = 10^{-20}$ was assumed for domain Ω_2 in the finite-element simulations.

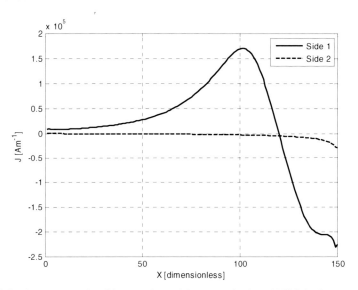

Fig. 4.18 Surface current densities set along sides 1 and 2 of the HTSC in Ω_2.

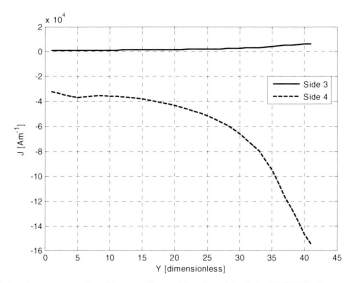

Fig. 4.19 Surface current densities set along sides 3 and 4 of the HTSC in Ω_2.

After the magnetic analysis in the domain $\Omega_1 \cup \Omega_2$, the current densities along $\Omega_1 \cap \Omega_2$, identifying the boundary conditions for the subsequent analysis in the domain Ω_1, have been obtained; they are shown in Fig. 4.18 and Fig. 4.19.

4.6.5 Case Study: Trapped-Flux Model of a Magnetic Bearing

For the sake of an application, attention is focused on HTSC magnetic bearings in which the superconductor operates in the trapped-flux state. This means that the conductor keeps memory of the inner distribution of flux lines in the activation position, *i.e.* its position when the transition to the superconducting state took place (operational-field cooling) [34]. Assuming small displacements, the inner field of the conductor does not change when it is moved from the activation to the operational position with respect to the exciting magnet. As a result, restoring forces acting on the conductor, as shown in Fig. 4.20, appear.

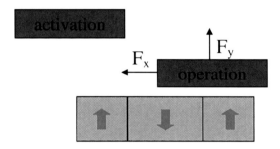

Fig. 4.20 Activation position and operational position for the HTSC.

The field region of interest is a doubly-connected domain, obtained after removing the HTSC and substituting it with a surface current distribution. The latter is identified by means of three field simulations based on finite elements, by varying state and position of the HTSC. Specifically, the following sub-problems are solved:

1) the conductor is modelled as a material with permeability μ_0, placed in the activation position (Fig. 4.21a);
2) the conductor is modelled as a perfectly diamagnetic material, placed in the operational position (Fig. 4.21b);
3) the conductor is modelled as a perfectly diamagnetic material, placed in the activation position (Fig. 4.21c).

The rationale is that sub-problem 1) captures the initial field, while 2) and 3) make the conductor non-permeable to the external magnetic field.

For each sub-problem, the current distribution along the HTSC boundary is identified by means of (5). Then, the superposition principle is exploited to determine the final current distribution. According to the equivalence principle stated in paragraph 6.1, a doubly-connected domain, corresponding to the outer field region is considered.

The procedure implemented offers a possible saving of runtime when various positions between HTSC and magnet are to be simulated; in fact, the field in the activation position (aforementioned step 1) is computed only once, while the field in the operational position is updated. The method here proposed to model the trapped-flux state (i.e. superposing surface current densities) is similar to that developed in [35], where the superposition of magnetic vector potential in the whole field region is considered.

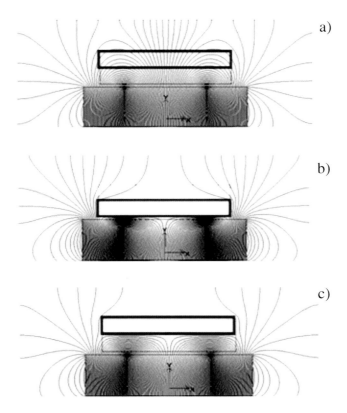

Fig. 4.21 Sub-problems to be solved for the superposition principle: a) conductor modelled as a material with permeability μ_0, placed in the activation position; b) conductor modelled as a perfectly diamagnetic material, placed in the operational position; c) conductor modelled as a perfectly diamagnetic material, placed in the activation position.

To compare results, a second method was used for field analysis, exploiting SC conformal maps of the external domain [1], paragraph 4.4, obtained after removing the HTSC to compute the fields in the above three conditions. After computing the field on the HTSC boundary in the activation position (step 1) by the Biot and

Savart's law, the external sub-domain and the current sheets (in fact, 80 discrete currents) are mapped into the SC upper half-plane, the real axis representing the HTSC boundary (see Fig. 4.22).

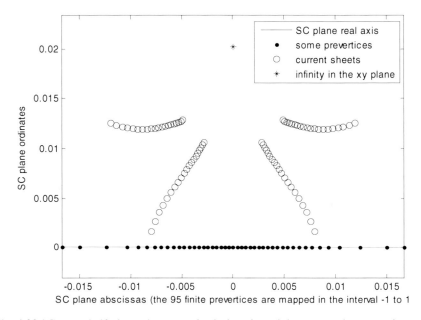

Fig. 4.22 SC upper half-plane: the external sub-domain and the current sheets are shown.

So, to perform the calculations involved in step 2 and step 3, it is sufficient to apply the Biot and Savart's law in the SC half plane, after setting homogeneous Neumann conditions on the real axis. The CPU times involved are of the order of one minute on a personal computer to map the sub-domain, and of the order of few seconds to map any new current sheet system, to simulate the relative movement between HTSC and exciting magnets in any direction. This way, fast calculations can be performed, with acceptable errors with respect to more refined FE results. Flexibility and accuracy of the SC mapping procedures is provided by modern computation methods [4]. The half-plane maps are obtained by first mapping the outer domain into the unit disc and, then, by mapping the disc into the half-plane; after defining the numerical transformation, the map of any current sheet system is almost immediately derived.

When the HTSC cross-section exhibits a simple rectangular shape, the above mapping procedure, useful to handle up to one-hundred vertices on a standard personal computer, can be replaced by the map of the external region into the SC half-plane presented in [8], see paragraph 9.3, equations (9.38) and (9.57). In this case, the whole three-step procedure of Fig. 4.21 is performed in about 5 seconds.

Fast Algorithms for the Design of Complex-Shape Devices in Electromechanics 81

The levitation device considered is shown in Fig. 4.23. It consists of a sample of HTSC excited by three gap-oriented permanent-magnets (NdFeB, coercive field H_c=827.6 kAm^{-1}). Typical dimensions are about 50 mm for height and 80 mm for width, while activation position g_{act} = 12 mm and operational position g_{op} = 2 mm are assumed.

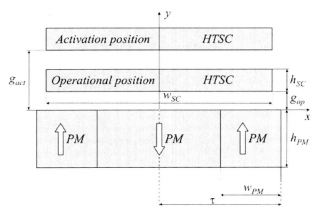

Fig. 4.23 Geometry of the HTSC with gap-oriented exciting magnets.

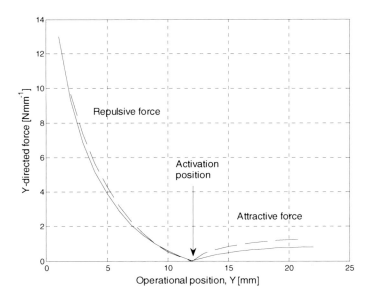

Fig. 4.24 Force-displacement curve using the substitution theorem (dashed line), and the Schwarz-Christoffel transformation (continuous line).

In Fig. 4.24 the results obtained for the trapped-flux state model are presented. In particular, the y-directed levitation force, acting on the HTSC in a set of operational positions, has been simulated by means of both analysis methods. The force has been evaluated by means of the Maxwell's stress tensor. In particular, it can be noted that the force-position curve exhibits two branches, corresponding to repulsive and attractive action, respectively: the zero of force is located just at the activation position.

Given a feasible geometry of conductor and magnet (h_{PM}=25 mm, w_{PM}=25 mm, τ=50 mm, w_{SC}=80 mm, h_{SC}=10 mm), assuming again g_{op}=2 mm and g_{act}=12 mm, the x-directed force component has been simulated by means of the SC transformation for several displacements of the magnet along the x axis. This way, the force-position curves represented in Fig. 4.25 have been obtained: for the sake of a comparison, the computation has been repeated for some positions by means of FE analysis.

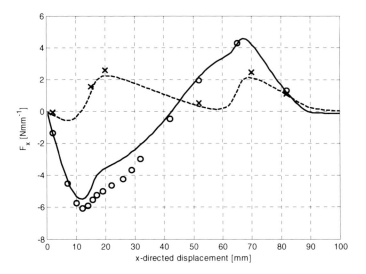

Fig. 4.25 F_x vs x. Dashed line: perfectly-diamagnetic material (SC), cross: perfectly-diamagnetic material (FE), continuous line: operational-field cooling (SC), circle: operational-field cooling (FE).

The various curves are in agreement; in particular, F_x=0 at x=0, *i.e.* when the magnet axis is coincident with the conductor axis, and F_x vanishes when the x displacement is large with respect to the magnet size. In Fig. 4.26 flux line maps for a given displacement are shown.

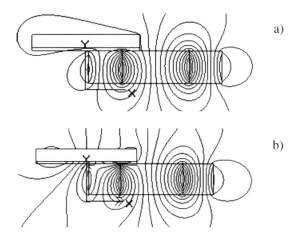

Fig. 4.26 Flux lines for x-displacement equal to 52 mm in the case of perfectly-diamagnetic material (a) and operational-field cooling (b).

4.6.6 An Application in Automated Optimal Design

Finally, to show the performance of SC transformations also in the case of an inverse problem, the following design problem has been formulated:

given a geometry of HTSC and a value of g_{op}, assuming (h_{PM}, w_{PM}, τ) as the design variables, find the magnet geometry such that

- the levitation force F_y at x=0 and y=g_{op} is maximum, and
- the absolute value of the force stiffness $\dfrac{\partial F_x}{\partial x}$, at x=0 and y=$g_{op}$, is maximum.

The latter objective accounts for a stable operational position of the HTSC against small perturbations along the x axis. For the sake of computation, the force stiffness has been approximated by a first-order finite difference.

The solution to the design problem was based on a Monte Carlo sampling of the three-dimensional design space, driven by a uniform probability density. It has been assumed w_{SC}=35 mm, h_{SC}=8 mm and g_{op}=2 mm. The resulting distribution of points in the objective space, obtained by means of a large number of SC analyses, is represented in Fig. 4.27.

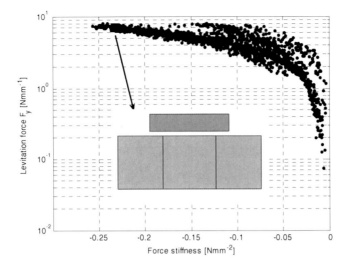

Fig. 4.27 Monte Carlo sampling of the objective space and geometry of a candidate optimum (square, h_{PM} = 28.3 mm, w_{PM} = 20 mm, τ = 31.7 mm) found by Nelder & Mead algorithm.

At a first glance, the two objective functions seem not to be in conflict, possibly identifying a unique optimum located at a corner of the point distribution.

Thanks to the short computing time, allowed by the conformal mapping procedures, these results have been assessed by means of a deterministic optimisation (Nelder & Mead algorithm), based on a preference function including both levitation and stiffness terms.

4.7 Conclusion

A deep investigation has been performed on a class of HTSC magnetic levitators. The effectiveness of the coordinated FE-SC approach has been proven; in the meantime, modern techniques for the numerical inversion of the SC formula have been found to be both fast and accurate.

References

[1] Driscoll, T.A., Trefethen, L.N.: Schwarz-Christoffel Mapping. Cambridge University Press, Cambridge (2002)
[2] Stoll, R.L.: Simple computational model for calculating the unbalanced magnetic pull on a two-pole turbogenerator rotor due to eccentricity. IEE Proc. - Electr. Power Appl. 144(4), 263–270 (1997)
[3] Binns, K.J., Lawrenson, P.J., Trowbridge, C.W.: The Analytical and Numerical Solution of Electric and Magnetic Fields. Wiley, New York (1992)

[4] Costamagna, E., Di Barba, P., Savini, A.: An effective application of Schwarz-Christoffel transformations to the shape design of permanent-magnet motors. Intl J. of Applied Electromagnetics and Mechanics 21, 21–37 (2005)
[5] Costamagna, E., Di Barba, P., Savini, A.: Synthesis of boundary conditions in a subdomain using the Schwarz-Christoffel transformation for the field analysis. The Intl J. for Computation and Mathematics in Electrical and Electronic Engineering (COMPEL) 25(3), 627–634 (2006)
[6] Costamagna, E., Di Barba, P., Savini, A.: Shape Design of a MEMS Device by Schwarz-Christoffel Numerical Inversion and Pareto Optimality. The Intl J. for Computation and Mathematics in Electrical and Electronic Engineering (COMPEL) 27(4), 760–769 (2008)
[7] Durand, E.: Electrostatique et Magnétostatique. Masson, Paris (1953)
[8] Binns, K.J., Lawrenson, P.S.: Analysis and Computation of Electric and Magnetic Field Problems. Pergamon Press, Oxford (1963)
[9] Trefethen, L.N.: Numerical computation of the Schwarz-Christoffel transformation. SIAM J. Sci. Stat. Comput. (1), 82–102 (1980)
[10] Henrici, P.: Applied and Computational Complex Analysis, vol. 1. Wiley, New York (1974)
[11] Howe, D.: The application of numerical methods to the conformal transformation of polygonal boundaries. J. Inst. Math. Appl. (12), 125–136 (1973)
[12] Costamagna, E., Maltese, U.: Unpublished works and FORTRAN programs, Marconi Italiana S.p.A., Genova (1970-1971)
[13] Costamagna, E.: On the numerical inversion of the Schwarz-Christoffel conformal transformation. IEEE Trans. Microwave Theory Tech. 35(1), 35–40 (1987)
[14] Chaudhry, M.A., Schinzinger, R.: Numerical computation of the Schwarz-Christoffel transformation parameter for conformal mapping of arbitrarily shaped polygons with finite vertices. The Intl J. for Computation and Mathematics in Electrical and Electronic Engineering (COMPEL) II(1), 263–275 (1992)
[15] Chuang, J.M., Gui, Q.Y., Hsiung, C.C.: Numerical computation of Schwarz-Christoffel transformation for simply connected unbounded domains Comput. Methods Appl. Mech. Engr. 105, 93–109 (1993)
[16] Costamagna, E.: Integration formulas for numerical calculations of the Schwarz-Christoffel conformal transformation. Microwave Opt. Technol. Lett. 15(4), 219–224 (1997)
[17] Dwigth, H.B.: Tables of Integrals and Other Mathematical Data, 3rd edn. MacMillan, New York (1957)
[18] Takahasi, H., Mori, M.: Double exponential formulas for numerical integration. Publ. RIMS, Kyoto Univ. (9), 721–741 (1974)
[19] Singh, R., Singh, S.: Efficient evaluation of singular and infinite integrals using the tanh transformation. IEE Proc. Microwave Antennas Propagat. 141(6), 464–466 (1994)
[20] Evans, G.A., Forbes, R.C., Hyslop, J.: The tanh transformation for singular integrals. Int. J. Comput. Math. 15, 339–358 (1984)
[21] Costamagna, E.: Error masking phenomena during numerical computation of Schwarz-Christoffel conformal transformations. Microwave Opt. Technol. Lett. 20(4), 223–225 (1999)
[22] Hu, C.: Algorithm 785: A software package for computing Schwarz-Christoffel conformal transformation for doubly connected polygonal regions. ACM Trans. Math. Soft. 24(3), 317–333 (1998)

[23] Schinzinger, R., Laura, P.A.A.: Conformal Mapping: Methods and Applications. Elsevier, Amsterdam (1991)
[24] Driscoll, T.A., Vavasis, S.A.: Numerical conformal mapping using cross-ratios and Delaunay triangulations. SIAM J. Sci. Comput. 19(6), 1783–1803 (1998)
[25] Howell, L.H., Trefethen, L.N.: A modified Schwarz-Christoffel transformation for elongated regions. SIAM J. Sci. Stat. Comput. 11(5), 928–949 (1990)
[26] Costamagna, E.: A new approach to standard Schwarz-Christoffel formula calculations, Microwave Opt. Technol. Lett. 32(3), 196–199 (1997)
[27] Driscoll, T.A.: Schwarz-Christoffel Toobox Userr's Guide, version 2.3, see [1] for website
[28] Bowman, F.: Notes on two-dimensional electric field problems. Proc. London Math. Soc. 39, 205–215 (1935)
[29] Papamichael, N., Kokkinos, C.A.: The use of singular functions for the approximate conformal mapping of doubly connected regions. SIAM J. Sci. Stat. Comput. (5), 685–700 (1984)
[30] Smythe, W.R.: Static and Dinamic Electricity, 3rd edn. MacGraw-Hill, New York (1968)
[31] Costamagna, E., Fanni, A.: Computing capacitances via the Schwarz-Christoffel transformation in structures with rotational symmetry. IEEE Trans. Magn. 34(5), 2497–2500 (1998)
[32] Alfonzetti, S., Costamagna, E., Fanni, A.: Computing capacitances of vias in multilayered boards. IEEE Trans. Magn. 37(5), 3186–3189 (2001)
[33] Costamagna, E., Di Barba, P., Savini, A.: Conformal mapping of doubly connected domains: an application to the modelling of an electrostatic micromotor. IET Science, Measurement & Technology 3(5), 334–342 (2009)
[34] Krabbes, G., Fuchs, G., Canders, W.R., May, H., Palka, R.: High Temperature Superconductor Bulk Materials. Wiley-VCH, Berlin (2006)
[35] May, H., Palka, R., Portabella, E., Canders, W.R.: Evaluation of the magnetic field – high temperature superconductor interactions. The Intl J. for Computation and Mathematics in Electrical and Electronic Engineering (COMPEL) 23(1), 286–304 (2004)
[36] Electronic sources: http://www.infolytica.com

Chapter 5
Optimization of Wound Rotor Synchronous Generators Based on Genetic Algorithms

Xavier Jannot, Philippe Dessante, Pierre Vidal, and Jean-Claude Vannier

SUPELEC Energy Department, 3 rue Joliot-Curie, F-91192 Gif-sur-Yvette Cedex, France
`{xavier.jannot,philippe.dessante,pierre.vidal,`
`jean-claude.vannier}@supelec.fr`

Abstract. The manufacturers are tempted to reduce the amount of active materials in the devices in order to lower the material bill. However, reducing the weight of the devices directly affects the energy efficiency. In this paper, a solution to this problem is proposed for the case of wound rotor synchronous generators. The trade-off between cost and efficiency is formulated as a constrained optimization problem and solved using a Genetic Algorithm. The cost optimization of three different machines is carried out through various design approaches. The proposed approach always gives better results than the classical approach concerning the global cost of the range.

5.1 Introduction

The recent rise in the cost of raw materials is motivating the manufacturers of electrical machines to improve the designs of their devices. The manufacturers are tempted to reduce the amount of active materials in the devices in order to lower the material bill. However, reducing the weight of the devices directly affects the energy efficiency, and therefore goes against the current trend and European suggestions [1], which recommend that the manufacturers increase the efficiency of their products. Manufacturers are thereby confronted with a constrained optimization problem: reducing the costs of devices and respecting the fixed efficiencies. In this paper we propose a solution to this problem for the case of a series of Wound Rotor Synchronous Generators (WRSG) using an optimization method based on Genetic Algorithm (GA).

The optimal design approach is used more and more often in the case of electromagnetic device design. This approach needs to accurately model various physical domains [2-3]. Magnetic, electric, thermal or mechanical models of the systems are often developed in order to design optimal devices respecting physical constraints from various fields of physics. The accuracy of the model and its execution time basically depend on the modeling complexity: 1D analytical model [3-6], electric circuit type model [3, 7-8] or Finite Element model [3, 9-10].

In this work, a 1D multiphysic modeling of the WRSG was adopted whose short execution time allowed coupling it with a stochastic optimization method. A GA optimization technique was employed because it is well suited to a mixed-variable constrained problem and it allows searching for a global optimum. The complexity of the treated objective function also leads to choose a zero-order derivation method.

The modeling of the machine and, then, the optimization method will be detailed and followed by the optimization results. Single and multi-objective optimizations of the range are performed. The single objective optimizations introduce a new procedure, different from the classical range design approach used in industry. The introduced method gives better results compared with the classical one for various test cases and permits rationalization of the optimal design of a series of machines from an industrial point of view. The multi-objective optimizations' goal is to propose a trade-off between saving money and saving energy when considering the whole range.

Finally, the originality of this work is to propose a global view of the design of a range of WRSG linked with an optimal design conception.

5.2 Wound Rotor Synchronous Generator Model

The geometry of the studied WRSG is shown in figure 5.1. The stator tooth width is constant except for the slots' bottom and back sides. The air gap is progressive and the rotor winding is ordered. Moreover the poles' number is fixed to four.

Fig. 5.1 Geometry of the studied WRSG with salient poles.

5.2.1 Electromagnetic Modeling of the WRSG

The goal of the following modeling is to determine the electromagnetic design of a WRSG that satisfy the specifications. The WRSG should be design in order to satisfy a fixed apparent power P, at a fixed line voltage V, and finally for a given power factor $\cos(\varphi)$. The three previously listed quantities constitute the specifications of the WRSG.

5.2.1.1 Air Gap Magnetic Flux Density Computation at No Load

The fundamental quantity used for the machine modeling and optimization is the line voltage. This voltage waveform is considered purely sinusoidal. The knowledge of this voltage, neglecting the winding resistance, allows determination, via integration, of the amplitude of the fundamental of total flux, due to rotor excitation, through the WRSG windings.

$$\phi_1 = \frac{V}{\pi\sqrt{2} N_s k_w f} \tag{1}$$

Where N_s is the winding turn linking the total flux, k_w the global winding factor related to the first harmonic and f the electrical frequency of the WRSG. From the fundamental of total flux, one can compute the fundamental of air gap magnetic flux density. Indeed the fundamental of the total flux is obtained from the integration of the fundamental of air gap magnetic flux density through the winding surface. If the axial end effects are neglected in the machine, the peak value of the fundamental of air gap magnetic flux density is computed, after integration, with:

$$B_1 = \frac{\phi_1}{R L_s}. \tag{2}$$

Where R is the stator inner radius and L_s the WRSG stack length. The first harmonic of air gap magnetic flux density, due to rotor magnetomotive force, is now known and the next step is to determine the exact air gap magnetic flux density waveform. The knowledge of this local quantity is useful in determining the magnetic flux density map of the WRSG. Because of salient rotor poles this quantity is to be computed under some hypothesis.

The rotor and the stator present permeance variations due to either the slots and to the rotor saliency. For the stator, an equivalent smooth armature is introduced. This simplification can be performed but only with the introduction of an equivalent larger air gap. The air gap increasing is obtained via the well know Carter coefficient, k_C.

With the hypotheses assumed on a smooth armature stator, the rotor permeance variations are included considering a simple air gap permeance function: the function is the inverse of the air gap length under the poles and between the poles the air gap permeance function is one-sixth of the air gap length under the poles. This consideration leads to the situation of figure 5.2 for the magnetic flux density produced by the rotor excitation. The plotted magnetic flux densities correspond to the case of a salient rotor with non uniform air gap.

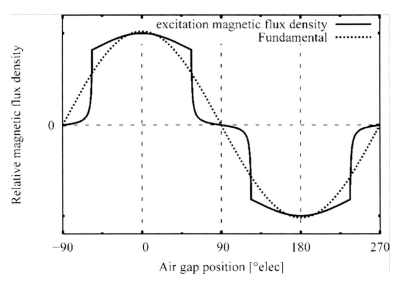

Fig. 5.2 Exact waveform of air gap magnetic flux density and its fundamental due rotor current excitation.

The non uniform air gap is designed in order to make the air gap magnetic flux density closer to the ideal sinus waveform. The air gap length can evolve according to the following law:

$$e = a - b.\cos(\theta). \qquad (3)$$

The reference for the angle θ is taken in the middle of the pole. Neglecting the magnetic field between the poles, one can compute the magnitude of the fundamental of air gap magnetic flux density from its "exact" waveform with:

$$B_1 = B_{ag} \frac{4}{\pi}(a-b)k_C \left(\frac{2a}{b\sqrt{a^2-b^2}} \operatorname{atan}\left(\tan\left(\frac{\beta\pi}{4}\right)\sqrt{\frac{a+b}{a-b}} \right) - \frac{\beta\pi}{2b} \right). \qquad (4)$$

Where B_{ag} is the peak value of the air gap magnetic flux density "exact" waveform and β the rotor pole opening factor. The goal is to obtain the relation giving the peak value of air gap magnetic flux density when knowing the peak value of its fundamental component. From the previous expression (4), one obtains:

$$B_{ag} = \frac{B_1}{\frac{4}{\pi}(a-b)k_C \left(\frac{2a}{b\sqrt{a^2-b^2}} \operatorname{atan}\left(\tan\left(\frac{\beta\pi}{4}\right)\sqrt{\frac{a+b}{a-b}} \right) - \frac{\beta\pi}{2b} \right)}. \qquad (5)$$

5.2.1.2 Computation of Magnetic Flux Density in the Iron Paths

From the "exact" air gap magnetic flux density, and its peak value, one can compute the total flux under one pole. This is performed by integrating the air gap magnetic flux density under the pole and neglecting the flux flowing between two consecutive magnetic poles. The total flux is:

$$\Phi_0 = B_{ag} k_C . 2R.L_s \frac{a-b}{\sqrt{a^2 - b^2}} \operatorname{atan}\left(\tan\left(\beta\frac{\pi}{4}\right)\sqrt{\frac{a+b}{a-b}}\right). \qquad (6)$$

Now that the total flux under the pole is computed, the magnetic flux density in the magnetic paths of the machine can be determined applying the Gauss's law. The magnetic flux density in the rotor pole is:

$$B_{pole} = \frac{\Phi_0}{L_s l_p}. \qquad (7)$$

Where l_p is the rotor pole width. Concerning the magnetic flux density in the stator teeth, a concentration factor must be taken into account. It characterizes the fact that the air gap magnetic flux under one stator tooth pitch is concentrated in the tooth. Moreover, the air gap magnetic flux density is not uniform under the pole, see figure 5.2. The magnetic flux density in a given tooth varies versus time according to the synchronous rotation of the rotor and the stator current reaction. The peak value of this quantity is obtained when the tooth is in front of the peak value of the air gap magnetic flux density. It leads to the expression of the magnetic flux density in the teeth:

$$B_{sth} = \frac{2\pi R}{N_{slot} l_t} B_{ag}. \qquad (8)$$

Where N_{slot} is the number of slots of the stator and l_t is the width of a stator tooth. Next is the computation of the stator yoke magnetic flux density. The flux flowing through the pole, crosses the air gap, the stator teeth, and then is divided. One half is flowing through one side of the stator yoke and the second half is flowing through the opposite side of the stator yoke. The resulting magnetic flux density in the stator yoke can be computed with:

$$B_{sbi} = \frac{\Phi_0}{2L_s h_{sbi}}. \qquad (9)$$

Where h_{sbi} is the height of the stator yoke.

5.2.1.3 Magnetic Field Circulation in the WRSG

The previous computations give the magnetic flux density in the various parts of the WRSG: inside the air gap, the rotor pole, the stator teeth, and the stator yoke.

These quantities allow determining the corresponding magnetic field strength. The material constitutive laws permit to link the magnetic flux density to the magnetic field strength. In the air gap, the relation is linear and leads to:

$$H_{ag} = \frac{B_{ag}}{\mu_0}. \tag{10}$$

Where μ_0 is the magnetic permeability of vacuum. Concerning the magnetic paths, the relation is much complicated because of the non linear relation between the magnetic flux density and the magnetic field strength. A typical relation is illustrated in figure 5.3. A relation, depending on the iron sheet quality, between the magnetic flux density and field strength, must be defined (even numerically). For the iron paths, it leads to:

$$H_{ir} = HB_{curve}(B_{ir}, iron\ quality). \tag{11}$$

The previous relation is applied to the rotor pole, the stator teeth, and the stator yoke and respectively gives H_{pole}, H_{sth}, and H_{sbi}.

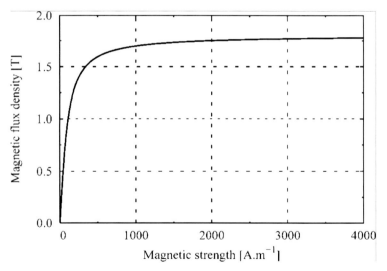

Fig. 5.3 Non linear relation between the magnetic flux density and the magnetic field strength in iron.

Knowing the magnetic field strength in every magnetic flux path, one can compute the field circulation on a contour. The field circulation is then evaluated in the rotor pole, in the air gap, in the stator teeth, and in the stator yoke. In the air gap, the field circulation is computed with:

$$A_{ag} = H_{ag} k_C (a-b). \tag{12}$$

In the rotor pole the field circulation is expressed:

$$A_{pole} = H_{pole} h_{pole}. \qquad (13)$$

Where h_{pole} is the radial height of the rotor pole. Concerning the stator teeth, the field circulation is:

$$A_{sth} = H_{sth} h_{sth}. \qquad (14)$$

With h_{sth} the stator tooth height. In the stator yoke the field circulation is estimated with:

$$A_{sbi} = H_{sbi} \frac{\pi}{4} \left(R_{ext} - \frac{h_{sbi}}{2} \right). \qquad (15)$$

Where R_{ext} is the stator outer radius of the WRSG.

5.2.1.4 Computation of Rotor Magnetomotive Force

The previous computations' goal is to go further in the rotor magnetomotive force determination. The magnetic field strength circulations in the machine permit to compute the total magnetomotive drop in the machine and under non linear magnetic behavior. When the power factor is around 0.8, the total rotor magnetomotive force ε to be supplied is roughly the sum of two mmf: ε_0 that should be supplied to obtain the nominal voltage at no-load and ε_{cc} that should be supplied to deliver the current I in short-circuit condition:

$$\varepsilon = \varepsilon_0 + \varepsilon_{cc}. \qquad (16)$$

The determination of ε_0 is a straight forward process when knowing the magnetomotive force consumption in the different parts of the WRSG:

$$\varepsilon_0 = A_{ag} + A_{pole} + A_{sth} + A_{sbi}. \qquad (17)$$

The determination of ε_{cc} is detailed as follows. The current flowing in the stator windings produces a mmf which will be called the armature current reaction. From this quantity, the air gap magnetic flux density due to the armature reaction is deduced. Then, according to the principle detailed before, the fundamental of this magnetic flux density, in d-axis, noted B_{1d} is determined. B_{1d} depends only on ε_{cc}.

The current flowing in the stator can be computed from the WRSG specifications. With the apparent power, the line voltage, and the power factor, one can deduce the stator line current:

$$I = \frac{P}{3V \cos(\varphi)}. \qquad (18)$$

The stator current produces a magnetomotive force along the air gap. Only the d-axis stator current reaction is taken into account. In this axis, the maximum of magnetomotive force is located in the middle of the rotor pole and the peak value is usually determined with:

$$mmf_d = \frac{3\sqrt{2}.N_s.k_w I}{2\pi}. \tag{19}$$

The peak value of the magnetic flux density reaction in the air gap is obtained where the air gap is minimal and its value is:

$$B_r = \mu_0 \frac{mmf_d}{k_C(a-b)}. \tag{20}$$

Because of the non uniform air gap and of the rotor saliency, the fundamental value of the air gap magnetic flux density due to d-axis stator current reaction, plotted in figure 5.4, is to be calculated with:

$$B_{1d} = Br \frac{4}{\pi}(a-b)k_C \left(\frac{2a^2 \mathrm{atan}\left(\tan\left(\frac{\beta\pi}{4}\right)\sqrt{\frac{a+b}{a-b}} \right)}{b^2\sqrt{a^2-b^2}} - \frac{a}{b^2}\frac{\beta\pi}{2} - \frac{\sin\left(\frac{\beta\pi}{2}\right)}{b} \right). \tag{21}$$

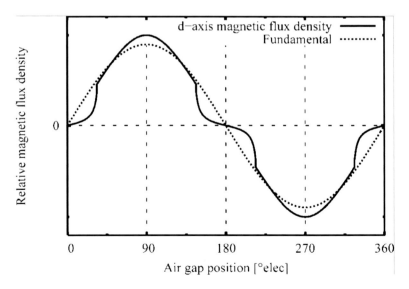

Fig. 5.4 Exact waveform of air gap magnetic flux density and its fundamental due to d-axis stator current reaction.

In short-circuit, the air gap magnetic flux density value is low, so that the machine is not saturated; on the contrary, it is working under linear conditions. As a consequence it can be stated that the mmf ε_{cc} needed to obtain the air gap magnetic flux density is directly proportional to this very value. With a simple proportional relation, one can deduce:

$$\varepsilon_{cc} = \varepsilon_0 \frac{B_{1d}}{B_1}. \tag{22}$$

Finally the total magnetomotive force in load (16), that must be supplied by the rotor excitation, is completely determined. It remains the computation of the current flowing in the rotor conductors. It is simply determined with:

$$I_r = \frac{\varepsilon}{N_r}. \tag{23}$$

Where N_r is the turn number of the rotor winding.

The electromagnetic modeling of the WRSG is completed. It starts with the definition of the specifications and, for a given geometry; it ends with the rotor current. By the way, the magnetic flux densities in the machine are computed and will be involved in the magnetic loss determination.

5.2.2 Losses and Thermal Model

To ensure normal working conditions of the machine, the thermal behavior has to be studied. First the different kinds of losses occurring in the machine must be evaluated to obtain the input parameters of the thermal model that yield the stator and rotor temperatures.

5.2.2.1 Losses' Estimation

There are various losses occurring in the WRSG. There are the classical copper and iron losses that are located both in the rotor and in the stator. In case of a fan cooled machine where the fan is mounted on the shaft, there are losses happening that should be taken into account. Still concerning the aerodynamic losses, another locus where they should be computed is the air gap. Indeed due to the relative motion between the rotor and the stator, the fluid in the air gap is submitted to a shear stress leading to a power loss. Another loss item is in the bearing, the mechanical losses. Whatever the kind of bearing, ball bearing or magnetic bearing, there are losses occurring in these parts. In the following the focus is put on the copper, iron, and windage losses.

The copper losses occur in stator and rotor windings. The rotor losses can be computed classically as follow:

$$P_{cu,r} = R_r . I_r^2 . \quad (24)$$

Where R_r is the rotor winding DC resistance. This parameter depends on the rotor temperature. This dependence is explicated latter in the text. The stator copper losses are computed in the same manner:

$$P_{cu,s} = 3R_s . I^2 . \quad (25)$$

Where R_s is the stator winding AC resistance at the nominal working frequency (usually 50Hz or 60Hz). This parameter also depends on temperature, as for the rotor.

Latter is the determination of the iron losses. These losses are due to the time variation of the magnetic flux density in the iron paths. They correspond to the hysteretic behavior of the materials: when injecting a certain quantity of energy, the materials restitute less energy. These losses vary with the time variation of the magnetic flux density. The time dependence approach can be turned into a frequency dependence of the losses. In the case of alternative machines, such as the WRSG, one can assume that the magnetic flux density varies sinusoidally. Moreover in the case of WRSG, the frequency is usually fixed (50Hz or 60Hz). It is well admitted that the magnetic losses can fall into three components: the hysteresis losses, the classical eddy-current losses, and the excess losses. This decomposition has been introduced by Bertotti. The model adopted is simpler, and based on the observations made by Steinmetz. The iron losses per kilogram are computed with the law:

$$P_{ir} = k_{ir} f \left(\frac{B}{1.5} \right)^\alpha . \quad (26)$$

Where k_{ir} is the specific loss factor – which depends on the magnetic material quality – at 50Hz and for a magnetic flux density of 1.5T, B the magnetic flux density in the material, and α the called Steinmetz coefficient. In the WRSG the rotation speeds of the field and the rotor are identical so that in first approximation the rotor magnetic field is constant. There is no time variation of the magnetic flux density so no iron losses occur in the rotor. Concerning the stator, the magnetic field is rotating at the electrical frequency generating some iron losses. The magnetic path of the stator is mainly composed of the teeth and the yoke. These two parts need to be considered separately since the magnetic flux density inside is different. Concerning the iron loss density in the teeth, it can be evaluated with:

$$P_{ir,sth} = k_{ir} f \left(\frac{B_{sth}}{1.5} \right)^\alpha . \quad (27)$$

Optimization of Wound Rotor Synchronous Generators

In the same way, the iron loss density in the stator yoke is:

$$P_{ir,sbi} = k_{ir} f \left(\frac{B_{sbi}}{1.5} \right)^{\alpha}. \tag{28}$$

Finally the total iron losses are:

$$P_{ir,s} = P_{ir,sbi} M_{sbi} + P_{ir,sth} M_{sth}. \tag{29}$$

Where M_{sbi} is the weight of the stator yoke and M_{sth} is the weight of the stator teeth.

The last item of losses concerning the fan is computed with a semi-empirical law. Admitting that the fan outer radius is the same as the WRSG outer radius, the mechanical losses are computed with:

$$P_{mec} = k_{mec} R_{ext}^{4.5}. \tag{30}$$

Where k_{mec} are the mechanical losses for a unitary stator outer radius.

Finally the efficiency, as defined by [11], of the WRSG can be determined with the following expression:

$$\eta = \frac{P.\cos(\varphi)}{P.\cos(\varphi) + P_{ir,sth} + P_{ir,sbi} + P_{cu,r} + P_{cu,s} + P_{mec}}. \tag{31}$$

5.2.2.2 Thermal Modeling

As written before, some losses greatly depend on the material temperature. It is the case of the copper losses. This is a reason to estimate the temperature of the WRSG. Another reason is to limit the temperature rise of the electric machine, especially because of the insulating materials of the copper wire. These materials are specified for a maximal working temperature and going beyond this limit will make their life time become shorter, and thus the life time of the machine.

As soon as the losses are determined the temperature rise in the stator and the rotor can be evaluated using a formulation derived from experimental correlations. It is assumed that the copper losses are the main losses so that they drive the global thermal behavior of the rotor and the stator. The temperature rise is then given by:

$$\Delta T = \frac{P_{cu}}{\text{eff}_{temp}(R_{ext}, L_s)}. \tag{32}$$

Where eff$_{temp}$ characterizes the ability of evacuating the heat generated by losses. This formulation is to be applied to the rotor:

$$\Delta T_r = \frac{P_{cu,r}}{\text{eff}_{temp,r}(R_{ext}, L_s)}. \tag{33}$$

Concerning the stator, it leads to:

$$\Delta T_s = \frac{P_{cu,s}}{\text{eff}_{temp,s}(R_{ext}, L_s)}. \tag{34}$$

The thermal efficiencies of the rotor and the stator are function of the stator outer radius as well as the machine stack length. This point must be underlined since the classical thinking that the machine length can be adapted to adjust the apparent power is now limited by the thermal considerations, themselves related to the machine length.

5.2.2.3 Thermal Loop

It has been explained that the copper losses depend on the copper temperature. This dependence consists in the electrical resistivity dependence with the temperature. The evolution law of the electrical resistivity with temperature is modeled by:

$$\rho_T = \rho_{T_0}[1 + \alpha_{cu}(T - T_0)]. \tag{35}$$

ρ_{T_0} is the electrical resistivity at the reference temperature T_0 and α_{cu} is the thermal coefficient of the copper for the electrical resistivity. With this law, the copper losses now evolve with temperature as:

$$P_{cu,T} = P_{cu,T_0}[1 + \alpha_{cu}(T - T_0)]. \tag{36}$$

Where P_{cu,T_0} are the copper losses at the reference temperature T_0.

The electro-thermal coupled modeling of the copper losses with the WRSG rotor and stator temperatures is solved with the fixed point method. The temperature is fixed, and then the copper losses are evaluated. With these copper losses, the temperature field of the WRSG is reevaluated. With this new temperature, the copper losses are recomputed. And so on until the temperature and the copper losses reach their equilibrium. This method is detailed in the flowchart of figure 5.5. It has to be noted that a stop criterion is introduced concerning either the electro-thermal equilibrium or the maximal execution number of the thermal loop. This latter point avoids staying infinitely in the thermal loop in case of a problem.

Optimization of Wound Rotor Synchronous Generators

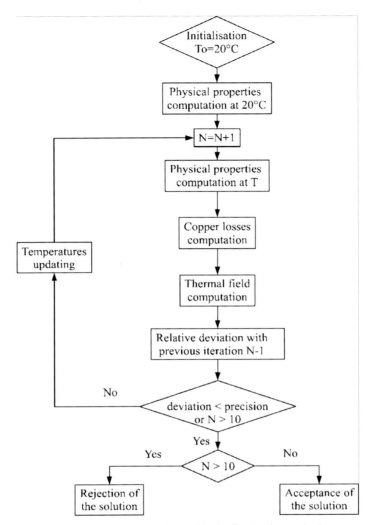

Fig. 5.5 Flowchart of the thermal loop solved with the fixed point method.

5.3 Optimization Problem

5.3.1 Interest of Designing with Optimization Procedure

Optimal design approach allows us to find the best system meeting the specifications of an application. The optimal design of electric machines often leads to minimize the weight, the cost or else to maximize the efficiency. In a way it permits to rationalize the design process. To perform this, the system has to be modeled with mathematical expressions as done in section 2. Generally the input parameters of the models can vary within specified ranges and the definition space of the output quantities is restricted due to constraints. These reasons lead the designer to define an optimization problem under constraints.

The optimization problem consists in determining the functions that have to be optimized, the variables of optimization and finally the constraints of optimization. When one single function is to optimize, it deals with single objective optimization, while when two or more objectives are to be optimized, it is the field of multi-objective optimization. The optimization constraints, in case of electric machine design, are often related to physical limitations, industrial and economical constraints, and to a minimal level of performances to satisfy.

In the following of the paper three single objective optimizations are presented: one taking individually the different machines of the range and two others that concern optimizations of a range of machines. Finally, a multi-objective optimization of the range efficiency and the average cost per machine of the range is carried out.

5.3.2 Choice of the Optimization Technique

WRSG design optimization problem is complex. It involves mixed variables, the objective functions are non-linear and have no simple analytical expressions. Moreover the global optimum is researched. These particular features are well treated by Genetic Algorithm (GA). Now concerning the solution representation in case of multi-objective optimization, the Pareto front determination can be performed using various techniques. Using a gradient-based algorithm, the Pareto front is determined in two steps. First, one objective is fixed and the other one is optimized. Then, this is done for several values of the first objective and it finally gives the Pareto front of the problem. This method requires as many optimization routines as the number of desired points in the Pareto front. The complexity in determining the Pareto front can be reduced using 'Multiobjective Optimization Evolutionary Approaches' (MOEA) [12]. These algorithms are based on the evolution theory of species and permit to converge toward the Pareto front into a single run of the optimization process. The definition of the optimization problem as well as the representation of the objectives justifies the choice of a MOEA based on GA. Among the various MOEA (NSGA II [13], NPGA II [14], SPEA 2 [15] ...) a NSGA II-based algorithm is used for solving the WRSG optimization problem.

Finally the single objective and multi-objective optimization routines will both involve GAs.

5.4 Optimization with Genetic Algorithm

5.4.1 Objectives of Optimization: Single and Multi-objective Optimization

In case of single objective optimization the goal is to maximize or minimize one or another of the characteristics of the system. In the case of a WRSG, the goals could be to minimize the cost, maximize the efficiency, the power density, or the ratio power/cost. The result is a set of optimization parameters optimizing the given objective.

In case of multi-objective optimization two main techniques are used: the aggregate function technique and the Pareto front technique [12, 16]. The philosophies of these two techniques are radically different. In the aggregate technique the multi-objective problem is transformed into a single-objective problem defining a new function:

$$F_{single} = \sum_{k} w_k F_k .\qquad(37)$$

Where w_k is the weighting factor associated to the objective F_k. This method supposes to set *a priori* the relative importance of the various objectives. On the contrary, the Pareto front philosophy suggests that the designer is not able to state clearly the relative importance of the various objectives. The outcome of this technique is a curve composed of different solutions where none of them are dominated; they are Pareto-optimal solutions. It means that, for the non dominated solutions, it is not possible to improve one single objective without decreasing at least another one. In the case of optimizing both the efficiency and the cost, the efficiency cannot be improved without making the cost increasing at the same time. Moreover the comparison between two solutions requires the introduction of a new comparison operator: solution A dominates solution B if all the objectives for solution A are superior to those of solution B [13, 16]. These concepts are illustrated in figure 5.6.

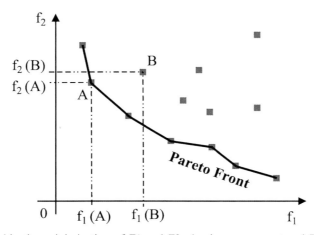

Fig. 5.6 Bi-objective minimization of F1 and F2: dominance concept and Pareto-optimal solutions.

Finally, using this tool the designer can make a choice *a posteriori* in a set of different solutions in order to privilege an objective or another.

5.4.2 Principle of Genetic Algorithm

The GA acts on a 'population of individuals'. Each individual is composed of a set of 'genes'. A gene, in our real-coded algorithm, corresponds to an optimization variable. The population, which must be initialized, evolves during the optimization process towards the global optimum. The objective functions are evaluated for each individual and then ranked. For the next population the elitism principle [12] is used as well as genetic operators. Only the crossover and the mutation operators are applied in this study. Finally the process is repeated until the algorithm stops, because of a maximum number of iterations or because of a stop criterion. The flow chart of the algorithm is described in figure 5.7.

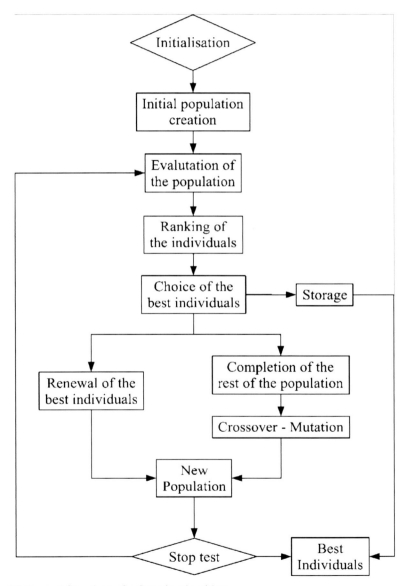

Fig. 5.7 General flowchart of a Genetic Algorithm.

5.4.2.1 Initialization

As for all algorithms, the process must be initialized. In this study a constrained optimization is performed which makes the choice of the initial population complex. Indeed, choosing randomly the initial population, the individuals' probability to satisfy the very strict constraints is low. To avoid such a behavior, which would slow the starting of the algorithm, some individuals satisfying the constraints are included in the random initial population.

5.4.2.2 Evaluation

This step simply corresponds to the computation of the objective functions such as the efficiency or the cost of the WRSG for all the individuals of the population.

5.4.2.3 Ranking

Single Objective Ranking

The individuals are ranked according to the value of their objective functions. Thus, the best individuals are on top of the rest of the population and the most interesting solutions are more likely to be involved in the generation of the next population.

Multi-objective Ranking

The implemented algorithm uses the ranking method developed by Deb in the NSGA II [13]. Two important concepts are used that must be explained: the particular sorting according to the front rank and the notion of 'crowding distance'. Once the whole population is evaluated, for each individual, the number of individuals dominating this individual is computed. This number corresponds to the front number of the individual. This way, all the individuals can be ranked according to their front number value. One issue in determining a Pareto front is to obtain a set of solutions well distributed all over the front. To solve this problem, Deb introduced the 'crowding distance'. This quantity is relative to the density of individuals on the front. For each individual, the Euclidian distance between each neighboring individual is computed, based on all the objectives. Finally inside each front number a second ranking is performed according to the crowding distance. These two concepts permit to accelerate the convergence of the algorithm towards the Pareto optimal front.

5.4.2.4 Selection

The selection involves the principle of elitism [12]. It is the inclusion of parents in the next population with their children and it allows making the convergence faster. The parents that are renewed in the next generation - whose proportion is tunable - are selected according to the performed ranking. The rest of the next population is created from the renewed parents, involving crossover and mutation.

5.4.2.5 Crossover

The crossover is an operation that allows creating a 'child' by combining the genes of existing individuals. This operation corresponds to an intensification of the research within one area of the research space. The crossover is linked to an occurrence probability. A random number is compared to the fixed crossover probability. Then the child is either a copy of one parent or a mix of several parents.

5.4.2.6 Mutation

The mutation is essential in the convergence of the algorithm. This operation enables to diversify the population and to leave a local optimum. The principle of mutation is to change randomly one gene of some individuals of the population. In our algorithm the mutation operator is applied once the whole population is completed using crossover. Only few individuals are changed with mutations.

5.4.2.7 Discrete Variable Handling

The crossover and mutation operators, in our algorithm, involve combining genes from parents including a small random dimension. This results in genes whose values are continuous. For the discrete genes, the round operator [17], relative to the discrete values allowed for the gene, is applied after the crossover and the mutation to ensure these particular genes to remain discrete.

Finally, all the steps of the GA flowchart are the same in the case of single and multiobjective optimization except the ranking.

5.5 Mono-objective Optimization of a WRSG Range

Several optimizations are run to minimize various objective functions. A set of three machines covering a power range is considered for optimization. From the initial 165kVA WRSG, whose characteristics are given in table 5.1, a range of three WRSGs has to be built. The three machines have power ratings of 125 kVA, 165 kVA, and 180 kVA. The goal of the consideration of a series of machine size is to show the interest of considering an ensemble of machine size when trying to optimize machines. First, the three WRSGs' costs are optimized separately. This approach leads to three different stator cross section geometries. This is unacceptable for industrial applications because it leads to a large cost of production tools. Then, a classical range design approach is applied to determine a series of machines in which the cross section geometry is identical for the three WRSGs. At last, a new objective function is introduced that integrates the specifications of all the machines at once.

The cost function, expressed in (38), is determined by the sum of the copper and iron weights multiplied by their respective material cost [5, 7, 9].

$$WRSG_{cost} = \sum weight \cdot material_{cost} \qquad (38)$$

It should again be noted here that there are eight design variables given in table 5.1 with their respective ranges to be considered in the optimization. The constraints to be considered are maximum temperature rise and minimum efficiency listed in table 5.2, and they are handled via penalty functions as in [5].

Table 5.1 WRSG's initial variable values with their ranges.

WRSG's parameters	125kVA Initial design	Range
Effective Power [kVA]	165	-
Outer stator diameter [mm]	390	[390,420]
Inner stator diameter [mm]	270.5	[255,290]
Slot diameter [mm]	317.5	[300,360]
Tooth width [mm]	10.2	[5,24]
Rotor pole width [mm]	86.5	[70,100]
Rotor pole opening factor	0.7045	[0.6,0.8]
Machine length [mm]	410	[210,420]
Conductor number	6	[5,12]
Relative cost	1	-

Table 5.2 Optimization constraints on output quantities.

Quantity	Constraint
Stator temperature rise	< 125K
Rotor temperature rise	< 125K
125kVA WRSG efficiency	$\eta > 91.6\%$
165kVA WRSG efficiency	$\eta > 91.7\%$
180kVA WRSG efficiency	$\eta > 91.7\%$

5.5.1 Independent Optimization of Three Different Machines

The three machines are first optimized independently. The optimizations are carried out on the eight variables of table 5.1. The cost is minimized respecting the minimum efficiencies of 91.6%, 91.7%, and 91.7% respectively for the 125, 165, and 180 kVA WRSGs. The results are given in table 5.3 and the convergence of the algorithm, in case of the 125kVA optimization, is presented in figure 5.8.

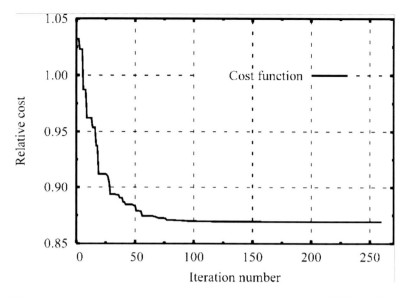

Fig. 5.8 Cost function evolution vs. iteration number for the 165 kVA WRSG. The results are normalized according the initial 165 kVA design.

The optimization is stopped according to two rules. The first criterion, for a single objective optimization, is the number of consecutive iterations where the objective function value does not vary. It is considered that if the objective function value is the same during n successive iterations, the algorithm has converged. The second criterion is a maximal number of iterations. When this number is reached, the algorithm stops. This allows limiting the computation time.

Table 5.3 Machine characteristics resulting of independent optimizations of the WRSGs.

variables	125 kVA	165 kVA	180 kVA
Outer stator diameter [mm]	393.7	420	420
Inner stator diameter [mm]	255.2	276.9	278.9
Slot diameter [mm]	311	326.1	331.5
Tooth width [mm]	11.1	12.9	12.4
Rotor pole width [mm]	89	100	96.9
Rotor pole opening factor	0.748	0.7415	0.7415
Machine length [mm]	285.2	331.7	349.9
Conductor number	8	6	6
Stator temperature rise [K]	83.0	125	125
Rotor temperature rise [K]	124.8	125	125
Efficiency [%]	91.60	91.79	92.34
Relative cost	0.7129	0.8693	0.9589

It can be noted in table 5.3 that for the 165 and 180 kVA machines the constraints on rotor and stator temperature rises are saturated as well as outer stator diameter. The 125 kVA WRSG has two saturated constraints: the rotor temperature rise and the efficiency. These results are in good agreement with usual remarks: larger machines are often very efficient but limited by temperature constraints. On the other hand smaller machines are known to have lower heating but poorer efficiencies.

5.5.2 Classical Design Approach Covering a WRSG Range

When designing a series of machines in an industrial environment, an important constraint is the production cost. To reduce investment costs, a strong constraint is introduced: all the machines must have the same cross section geometry. This means that the three machines of our power range differ only in their stack length and their conductor number. Usually the largest machine cost is optimized and then the stack length is adjusted as well as the conductor number in order to obtain the two smaller power machine designs.

Thus starting from the 180kVA optimal design of table 5.3, two-variable optimizations (stack length and conductor number) are carried out for the powers of 125kVA and 165kVA, according to the constraints of table 5.2. The results are given in table 5.4.

Table 5.4 Machine characteristics resulting from the classical design approach: adapting the largest machine to obtain the two smaller power machines.

variables	125 kVA	165 kVA	180 kVA
Outer stator diameter [mm]		420	
Inner stator diameter [mm]		278.9	
Slot diameter [mm]		331.5	
Tooth width [mm]		12.4	
Rotor pole width [mm]		96.9	
Rotor pole opening factor		0.7415	
Machine length [mm]	261.8	349	349.9
Conductor number	8	6	6
Stator temperature rise [K]	87.1	97.8	125
Rotor temperature rise [K]	90.3	106.9	125
Efficiency [%]	91.61	92.53	92.34
Relative cost	0.7526	0.9568	0.9589

One can think of starting from the cross section design of the smallest optimized machine and then having the stack length increased and the conductor number changed in order to reach the specified powers of 165kVA and 180kVA. An optimization was performed according to this idea but the algorithm was unable to find any solution. Indeed a feasible solution for the largest machine would be to enlarge the stack length beyond the upper fixed limit: the optimal stack length of the largest machine would be 0.478 m. This confirms the approach adopted by manufacturers: designing the largest machine and then having the stack length shortened. This way, one can be sure to keep all the machines of the series within a fixed volume. But fixing the cross section geometry from the largest optimized WRSG may not be the most optimal solution when considering the cost of the entire range.

5.5.3 Simultaneous Optimization of a Series of Machines

Here, an optimization of the cost of the entire range, considering the three machines at the same time, is proposed. The objective function to minimize is then a weighted sum (39) of the individual costs of the three machines. This method is similar to the annual energy production (AEP) optimization when considering a wind generator [5, 10].

$$range_{cost} = \sum_{i=1}^{3} \alpha_i \cdot WRSG_{cost,i} \qquad (39)$$

The weighting factors α_i correspond to the repartition of the sales inside the considered range. Thus the cost function to minimize is the average cost per machine.

The design variables are therefore changed. Instead of one stack length and one conductor number, there are now three stack lengths and three conductor numbers, one stack length and one conductor number per machine. Moreover, the geometric variables of table 5.1 are still involved in the optimization. This leads to a 12 variables optimization. The constraints of table 5.2 should still be satisfied.

The steps are as follows. For a given individual (vector of given parameters), the geometric variables plus the stack length and conductor number related to the 125kVA are used to computed the 125kVA WRSG cost and to verify the constraints. If the variables do not allow satisfying the 125kVA constraints, the penalty function is applied to the 125kVA WRSG cost. The same is done for the two other machines. Finally, one gets three costs, eventually penalized if some constraints are not satisfied, that now must be summed weighted to build the average cost per machine of the global range.

Three different cases are studied, corresponding to three different sets of weighting factors given in table 5.5.

Table 5.5 Weighting factors for the three cases under study.

	125 kVA	165 kVA	180 kVA
Case I	8/10	1/10	1/10
Case II	1/10	8/10	1/10
Case III	1/10	1/10	8/10

Optimization of Wound Rotor Synchronous Generators

The optimization results of the three cases are given in tables 5.6, 5.7, and 5.8 associated with the considered constraints. Moreover the savings due the new proposed method are given.

Table 5.6 Optimal Machines' characteristics in case of non uniform sales, case I.

	Case I		
	125 kVA	165 kVA	180 kVA
Outer stator diameter [mm]		412	
Inner stator diameter [mm]		267.6	
Slot diameter [mm]		324.1	
Tooth width [mm]		12.1	
Rotor pole width [mm]		94.7	
Rotor pole opening factor		0.7472	
Machine length [mm]	266.2	360.7	394.8
Conductor number	8	6	6
Stator temperature rise [K]	82.3	93.3	125
Rotor temperature rise [K]	110.4	124.9	124.9
Efficiency [%]	91.60	92.53	92.40
Relative cost	0.726	0.9372	1.0135
Relative average cost per machine		0.7759	
Saving compared to classical approach		- 2.24 %	

Table 5.7 Optimal Machines' characteristics in case of non uniform sales, case II.

	Case II		
	125 kVA	165 kVA	180 kVA
Outer stator diameter [mm]		420	
Inner stator diameter [mm]		276.9	
Slot diameter [mm]		332.6	
Tooth width [mm]		12.3	
Rotor pole width [mm]		95.1	
Rotor pole opening factor		0.7472	
Machine length [mm]	266.1	306.6	355
Conductor number	8	7	6
Stator temperature rise [K]	78.6	125	111.7
Rotor temperature rise [K]	89.5	125	125
Efficiency [%]	91.89	92.00	92.59
Relative cost	0.7753	0.8712	0.9857
Relative average cost per machine		0.8731	
Saving compared to classical approach		- 6.78 %	

Table 5.8 Optimal Machines' characteristics in case of non uniform sales, case III.

	Case III		
	125 kVA	165 kVA	180 kVA
Outer stator diameter [mm]		420	
Inner stator diameter [mm]		277.6	
Slot diameter [mm]		330.6	
Tooth width [mm]		12.4	
Rotor pole width [mm]		96.8	
Rotor pole opening factor		0.7427	
Machine length [mm]	261.3	348	352.1
Conductor number	8	6	6
Stator temperature rise [K]	86.7	97.3	125
Rotor temperature rise [K]	92.5	110.1	125
Efficiency [%]	91.60	92.51	92.35
Relative cost	0.7478	0.95	0.9596
Relative average cost per machine		0.9381	
Saving compared to classical approach		- 0.07 %	

Whatever the chosen weighting factors may be, the proposed cost function gives better results than the classical design approach. This will give a non-negligible cost reduction for manufacturers. When analyzing the cost of the optimized WRSGs, it can be observed that the heaviest weighted machine is the one which is privileged during the optimization process. This also means that the saturated constraints are often those of the privileged WRSG as can be seen in tables 5.6, 5.7, and 5.8. In case I the minimum efficiency is saturated for the smallest machine and the larger ones are constrained by their temperature rises. In case II the 165 kVA and 180 kVA WRSG are both limited by temperature rise. At last in case III the largest WRSG is thermally constrained and the smallest one by its minimal efficiency.

As expected the cross section design resulting from the simultaneous optimization of the whole series depends on the sales' distribution. The differences in the results confirm the adopted approach.

The global range consideration has been included into a single objective optimization concerning the global range cost. An extension of the method to multi-objective optimizations, involving efficiencies' considerations, is proposed in the following section.

5.6 Multi-objective Optimization of a WRSG Range

In the previous sections, the optimizations were conducted with the goal of minimizing the global cost of the range. Three cases of sale repartitions have been considered. For each case, one of the WRSG dominates the sales. It can be interesting to propose a trade-off between the global cost of the range and the efficiency of the most sold WRSG.

The twelve optimization variables are the same as for the simultaneous optimization of a series of machines at once (paragraph 5.3.). The constraints of table 5.2 are still of concern.

The following multiobjective optimization is the minimization of the relative average cost per machine and the maximization of the 125kVA WRSG efficiency; it corresponds to the sales' distribution of case I. This is performed respecting all the optimization constraints of table 5.2. Moreover the constraints on the minimal efficiency concerning the 165kVA and the 180kVA WRSG are considered. The resulting Pareto front of the machine mean cost vs. the 125kVA WRSG efficiency is plotted in figure 5.9. For each solution of the front, the efficiencies of the two others machines of the range are computed and reported in figure 5.9.

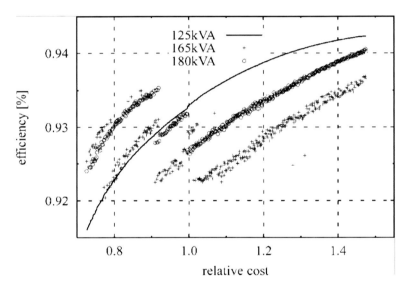

Fig. 5.9 Pareto front of the machine mean cost vs. the 125kVA WRSG efficiency. Efficiencies of the 165kVA and 180kVA WRSGs are reported as output values.

One can observe that the efficiency grows with the relative average cost per machine. The ratio between the less expensive solution and the most expensive one is around 2. Concerning the efficiency, the maximal efficiency is roughly 0.25 point larger than the minimal efficiency. It is to be noted that the constraint on minimal efficiency is satisfied whatever the solution of the front may be.

In some cases it would be preferable to choose a 125kVA WRSG whose efficiency is about 0.93 instead of one of 0.935. Indeed in the first case, one can see that the efficiencies of the other machines of the range are larger than in the second case. The same observation can be made concerning the 125KVA WRSG whose cost is around 1. It should be better to chose a solution whose cost is slightly lower than 1 because the efficiencies of the other machines of the range are better than in the case of solution whose price is slightly higher than 1.

The choice of a solution in a Pareto front is not always an easy task. A tool to help in the final choice is to represent other output quantities according to the front. That is what is performed in figure 5.9 where the Pareto front is superimposed with the efficiencies of the other machines of the range.

In this section, the global mean cost of the range has been considered as an optimization goal and the efficiency of the most sold WRSG of the range as the other optimization goal. To go further, the global efficiency of the range can be considered instead of the efficiency of the most sold WRSG of the range.

5.7 Conclusion

A GA optimization process has been applied to wound rotor synchronous generators' design aiming at minimizing the cost of the machine. First, three different WRSGs, differentiated by their power, have been independently optimized under geometrical, thermal and efficiency constraints. Then the three WRSGs' designs were considered from an industrial point of view which introduces a new geometric constraint: the cross section dimensions of the three WRSGs must be the same in order to avoid excessive investments in the production tool. Using this consideration a classical design approach is carried out and its results are compared with the results from the minimization of the newly proposed objective function. As shown, the new approach gives better results for every studied case. Very few adjustments are needed to optimize the whole range at once and the results show that the global optimization of the series can bring an important reduction of the global range cost.

The method is extended to multi-objective optimization of the global cost of the range and of the efficiency of the most sold machine. One can think about an extension of this multi-objective optimization by changing the efficiency objective into a function representing the ability of the range in converting the mechanical energy into electrical energy. These considerations will be presented in a further paper.

The method can be extended to series of any kind of electrical equipment through a straightforward process.

References

[1] Belmans, R., De Keulenaer, H., Blaustein, E., Chapman, D., De Almeida, A., De Wachter, B., Radgen, P.: Energy Efficient Motor Driven Systems. Technical report, Motor Challenge (2004)
[2] Kreuawan, S., Gillon, F., Brochet, P.: Optimal design of permanent magnet motor using Multidisciplinary Design Optimization. In: Proceedings of the 2008 International Conference on Electrical Machines (2008)

[3] Legranger, J., Friedrich, G., Vivier, S., Mipo, J.-C.: Combination of Finite-Element and Analytical Models in the Optimal Multidomain Design of Machines: Application to an Interior Permanent Magnet Starter Generator. IEEE Trans. on Energy Convers (2010) doi:10.1109/TIA.2009.2036549
[4] Bellegarde, N., Dessante, P., Vidal, P., Vannier, J.-C.: Optimisation of a drive system and its epicyclic gear set. Stud. in Comput. Intell (2008), doi:10.1007/978-3-540-78490-6_29
[5] Li, H., Chen, Z., Polinder, H.: Optimization of Multibrid Permanent-Magnet Wind Generator Systems. IEEE Trans. on Energy Convers (2009) doi:10.1109/TEC.2008.2005279
[6] Dessante, P., Vannier, J.-C., Rippol, C.: Optimization of a linear brushless DC motor Drive. Recent Dev. of Electr. Drives (2006), doi:10.1007/978-1-4020-4535-6_12
[7] Candela, C., Morín, M., Blázquez, F., Platero, C.A.: Optimal Design of a Salient Poles Permanent Magnet Synchronous Motor Using Geometric Programming and Finite Element Method. In: Proceedings of the 2008 International Conference on Electrical Machines (2008)
[8] Kano, Y., Matsui, N.: A Design Approach for Direct-Drive Permanent-Magnet Motor. IEEE Trans. on Industry Appl (2008), doi:10.1109/TIA.2008.916600
[9] Schätzer, C., Müller, W., Binder, A.: Vector optimization of two-dimensional numerical field problems applied to the design of a wind turbine generator. Math. and Comput. in Siml (1999), doi:10.1016/S0378-4754(99)00012-9
[10] Jung, H., Lee, C.-G., Hahn, S.-C., Jung, S.-Y.: Optimal Design of Direct-Driven PM Wind Generator Applying Parallel Computing Genetic Algorithm. In: Proceeding of the 2007 International Conference on Electrical Machines and Systems (2007)
[11] Methods for Determining Losses and Efficiency of Rotating Electrical Machinery from Tests (Excluding Machines for Traction Vehicles), IEC 34-2 (1972)
[12] Zitzler, E., Deb, K., Thiele, L.: Comparison of multiobjective evolutionary algorithms: Empirical results. Evol. Comput (2000), doi:10.1162/106365600568202
[13] Deb, K., Pratab, A., Agrawal, S., Meyarivan, T.: A fast elitist non-dominated sorting genetic algorithm for multi-objective optimization: NSGA II. IEEE Trans. On Evol. Comput. (2002), doi:10.1109/4235.996017
[14] Erickson, M., Mayer, A., Horn, J.: The niched pareto genetic algorithm 2 applied to the design of groundwater remediation systems. In: Evol. Multi-Criterion Optim. Springer, Heidelberg (2001), doi:10.1007/3-540-44719-9_48
[15] Zitzler, E., Laumanns, M., Thiele, L.: Spea2: Improving the strength pareto evolutionary algorithm, Tech. Rep., Swiss Federal Institute of Technology (ETH), Zurich, Switzerland (2001)
[16] Dréo, J., Petrowski, A., Taillard, E., Siarry, P.: Metaheuristics for Hard Optimization. Springer Editions, Paris (2006)
[17] Lebesnerais, J., Lanfranchi, V., Hecquet, M., Brochet, P.: Mixed-variable multi-objective optimization of induction machines including noise minimization. IEEE Trans. on Magn (2008), doi:10.1109/TMAG.2007.916173

Chapter 6
Simple and Fast Algorithms for the Optimal Design of Complex Electrical Machines

Kazumi Kurihara[1], Tomotsugu Kubota[1], and Yuki Imaizumi[2]

[1] Department of Electrical and Electronic Engineering, Ibaraki University,
4-12-1Nakanarusawa Hitachi 316-8511, Japan
`kurihara@mx.ibaraki.ac.jp`
[2] Panasonic Electric Works Obihiro Co., Ltd. 1-2-1 Nishi25jyo-kita, Obihiro,
080-2493, Japan
`zumi1@ezweb.ne.jp`

Abstract. This paper presents the simple and fast algorithms for the optimal design of complex electrical machines. A single-phase capacitor-run permanent-magnet (PM) motor whose auxiliary winding is supplied through a capacitor is a kind of the complex electrical machine. The exact motor performance analysis using two-axis theory is very difficult because the operation has been complicated by the imbalance between the main and auxiliary winding voltages. Besides, this motor has the pulsating torque component corresponding to double fundamental frequency of the supply voltage due to the backward stator rotating field. The successful combined method of the time-stepping finite element analysis (FEA) and response surface methodology (RSM) have been proposed to minimize the torque ripple and maximize the efficiency. The optimum values of the capacitance for running and stator-slot skew pitch to minimize the torque ripple and maximize the efficiency are obtained respectively.

6.1 Introduction

The single-phase capacitor-run PM motor for application in home appliances, such as refrigerator compressors [1] is one of the typical examples of complex electrical machines. This motor is the single-phase version of a three-phase line-start PM motor (LSPM) [2]. However, in single-phase motors, where the auxiliary winding is supplied through a capacitor, the operation has been further complicated by the imbalance between the main and auxiliary winding voltages. Because of this, the backward stator rotating field occurs and increases the pulsating torque, losses, vibrations, and noises compared to those of the three-phase LSPMs. In particular, it is well-known that the backward stator rotating field causes the pulsating torque component corresponding to double fundamental frequency of the supply voltage source [3]. Herein, it is emphasized that the pulsating torque must be reduced by the optimal choice of the capacitance for running. Besides, the single-phase PM motors as well as three-phase PM motors have the pulsating torque component by the space harmonics. However, it is well-known that the latter component has been reduced drastically by the suitable skewing [4]. It is described that the optimum values of the capacitance for running and stator-slot skew pitch to minimize the torque ripple at the rated load torque are obtained.

In this paper, the 2-D time-stepping FEA to obtain the accurate line-currents and torque is used, where the multislice model taking the skew effect into account is used to reduce the computing time [5]. The fundamental equations for the magnetic field, the voltage and current equations for the circuit of the motor, and the dynamic equation are given respectively. Next, the calculation process to obtain the line-currents and rotational angle is shown. Then, the suitable time step and number of the multislice will be shown. The validity of the simulation results by the FEA is confirmed by the comparison with the experimental results. Besides, it is shown that the agreement between the computed and experimental results in starting is also very good. Further, the RSM [6]-[8] has been used to determine the values of the running capacitance and stator-slot skew pitch to minimize the torque ripple of single-phase capacitor-run PM motor at the rated voltage and load torque from the simulation results by the FEA. The response surface of the computed torque ripple at the rated voltage and load torque has been shown. The optimum values of the capacitance for running and stator-slot skew pitch to minimize the torque ripple are obtained respectively. Further, the response surface of the computed efficiency at the rated voltage and load torque has been shown. The proposed method is quite useful for the design and analysis of the single-phase PM motors using both capacitors for running and starting.

6.2 Experimental Motor and Circuit

Fig.6.1 shows the cross section of the experimental rotor [5] used in this paper. It is composed of aluminium cage bars, arc-shaped interior Neodymium-Boron-Iron magnets, and flux barriers. A 50-Hz, 100-V, two-pole single-phase capacitor-run induction motor is used for testing the experimental rotor. The rated torque and output power are 0.225N·m and 70.7W, respectively. The rotor slots are skewed by 1.15 times of one stator slot pitch (or 1.34 times of one rotor slot pitch).

Fig. 6.2 shows the experimental circuit of the experimental motor. C_r and C_s are the capacitances for running and starting, respectively. Each value of C_r and C_s is 14μF and 150μF, respectively. e_m and e_a are the induced voltages for main and auxiliary windings, respectively. i_m is the main winding current. i_a is the auxiliary winding current. i_{a1} and i_{a2} are the currents across the capacitance C_r and C_s, respectively. PTC is the resistance of a positive temperature coefficient resistor.

Simple and Fast Algorithms for the Optimal Design

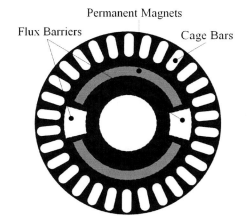

Fig. 6.1 Cross section of experimental rotor

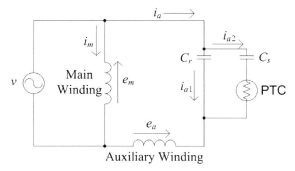

Fig. 6.2 Circuit of a single-phase capacitor-run PM motor

6.3 Method for Analysis

6.3.1 Finite Element Analysis (FEA)

The fundamental equations for the magnetic field are represented in the two-dimensional rectangular coordinates as
The following assumptions for this analysis are made.

1) The eddy currents in the iron cores are neglected because the iron cores are laminated.
2) The skewing effect is taken into account by skewing stator slots equivalently. The stator in the axial direction is considered as composing of multislices [4].
3) The effect of the eddy current for the rotor ends is taken into account by multiplying the conductivity of the rotor bars by the coefficient [5],[10].

These assumptions reduce the analysis to a two-dimensional problem.

The fundamental equations for the magnetic field are represented in the two-dimensional rectangular coordinates as

$$\frac{\partial}{\partial x}\left(\nu \frac{\partial A}{\partial x}\right) + \frac{\partial}{\partial y}\left(\nu \frac{\partial A}{\partial y}\right) = -J_0 - J_e - J_m \quad (1)$$

$$J_e = -\sigma \frac{\partial A}{\partial t} \quad (2)$$

$$J_m = \nu_0 \left(\frac{\partial M_y}{\partial x} - \frac{\partial M_x}{\partial y}\right) \quad (3)$$

where A is z component of magnetic vector potential $\mathbf{A}$, J_0 is the stator-winding current density, J_e is the eddy-current density, J_m is the equivalent magnetizing current density, M_x, M_y are the x and y components of the magnetization $\mathbf{M}$, respectively [5], σ is the conductivity, and ν is the reluctivity. The value of ν in the PM is assumed the same as the reluctivity of free space ν_0. J_m is assumed zero outside the PM.

The voltage and current equations are given as

$$e_m + r_m i_m + L_m \frac{\partial i_m}{\partial t} = v \quad (4)$$

$$e_a + R_a i_a + L_a \frac{\partial i_a}{\partial t} + \frac{1}{C_r} \int i_{a1} dt = v \quad (5)$$

$$i_a = i_{a1} + i_{a2} \quad (6)$$

$$\frac{1}{C_r} i_{a1} = \frac{1}{C_s} i_{a2} + r_s \frac{\partial i_{a2}}{\partial t} \quad (7)$$

where v is the terminal voltage, r_m and L_m are the resistance and end-winding leakage inductance of the main winding, r_a, L_a are the resistance and end-winding leakage inductance of the auxiliary winding, r_s is the resistance of PTC. e_m is given by the line integral of the vector potential round c_m which is along the main winding similarly; e_a is given by the line integral of the vector potential round c_a which is along the auxiliary winding

$$e_m = \oint_{c_m} \frac{\partial A^t}{\partial t} ds = \sum_{k=1}^{N} \left(\oint_{c_m^{(k)}} \frac{A^{t(k)} - A^{t-\Delta t(k)}}{\Delta t} ds \right) \quad (8)$$

$$e_a = \oint_{c_a} \frac{\partial A^t}{\partial t} ds = \sum_{k=1}^{N} \left(\oint_{c_a^{(k)}} \frac{A^{t(k)} - A^{t-\Delta t(k)}}{\Delta t} ds \right) \quad (9)$$

where A^t is A at time t. Δt is the time step. N is the number of slices. $c_m^{(k)}$ and $c_a^{(k)}$ are the vector potential rounds of the kth slice, respectively.

Simple and Fast Algorithms for the Optimal Design

One can obtain the vector potential and currents by solving (1), and (4)-(7) using the time-stepping finite element technique [5]. After that, the torque T can be obtained by

$$T = \sum_{k=1}^{N} T^{(k)}. \tag{10}$$

where $T^{(k)}$ is the torque in the kth slice and calculated by using the *Bil* rule [9].

Next, the calculation steps for this analysis are shown in Fig. 6.3.

1) First, the terminal voltage V, the load angle δ, and Δt, and the rotational step angle θ_s are set, respectively. The instantaneous value v^t of the terminal voltage at time t can be represented by

$$v^t = \sqrt{2}V \cos(\omega t + \phi_0) \tag{11}$$

where ϕ_0 is the initial phase angle of v^t and ω is the angular frequency.

$$\phi_0 = \delta \tag{12}$$

θ_s is the constant value at synchronous speed and given by

$$\theta_s = \frac{\omega \cdot \Delta t}{p} \tag{13}$$

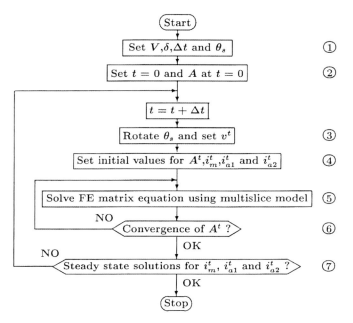

Fig. 6.3 Flowchart

where p is the pole pair number.
2) The vector potential A at $t = 0$ is set, where the static field caused by only PMs is given as the initial value.
3) At $t = t+\Delta t$, the terminal voltage at new t is set after the rotation of the rotor mesh by θ_s.
4) The initial values for A^t, i_m^t, i_{a1}^t, and i_{a2}^t are set.
5) The matrix equation constructed by the time-stepping finite element technique using multislice model is solved [5].
6) The convergence of A^t is tested. Unless A^t converges, the process returns to step 5).
7) After the convergence of A^t, i_m^t, i_{a1}^t, and i_{a2}^t, T^t can be calculated.
8) The calculation process from step 3) to step7) continues until the steady-state currents are obtained.

6.3.2 Response Surface Methodology (RSM)

In this paper, the RSM is used to determine the values of the running capacitance and stator-slot skew pitch to minimize the torque ripple of single-phase capacitor-run PM motor the at the rated voltage and load torque from the simulation results by the FEA. The polynomial approximation model for a second-order fitted response u can be represented by [8]

$$u = \beta_0 + \sum_{j=1}^{k} \beta_j x_j + \sum_{j=1}^{k} \beta_{jj} x_j^2 + \sum_{i \ne j}^{k} \beta_{ij} x_i x_j + \varepsilon \tag{14}$$

where β is regression coefficient. x is variable. ε is random error.

The variables are x_1 for the value of the stator-slot skew pitch and x_2 for the value of the capacitance for running. The approximation function $f(x_1, x_2)$ is represented as follows

$$u = f(x_1, x_2) + \varepsilon = \beta_0 + \beta_1 x_1 + \beta_2 x_2 + \beta_3 x_1^2 + \beta_4 x_2^2 + \beta_5 x_1 x_2 + \varepsilon \tag{15}$$

The least squares method is used to estimate unknown coefficients.

6.4 Steady-State Performance Characteristics

The validity of the simulation results by the FEA is confirmed by the comparison with the experimental results in this section. Table 6.1 shows the parameters for the simulation. The number of slices in the multislice model was 5 [5] and the time step was 69 μs. The value of the time step is directly concerned with the number of the stator and rotor slots [10]. This value was the suitable time step.

Simple and Fast Algorithms for the Optimal Design

6.4.1 EMF Due to PMs

Fig. 6.4 shows the flux distribution caused by PMs.

Fig. 6.5 shows the terminal voltage waveform generated by PMs in driving the experimental motor at 3000 r/min by the external motor. It is shown that the agreement between the computed and measured values of the generated voltage is excellent.

Table 6.1 Parameters for the simulation

Number of slices in multislice model	5
Time Step Δt	69(μs)

6.4.2 No-Load Performance Characteristics

Fig. 6.6 shows the main- and auxiliary-winding currents versus the terminal voltage at no-load under a single-phase voltage source. It is shown that the agreement between computed and measured values of the no-load currents is good.

Fig. 6.7 shows the no-load main-winding current waveform at 100V. It is shown that the agreement between computed and measured values of the main winding current is excellent.

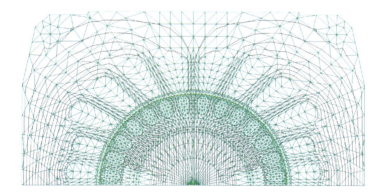

Fig. 6.4 Flux distribution caused by PMs

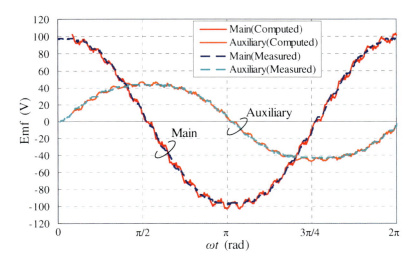

Fig. 6.5 EMF generated by PMs

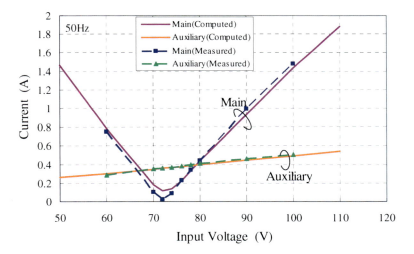

Fig. 6.6 No-load currents versus terminal voltage

Simple and Fast Algorithms for the Optimal Design

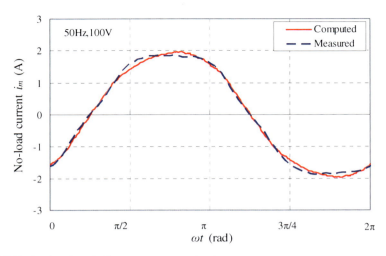

Fig. 6.7 No-load main-winding current at rated voltage

6.4.3 Load Performance Characteristics

Fig. 6.8 shows the computed and measured results of the main- and auxiliary-winding currents versus the output power at 100V. It is shown that the agreement between computed and experimental results is good.

Fig. 6.9 shows the computed and measured results of the efficiency versus the output power at 100V. It is shown that the agreement between computed and measured results is good. The measured efficiency at rated output was 72.6%. The measured value of the stator iron-loss is included in computed values only in computing the efficiency. Its value is 19.7[W].

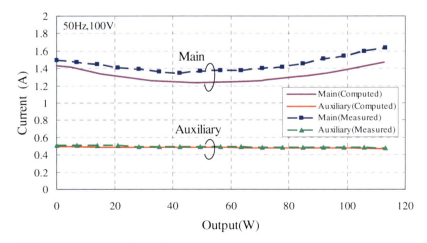

Fig. 6.8 Computed and experimental results of currents versus output

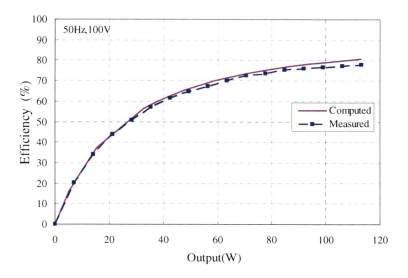

Fig. 6.9 Computed and experimental results of efficiency versus output

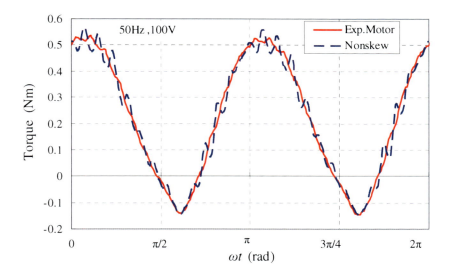

Fig. 6.10 Computed torque under rated load torque

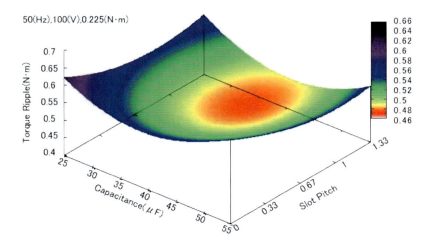

Fig. 6.11 Response surface of torque ripple at rated load torque

6.4.4 Torque Ripple and Efficiency

Fig. 6.10 shows the computed torque waveform under rated load torque. It is seen from the figure that the main component of the pulsating torque is due to the negative sequence field and the frequency is 100Hz. Besides, it is seen that higher harmonic components become small due to the skewing effect.

Fig. 6.11 shows the response surface of the computed torque ripple at the rated voltage and load torque. The optimum values of the capacitance for running and stator-slot skew pitch to minimize the torque ripple were 37.5µF and 0.97 times of one stator slot pitch, respectively. The minimum value of the torque ripple was 0.50N·m. 37.5µF was 2.68 times of C_r and 0.25 times of C_s used in the experiment. It was found that the RSM is quite useful for minimizing the torque ripple of the single-phase capacitor-run PM motor. Fig. 6.12 shows the response surface of the computed efficiency at the rated voltage and load torque. The optimum values of the capacitance for running and stator-slot skew pitch to maximize the efficiency were 7.25µF and 0.51 times of one stator slot pitch, respectively. The maximum value of the efficiency was 73.8%. 7.25µF is 0.51 times of C_r and 0.048 times of C_s used in the experiment. It was found from Fig. 6.11 and 6.12 that the value of C_r for minimizing the torque ripple is different from that for maximizing the efficiency.

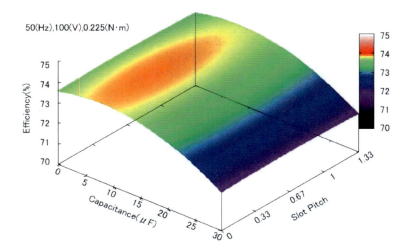

Fig. 6.12 Response surface of efficiency at rated load torque

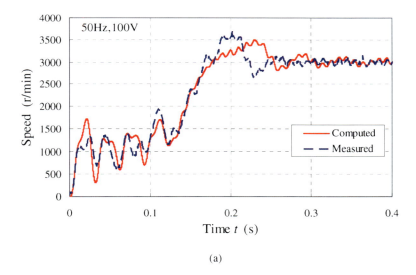

(a)

Fig. 6.13 Computed and measured results of starting performance. (a) Speed-time responses (b) Current i_m versus time.

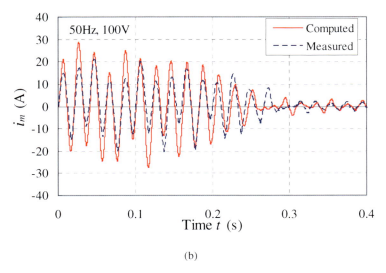

(b)

Fig. 6.13 (*continued*)

6.5 Starting Performance Characteristics

The same value as the time step used in the steady-state performance analysis has been used in the starting performance analysis.

Fig. 6.13 shows the computed and measured speed-time responses with time during run up and synchronization period when ϕ_0 is zero. It can be seen that good agreement between the computed and measured results exists. This is due to good correspondence with simulation and experimental conditions.

6.6 Conclusions

This paper presents the simple and fast algorithms for the optimal design of complex electrical machines. The numerical method for the analysis to minimize the torque ripple of the single-phase capacitor-run PM motor using the time-stepping FEA and RSM has been illustrated. The multislice model taking the skew effect into account reduced the analysis to a 2-D problem and the computing time for the steady-state and transient performance analysis using the time-stepping FEM. The optimum values of the capacitance for running and stator-slot skew pitch to minimize the torque ripple at the rated load torque have been obtained from the response surface of the computed torque ripple. Further, the optimum values of those to maximize the efficiency at the rated voltage and load torque has been shown from the response surface of the computed efficiency. It was found that the value of the capacitance for minimizing the torque ripple is different from that for maximizing the efficiency. The optimum design for realizing both maximum efficiency and minimum torque ripple concurrently will be presented in the future works.

It is considered that the proposed method is quite useful for the design and analysis of the complex electrical machines.

Acknowledgments

The authors wish to thank Dr. Ide, Dr. Mikami, S. Wakui, and M. Hori of the Hitachi Research Laboratory, Hitachi Ltd. for technical support and T. Ito, T. Ohtsu, H. Ishige of the Ibaraki University for experimental support.

References

1. Popescu, M., Miller, T.J.E., McGilp, M., Strappazzon, G., Trivillin, N., Santarossa, R.: Asynchronous performance analysis of a single-phase capacitor-start, capacitor-run permanent magnet motor. IEEE Trans. Energy Convs. 20(1), 142–149 (2005)
2. Miyashita, K., Yamashita, S., Tanabe, S., Shimozu, T., Sento, H.: Development of a high-speed 2-pole permanent magnet synchronous motor. IEEE Trans.Power App. Syst. 99(6), 2175–2181 (1980)
3. Hsu, J.S.: Monitoring of defects in induction motors through air-gap torque observation. IEEE Trans. Ind. Appl. 31(5), 1016–1021 (1995)
4. Kurihara, K., Monzen, T., Hori, M.: Steady-State and Transient Performance Analysis for Line-Start Permanent-Magnet Synchronous Motors with Skewed Slots. In: Proc. Int. Conf. Elect. Machines, Crete Island, Greece, September 2-5 (2006)
5. Hori, M., Kurihara, K., Kubota, T.: Starting performance analysis for a single-phase capacitor run permanent magnet motor with skewed rotor slots. In: Proc. Int. Conf. Elect. Machines, Vilamoura, Portugal, September 6-9 (2008)
6. Park, J.M., Kim, S.I., Hong, J.P., Lee, J.H.: Rotor design on torque ripple reduction for a synchronous reluctance motor with concentrated winding using response surface methodology. IEEE Trans. on Magn. 42(10), 3479–3481 (2006)
7. Choi, Y.C., Park, M.H., Lee, M.M., Lee, J.H., Chun, J.S.: Optimum design solutions to minimize torque ripple of concentrated winding synRM using response surface methodology. In: presented at the 16th Int. Conf. on the Computation of Electromagnetic Fields (Compumag 2007), Aachen, Germany, June 24-28 (2007)
8. Hong, D.K., Kim, J.M.: Optimum design of Maglev lift system's electromagnet for weight reduction using response surface methodology. COMPEL 27(4), 797–805 (2008)
9. Binns, K.J., Riley, C.P., Wong, M.: The efficient evaluation of torque and field gradient in permanent-magnet machines with small air-gap. IEEE Trans. Magn. 21, 2435–2438 (1985)
10. Kurihara, K., Rahman, M.A.: High-efficiency line-start interior permanent-magnet synchronous motors. IEEE Trans. Ind. Applicat. 40(3), 789–796 (2004)

Chapter 7
The Flock of Starlings Optimization: Influence of Topological Rules on the Collective Behavior of Swarm Intelligence

Francesco Riganti Fulginei and Alessandro Salvini

Department of Applied Electronics - Roma Tre University

Abstract. This chapter presents an algorithm, the flock of starlings optimization that is inspired both to the famous Particle Swarm Optimization (PSO) and to recent naturalistic observations on collective animal behaviour, performed by M. Ballerini et al. The presented algorithm implements a virtual flock governed by topological interactions between its members. The proposed approach has been validated by using classical benchmarks and compared with different versions of PSO. Results have shown that the algorithm has high exploration capability, avoids local minima entrapments and is particularly suitable for multimodal optimizations.

7.1 Introduction

The Particle Swarm Optimization (PSO) is one of the most used and studied optimization methods first proposed by James Kennedy and Russell Eberhart in 1995 [1], starting from the works of Reynolds [2] and Heppner and Grenander [3]. In particular, in [2] is presented an algorithm for simulating the flocks of birds collective flight, by using computer graphics and by adopting simple metric rules. Similarly, but from a different point of view, Heppner and Grenander, in [3], have focused their attention on the rules that enable large numbers of birds to fly together harmoniously and synchronously.

The aim of the metric approach is to describe the behavior of a bird which is maintaining a distance from its neighbors within a fixed interaction range, i.e., all birds are maintaining alignments of velocity among flock members. Even if the metric rule allowed the simulation of a collective movement of animals, quite large differences with the real behavior still remained. However, on the basis of this paradigm, Kennedy and Eberhart [1], began their works by using a one-to-one correspondence between the motion of a (metric) flock searching for food and the iterative steps of an algorithm searching the best solution for optimization. As second step, they found that some metric rules of the paradigm were an obstacle for multi-objective optimization tasks. Thus, they improved the algorithm by removing some parts. For example, the nearest neighbor-velocity matching was removed, and so on. These variations altered the virtual collective movement and the final algorithm simulates a collective animal behavior more similar to a swarm of insects than to a flock of birds and it was therefore called PSO. Thus, we can say that the original canonical PSO started from a flock simulation for arriving to a swarm representation.

In the present paper we propose an opposite path: from the swarm to the (topological) flock. The idea came from recent scientific observations, published in [4], on the collective behavior of the European Starlings (Sturnus Vulgaris).

The authors of the paper [4] discovered a topological interaction among members of the same flock: the relevant quantity is how many intermediate birds separate two starlings, not how far apart they are. This means that the main property of the topological interaction is that each starling interacts with a fixed number of neighbors, i.e. their metric distance is not crucial. Then, real flocks of starlings have a behavior that can be numerically simulated by assuming topological rather than metric rules. In fact, the topological approach is able to describe the density changes typical of flock of birds, while the metric approach cannot. This collective animal behavior can be taken as a model for an optimization algorithm and it is indeed the starting point of the proposed algorithm, which we will call Flock of Starlings Optimization (FSO). We will show that the most significant advantage provided by FSO is that it allows full exploration of the search space and can prove very useful in optimization problems where the objective function is multi-modal.

7.2 Standard PSO Overview

Let us recall the standard PSO algorithm. It is an intriguing but simple algorithm, easy to implement and with few parameters to manage. It can be summarized with the following pseudo-code [5] for a generic function, $f(x_1...x_D)$, to be minimized in the search space having dimension $\mathbb{R}^D$:

0 - Define:
- Dimension of the search space, $\mathbb{R}^D \equiv (x_1...x_D)$: $x_k^{min} \leq x_k \leq x_k^{max}$ with $k = 1...D$;
- the values of the main parameters for each j-th particle $p_j \equiv \left(x_1^j...x_D^j\right)$, with $j = 1...n_{particles}$,
- Maximum number of iterations, T_{max};
- Fitness function $f(x_1...x_D)$;
- Maximum value of each velocity component V_{max};
- Initialization of velocities $v^j{}_k(t=0) = random(0,1) \cdot V_{max}$;
- Initial personal fitness $f_{p_j}(0) = \infty$;
- Initial global fitness $g(0) = \infty$;

The Flock of Starlings Optimization

- Initial position $p_j \equiv \left(x_1^j(0)...x_D^j(0)\right)$ of each j-th particle : $x_k^j(0) = random(0,1) \cdot \left(x_k^{max} - x_k^{min}\right) + x_k^{min}$;
- Inertial coefficient ω^j ;
- Maximum value of cognize coefficient, λ_{max} ;
- Maximum value of social coefficient, γ_{max} ;
- Fitness threshold $goal\_fitness = arbitrary\ small$

1 - For each j-th particle, for each step t, with t = 0... T_{max} :
- evaluate the fitness $f_j(t) = f(x_1^j(t)...x_D^j(t))$;
- If $f_j(t)$ is better than the personal best fitness of the j-th particle $f_{p_j}(t)$, then assign current position as personal best position and update the personal best fitness:

$$p\_best_k^j = x_k^j(t) \quad \forall k \qquad (1)$$

$$f_{p_j}(t) = f_j(t) \qquad (2)$$

- If $f_j(t)$ is better than global best fitness, then assign current position as global best position and update the global best fitness:

$$g\_best_k = x_k^j(t) \quad \forall k \qquad (3)$$

$$g(t) = f_j(t) \qquad (4)$$

- Update, for each particle p_j, the vector velocity components:

$$v_k^j(t+1) = \omega^j v_k^j(t) + \lambda^j(p\_best_k^j - x_k^j(t)) + \gamma^j(g\_best_k - x_k^j(t)) \quad \forall k \quad (5)$$

- where ω^j is the inertial coefficient, λ^j is the cognize coefficient and γ^j is the social coefficient. They are defined by the following relations: $\omega^j = \omega_{max}$, $\lambda^j = \lambda_{max} \cdot random(0,1)$, $\gamma^j = \gamma_{max} \cdot random(0,1)$
- Update, for each j-th particle, the position:

$$x_k^j(t+1) = x_k^j(t) + v_k^j(t+1) \qquad (6)$$

Figure 7.1 can help to understand how the PSO movement is. The collective movement of the swarm, indicated by the large grey arrow, results quite slow because the single particle (bird) has a non-synchronized movement with others and many iterations are necessary to observe a significant displacement of the whole swarm (group).

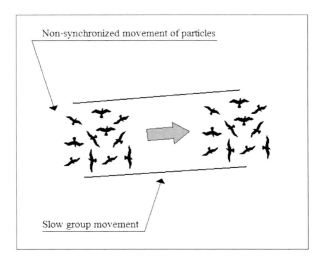

Fig. 7.1 Example of the PSO collective movement. The grey arrow indicates the collective movement of the swarm while the single particle (bird) has a non-synchronized movement with others.

7.3 Influence of Topological Rules: From Particle Swarm to Flock of Starlings Optimization

The introduction of topological rules into PSO is a new approach followed by some authors. Just to cite a more recent approach, [6] tests the implementation of three population topologies commonly used to modify standard PSO: the fully connected topology, in which every particle is a neighbor of any other particle in the swarm; the von Neumann topology, in which each particle is a neighbor of four other particles; and the ring topology, in which each particle is a neighbor of another two particles. The proposed FSO, adopts a different, although similar approach based on recent naturalistic observations [4], on the collective behavior of the European Starlings (Sturnus Vulgaris). The authors of [4] discovered a topological interaction among members of the same flock: the relevant quantity is how many intermediate birds separate two starlings, not how far apart they are. This means that the main property of the topological interaction is that each starling interacts with a fixed number of neighbors, i.e. their metric distance is not crucial. Then, real flocks of starlings have a behavior that can be numerically simulated by assuming topological rather than metric rules. In fact, the topological approach is able to describe the density changes typical of flock of birds, while the metric approach cannot. In real flocks each generic k-th bird controls and follows the flight of a number, N_{ctrl_birds}, no matter what their positions inside the flock are. A

The Flock of Starlings Optimization

typical value is $N_{ctrl\_birds} = 7$. Thus, the FSO algorithm adds to (5), written for PSO, a new term due to the previous naturalistic observation as it is described in the following pseudo-code:

0 - Define in addition to the previous PSO code:
- n_{birds} (from now on we use the term birds instead of particles) is the total number of birds into the flock;
- Maximum value of the topological coefficient, δ_{max};
- Number of birds into the flock controlled by one single bird, N_{ctrl_birds} (topology rule). At each j-th bird is associated a group composed of randomly chosen N_{ctrl_birds} of other members of the flock;
- A maximum value of the escape velocity under attack, $V_{esc\,max}$;

1 - For each j-th birds, for each step t, with $t = 0 \ldots T_{max}$:
- Update, in addition to the steps already written for PSO, the vector velocity components of each j-th bird:

$$v_k^j(t+1) = \omega^j v_k^j(t) + \lambda^j(p\_best_k^j - x_k^j(t)) + \gamma^j(g\_best_k - x_k^j(t)) + \delta^j \cdot Mccb_k^j(t) \tag{7}$$

The equation (7) differs from (5) for the presence of a new term: $\delta^j \cdot Mccb_k^j$ that introduces the topological rule according to [4]. It is constituted by the product between a topological coefficient, which is defined by the relation $\delta^j = \delta_{max} \cdot random(0,1) \quad \forall j$, and the quantity

$$Mccb_k^j(t) = \frac{1}{N_{crl\_birds}} \sum_{h=1}^{N_{crl\_birds}} v_k^{h,j}(t) \tag{8}$$

that is the mean value of the k-th velocity components, $v_k^{h,j}(t)$, of each h-th controlled starling by the j-th birds, $k = 1\ldots D$. Figure 7.2 and figure 7.3 can help to understand how (8) makes a deep change into the particle movement compared with PSO movement. In particular in figure 7.2 is shown an example about the control that the j-th bird makes on seven other members of the flock. Since each member has the same behavior by controlling seven different birds, after few iterations the velocities of each bird of the whole flock have parallel directions as shown in figure 7.3. Thus, the collective movement can be schematized as shown in figure 7.4. The collective movement of the swarm, indicated by the large grey arrow, results now quite fast because each single bird has a synchronized movement with others and a small number of iterations is necessary to observe a significant displacement of the whole flock (group).

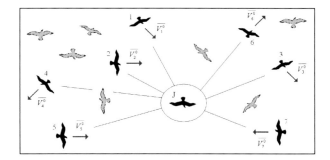

Fig. 7.2 Connections between the J-th bird and other seven birds at the starting iteration.

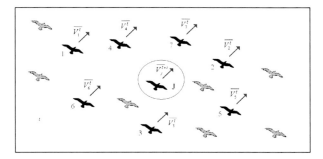

Fig. 7.3 Synchronization between the J-th bird and other seven birds at the t-th iteration due the effect of the term (8) present in (7).

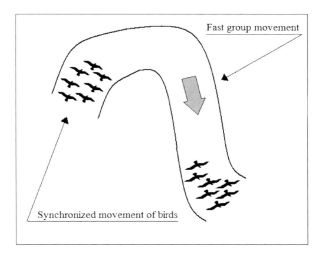

Fig. 7.4 Example of the FSO collective movement. The arrow indicates the collective movement of the flock. It is synchronous due to the toplogic rule (8) appering in (7).

7.4 FSO Validation and Comparison with the PSO Results

In this section the validation of the FSO is presented. The following listed benchmark functions have been tested and the obtained results have been compared with those reported in [7] (and in the references therein), for the same benchmarks evaluated with several different versions of PSO. All proposed benchmarks functions have a global minimum $f_{min} = 0$, with the exception of the two benchmarks that are called Bird function, $f_B(\overline{x})$,

Table 7.1

Benchmark Function	PSO best performance	FSO best performance
1 - Sphere function	$< 10^{-6}$ from [7]	$< 10^{-6}$
2 - Rosenbrock variant	$< 10^{-6}$ from [7]	$< 10^{-6}$
3 - Ackley function	< 0.104323 from [7]	< 0.005000
4 - De Jong's f4- no noise	$< 10^{-6}$ from [7]	$< 10^{-6}$
5 – Shaffer function	< 0.000155 from [7]	< 0.000071
6 - Griewank function	< 0.002095 from [7]	< 0.001170
7 - Rosenbrock function	< 39.118488 from [7]	< 29.337500
8 - Rastrigin function	< 46.468900 from [7]	< 3.765900
9 - Bird Function (bi-modal)	The implemented PSO found just one only minimum depending on the initial values used (100 different tests made)	≈ -106.72 and ≈ -106.24 i.e., the FSO finds each minimum at each test (100 different tests made)
10 - Decoy Function	The implemented PSO can finds or cannot the global minimum depending on the initial values used (100 different tests made)	≈ -2 i.e., the FSO finds the global minimum at each test (100 different tests made)

and Decoy function, $f_{Decoy}(\overline{x})$: $f_{B\,min} \approx -106.764537$ and $f_{Decoy\,min} = -2$. In particular, while all the presented benchmarks are well known in literature, the Decoy Function is a novel benchmark that has been ideated by the authors for emphasizing the exploration capability of the FSO. The reported results are referred to mean best evaluations after 2000 iterations of 100 different tests per benchmark with the exception of $f_B(x)$ and $f_{Decoy}(x)$. In these last two cases, due to the hard typology of the benchmark, it was impossible to fix a priori the maximum number of iterations. In particular, the FSO as well as the PSO were let to run till the finding of global minima. In table 7.1 all obtained results are shown for comparison. The ten different used benchmarks are listed and described separately in the following. For the first eight comparative analyses we have used the same values proposed in [7] for all parameters that are common to PSO (5) and FSO (7). For the last two benchmarks $f_B(x)$ and $f_{Decoy}(x)$, which have not been addressed in [7], the comparison is performed with parameter values just proposed in this work. The parameters of the topological rule (8) are separately indicated each time.

1. **Sphere function:** $f_{Sph}(\overline{x}) = \sum_{i=1}^{n} x_i^2$, for the tests shown in table 7.1, we used n = 30 in the range ±20 and the values of parameters appearing in (5) and (7) were: $V_{max} = 0.001$, $V_{esc\,max} = 1$, $\omega^j = 1$, $\lambda_{max} = 1$, $\gamma_{max} = 1$, $\delta_{max} = 0.3$, $N_{ctrl\_birds} = 7$.

With the aim of giving an idea about the type of this benchmark it has been plotted in Fig. 7.5 for the case n=2 in the range ±200.

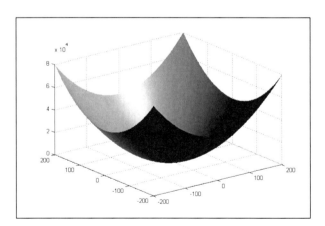

Fig. 7.5 Sphere function (for the case n=2).

2. **Rosenbrock variant :** $f_{Ros-Var}(\overline{x}) = 100(x_1^2 - x_2)^2 + (1 - x_1)^2$, for the tests shown in table 7.1, we used the range ±50 and the values of parameters

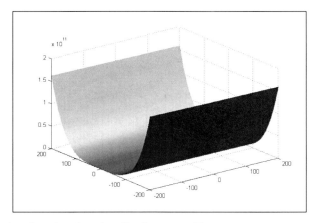

Fig. 7.6 Rosenbrock variant function (for the case n=2).

appearing in (5) and (7) were: $V_{max} = 0.001$, $V_{esc\,max} = 1$, $\omega^j = 1$, $\lambda_{max} = 1$, $\gamma_{max} = 1$, $\delta_{max} = 0.3$, $N_{ctrl\_birds} = 7$.

With the aim of giving an idea about the type of this benchmark it has been plotted in Fig. 7.6 in the range ±200.

3. **Ackley function :** $f_{Ak}(\overline{x}) = -20e^{-0.2\sqrt{\frac{1}{n}\sum_{i=1}^{n}x_i^2}} - e^{\frac{1}{n}\sqrt{\sum_{i=1}^{n}\cos(2\pi x_i)}} + 20 + e$, for the tests shown in table 7.1, we used n = 30 into the range ±32; the values of parameters appearing in (5) and (7) were: $V_{max} = 0.01$, $V_{esc\,max} = 1$, $\omega^j = 1$, $\lambda_{max} = 1$, $\gamma_{max} = 0.1$, $\delta_{max} = 0.3$, $N_{ctrl\_birds} = 7$.

With the aim of giving an idea about the type of the benchmark, it has been plotted in Fig. 7.7 for the case n = 2, in the range ±20.

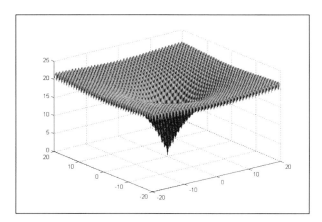

Fig. 7.7 Ackley function (for the case n=2).

4. **De Jong's f4- no noise:** $f_{f4}(\overline{x}) = \sum_{i=1}^{n} ix_i^4$, for the tests shown in table 7.1, we used n=D =30 into the range ±20; the values of parameters appearing in (5) and (7) were: $V_{max} = 0.001$, $V_{esc\,max} = 1$, $\omega^j = 1$, $\lambda_{max} = 1$, $\gamma_{max} = 1$, $\delta_{max} = 0.3$, $N_{ctrl\_birds} = 7$. With the aim of giving an idea about the type of the benchmark, it has been plotted in Fig. 7.8 for the case n = 2.

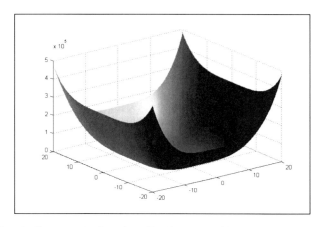

Fig. 7.8 De Jong's f4 – no noise function (for the case n=2).

5. **Shaffer function:** $f_{Shaf}(\overline{x}) = 0.5 + \dfrac{\sin^2 \sqrt{x_1^2 + x_2^2} - 0.5}{(1.0 + 0.001(x_1^2 + x_2^2))^2}$, for the tests shown in table 7.1, we used the range ±100; the values of parameters appearing

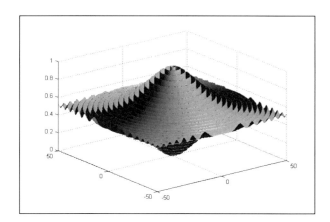

Fig. 7.9 Shaffer function

in (5) and (7) were: $V_{max} = 0.1$, $V_{esc\,max} = 1$, $\omega^j = 1$, $\lambda_{max} = 0.0001$, $\gamma_{max} = 0.01$, $\delta_{max} = 0.3$, $N_{ctrl\_birds} = 7$.

With the aim of giving an idea about the type of the benchmark, it has been plotted in Fig. 7.7 for the case n = 2 in the range ±50.

6. **Griewank function :**

$$f_{Griew}(\overline{x}) = \frac{1}{4000}\sum_{i=1}^{n}(x_i - 100)^2 - \prod_{i=1}^{n}\cos(\frac{x_i - 100}{\sqrt{i}}) + 1$$, for the tests shown in table 7.1, we used n = 30 in the range ±300; the values of parameters appearing in (5) and (7) were: $V_{max} = 0.1$, $V_{esc\,max} = 10$, $\omega^j = 1$, $\lambda_{max} = 1$, $\gamma_{max} = 1$, $\delta_{max} = 0.3$, $N_{ctrl\_birds} = 7$. With the aim of giving an idea about the type of the benchmark, it has been plotted in Fig. 7.10 for the case n = 2 in the range ±200.

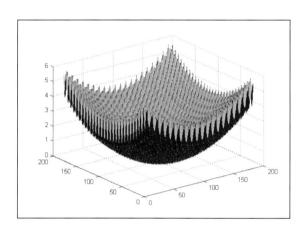

Fig. 7.10 Griewacnk function (for the case n=D=2)

7. **Rosenbrock function:** $f_8(\overline{x}) = \sum_{i=1}^{n}100(x_i^2 - x_{i+1})^2 + (x_i - 1)^2$, for the tests shown in table 7.1, we used n = 30 in the range ±10; the values of parameters appearing in (5) and (7) were: $V_{max} = 0.001$, $V_{esc\,max} = 1$, $\omega^j = 1$, $\lambda_{max} = 1$, $\gamma_{max} = 0.001$, $\delta_{max} = 0.3$, $N_{ctrl\_birds} = 7$. With the aim of giving an idea about the type of the benchmark, it has been plotted in Fig. 7.11 for the case n=D=2 in the range ±0.5.

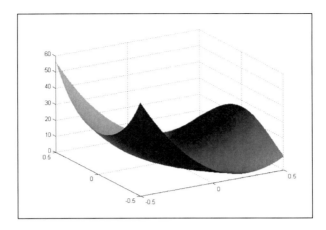

Fig. 7.11 Rosenbrock function (for the case n=D=2)

8. **Rastrigin function:** $f_{Ros}(\overline{x}) = \sum_{i=1}^{n}(x_i^2 - 10\cos(2\pi x_i) + 10)$, for the tests shown in table 7.1, we used n = 30 in the range ±5.12; the values of parameters appearing in (5) and (7) were: $V_{max} = 0.01$, $V_{esc\,max} = 1$, $\omega^j = 1$, $\lambda_{max} = 1$, $\gamma_{max} = 0.001$, $\delta_{max} = 0.3$, $N_{ctrl\_birds} = 7$. With the aim of giving an idea about the type of the benchmark, it has been plotted in Fig. 7.12 for the case n=2 in the range ± 6.

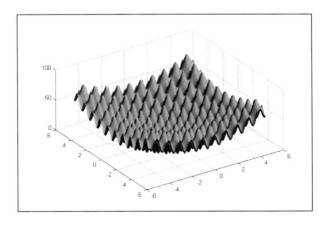

Fig. 7.12 Rastrigin function (for the case n=D=2)

Then, next function (whose minimization has not been addressed in [7]) is here proposed with the aim to show the effectiveness of the FSO both for multi-modal functions and for functions with tricky local-minima.

9. **Bird Function (bi-modal function):**

$f_B(\bar{x}) = \sin(x_1)e^{(1-\cos(x_2))^2} + \cos(x_2)e^{(1-\sin(x_1))^2} + (x_1 - x_2)^2$, for the tests shown in table 7.1, we used the range $\pm 2\pi$; the values of parameters appearing in (5) and (7) were: $V_{max} = 0.2$, $V_{esc\,max} = 2$, $\omega^j = 1$, $\lambda_{max} = 0.005$, $\gamma_{max} = 0.01$, $\delta_{max} = 0.3$, $N_{ctrl\_birds} = 7$. The used Bird function is shown in Figure 7.13, in the range ± 10.

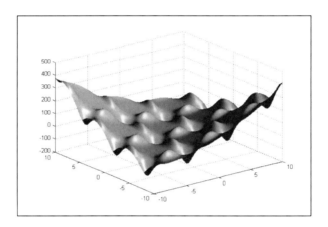

Fig. 7.13 Bird function (bimodal function)

The next benchmark is a novel function here proposed with the aim to emphasize the high exploration capability of the FSO.

10. **Decoy Function:**

11. $f_{Decoy}(\bar{x}) = \sin(\frac{2\pi}{10}x_1) \cdot \cos(\frac{2\pi}{10}x_2) - 3e^{-\left[\frac{(x_1-5)^2 + (x_2-5)^2}{10}\right]}$, for the tests shown in table 7.1, we used the range ± 100, the values of parameters appearing in (5) and (7) were: $V_{max} = 0.2$, $V_{esc\,max} = 2$, $\omega^j = 1$, $\lambda_{max} = 0.005$, $\gamma_{max} = 0.01$, $\delta_{max} = 0.3$, $N_{ctrl\_birds} = 7$. The peculiarity of the Decoy function, making its minimization difficult, is to have a very narrow global minimum (Decoy), hidden among a lot of local minima (see Fig. 7.14 and Fig. 7.15):

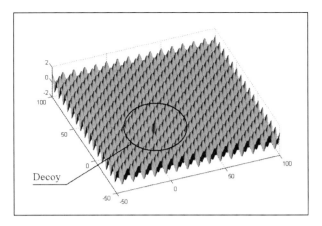

Fig. 7.14 Top view of Decoy function.

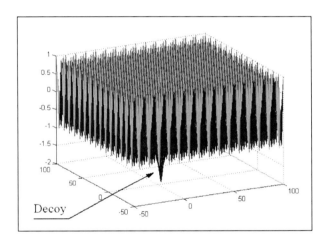

Fig. 7.15 Global minimum of Decoy function

7.5 Remarks

From the results shown in table 7.1 it is possible to observe that FSO outperforms PSO in seven out of ten benchmarks analyzed. For the other three cases the best performances of the two algorithms are similar. For the last two benchmarks different strategies are followed by the two algorithms. In particular, for the bimodal function, the implemented PSO found just one single minimum depending on the initial values used (i.e. it finds the minimum closest to the guess values); while the FSO finds all minima without the need of being launched more than once (i.e. it is immune to the initialization). For the Decoy function the implemented PSO can find or not the global minimum depending on the initial values while the FSO always finds the global minimum. The capability of PSO to find global minimum

The Flock of Starlings Optimization

depends on initial values of algorithm. On the contrary, the FSO is able to find the decoy position with a finite number of iterations starting from an arbitrary initialization of vector the variables $\bar{x}$. The analysis of the obtained results shows that the FSO algorithm has an exploration capability much greater than PSO. This behavior can be empirically justified by displaying for comparison the trajectory of a particle versus the trajectory of a bird in the search space for the same benchmark. Let us show in figure 7.16 and figure 7.17 one example referred to the Bird function (bimodal case). Both PSO and FSO have been initialized with the same starting positions (top right corner of the solution space) and with the same number of iterations (50.000). As it is evident from figure 7.16, the observed particle (PSO) is not able to exit from a limited region close to the initialization. On the contrary, figure 7.17 proves that the monitored bird (FSO) is able to explore regions definitely far from initialization.

A further important aspect should be point out discussing results shown in table 7.1. Let us consider the Rosenbrock function, for which the FSO returned an unsatisfactory result in 2000 iterations (even if it is slightly better than the PSO in [7]). The FSO was able to find a value $f_{Ros\,min} \approx 10^{-8}$ (practically the global minimum) in more or less 80000 iterations (remind that table 7.1 refers to results obtained in 2000 iterations). On the other hand, the PSO even if it was let to run for 80000 iterations was not able to improve the results achieved in 2000s.

Finally, it is important to remind the Theorem of Wolpert and Macready ("the no free lunch Theorems") [10] *"..all algorithms that search for an extremum of a cost function perform exactly the same, when averaged over all possible cost functions. In particular, if algorithm A outperforms algorithm B on some cost functions, then loosely speaking there must exist exactly as many other functions where B outperforms A"*. In other words, the average performance of all search algorithms applied over all problems is equal. Thus it will be also interesting to test the FSO in hybrid configurations with other optimizers (as suggested for other

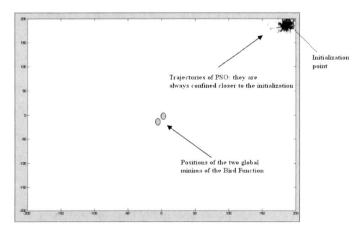

Fig. 7.16 Mapping of a particle movement of the PSO on the Bird function (bimodal). The trajectories of PSO remain confined closer to initialization point.

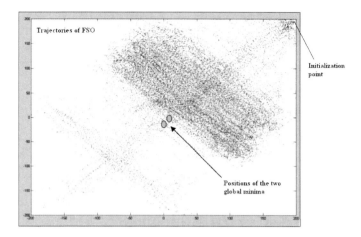

Fig. 7.17 Mapping of a bird movement of the FSO on the Bird function (bimodal). The trajectories of FSO are able to cover all the search space.

heuristics in [11]) for particular optimization tasks. For example, a hybrid configuration with FSO and Bacterial Chemotaxis Algorithm (BCA) has been tested in [8]: this further algorithm, which belongs to local search family methods, takes the place of FSO as a final step to complete the convergence. Practically, the hybrid algorithm working criterion is to employ the good exploration capability of FSO until the fitness does not reach a prefixed value; then the heuristics switched to the BCA since now the searching space is sufficiently reduced and the BCA is much faster than FSO in this context.

7.6 Conclusions

The standard PSO performs is a powerful heuristic that is able to perform very well in many optimization cases. But it has been noted that, especially when the search space is quite large, local minimum entrapments can occur. Furthermore, for multi-modal functions, the algorithm is not comprehensive. In fact, it is important to note that the standard PSO has undergone many changes. Many researchers have derived new versions and published theoretical studies of the effects of the various parameters (e.g. ω^j, λ^j, γ^j and so on) and aspects of the algorithm. In this work, we have proposed a new optimization algorithm, FSO, inspired both to PSO and to the topological behavior of real flocks of starlings recently discovered; FSO is very simple to implement; in fact it requires just some additions to the standard PSO code. Thus, the main characteristic that differentiates the FSO from the PSO is the topological interactions among members of the flock. The proposed validations show very good results returned by the proposed FSO. This algorithm seems to be competitive in comparison with the PSO standard versions. In the case of the bi-modal Bird Function, the FSO finds the two correct solutions without the necessity of being launched more than once while a standard PSO is able to find just one minimum depending on the initial values used. It is important to remark that in the presented validations and comparisons, the values of the main algorithm parameters (i.e. cognize, social, and topological coefficients, as well as the number of controlled birds by a single bird) have been empirically fixed. Obviously, the main open problem for future works is the determination of the "optimal parameter values" to be used in FSO formulas.

References

[1] Kennedy, J., Eberhart, R.: Particle swarm optimization. In: Proceedings of the IEEE International Conference on Neural Networks, Perth, Australia, vol. IV, pp. 1942–1948 (1995)
[2] Reynolds, C.W.: Flocks, herds and schools: a distributed behavioral model. Computer Graphics 21(4), 25–34 (1987)
[3] Heppner, F., Grenander, U.: A stochastic nonlinear model for coordinated bird flocks. In: Krasner, S. (ed.) The Ubiquity of Chaos. AAAS Publications, Washington (1990)
[4] Ballerini, M., Cabibbo, N., Candelier, R., Cavagna, A., Cisbani, E., Giardina, I., Lecomte, V., Orlandi, A., Parisi, G., Procaccini, M., Viale, M., Zdravkovic, V.: Interaction ruling animal collective behavior depends on topological rather than metric distance: Evidence from a field study. Proceedings of the National Academy of Science, 1232–1237 (2008)
[5] Engelbrecht, A.P.: Computational Intelligence: An Introduction. Wiley, Chichester (2002)
[6] Montes de Oca, M.A., Stutzle, T., Birattari, M., Dorigo, M.: Frankenstein's PSO: a composite particle swarm optimization algorithm. IEEE Transactions on Evolutionary Computation 13, 1120–1132 (2009)
[7] Clerc, M., Kennedy, J.: The particle swarm: explosion stability and convergence in a multi-dimensional complex space. IEEE Transactions on Evolutionary Computation 6(1), 58–73 (2002)
[8] Riganti Fulginei, F., Salvini, A.: Hysteresis model identification by the Flock-of-Starlings Optimization. International Journal of Applied Electromagnetics and Mechanics 30(3-4), 1383–5416 (2009)
[9] Ali, M.M., Kaelo, P.: Improved particle swarm algorithms for global optimization. Applied Mathematics and Computation, pp. 578–593. Elsevier, Amsterdam (2008)
[10] Wolpert, D.H., Macready, W.G.: No free lunch theorems for optimization. IEEE Transactions on Evolutionary Computation 1, 67–83 (1997)
[11] Riganti Fulginei, F., Salvini, A.: Comparative Analysis between Modern Heuristics and Hybrid Algorithms. COMPEL 26(2), 264–273 (2007)

Chapter 8
Multilevel Data Classification and Function Approximation Using Hierarchical Neural Networks

M. Alper Selver and Cüneyt Güzeliş

Department of Electrical and Electronics Engineering, Dokuz Eylül University, Izmir, Turkey
{alper.selver,cuneyt.guzelis}@deu.edu.tr

Abstract. Combining diverse features and multiple classifiers is an open research area in which no optimal strategy is found but successful experimental studies have been performed depending on a specific task at hand. In this chapter, a strategy for combining diverse features and multiple classifiers is presented as an exemplary new model in multilevel data classification using hierarchical neural networks. In the proposed strategy, each feature set and each classifier extracts its own representation from the raw data which results with measurements extracted from the original data (or a subset of original data) that are unique to each level of approximation/classification. Later on, the results of each level are linearly combined in function approximation or merged in classification. It is shown by advanced signal and image processing applications that proposed model of combining features/classifiers is especially important for applications that require integration of different types of features and classifiers.

8.1 Introduction

Neural Network (NN) (Haykin 1999) based approaches are one of the most powerful tools in the fields of function approximation and classification considering their generalization capability, robustness and adaptivity. The aim of these systems is to achieve the best possible approximation or classification performance which is sometimes limited with, two major drawbacks of NN based systems that are slow learning and over-fitting. To overcome these drawbacks and improve the performance of an NN based system, presentation of the data to the NN and the NN design strategy are very two very important issues.

Presentation of the data to an NN can simply be done using raw (original) data. However, NN based systems are often used together with feature extraction step which may include various methods that produce different information embedded inside several diverse features extracted from the same raw data. Each feature can independently be used to represent the original data, but in practice mostly a combination of them is needed to achieve a more complete representation. Unfortunately, measurements on effectiveness of features at hand are not an easy task. Since optimal combination and number of features is unknown, diverse features are often needed to be jointly used in order to achieve robust performance.

In a broad sense, two methodologies exist for combining diverse features. The first one is the use of a composite feature that is constructed by lumping diverse features together. However, as shown in the literature (Chen and Chin 1998), (Chen 2005), using a composite feature has several disadvantages including computational complexity, curse of dimensionality, formation difficulty, increased processing time and redundancy. Thus, instead of using all features as a composite vector and giving it as an input of a single and essentially very complex classifier, which eventually loses its generalization capacity, a second approach can be used that is based on combining multiple classifiers with diverse feature subset vectors (Chen 2005). Obviously, at the second approach, another important issue arises: using multiple classifiers and combining them. Although, it is stated in (Kuncheva 2003) that there is no optimal strategy for combining multiple classifiers with diverse features, several studies present advantages of the second approach.

It is shown in (Cao et al 1995); (Kittler et al. 1998), (Kong and Cai 2007), (Kumar et al. 2002); (Kuncheva 2003), (Valdovinos et al. 2005), (Xu et al. 1992), (Suen et al. 1993), (Ho et al. 1994); (Huang and Suen 1995), (Deng and Zhang 2005) and (Partridge and Griffith 2002) that the combination of classifiers outperforms the performance single classifiers when a single feature set is used. The reason behind this higher performance is introduced to be due to different regions of errors produced by each classifier in the input space (Alexandre et al. 2001). Moreover, it is shown that better performance can be achieved by combining multiple classifiers with diverse features (Xu et al. 1992); (Perrone 1993) (Chen et al. 1997) instead of using a combination of multiple classifiers with the same feature. In all these studies, several classifiers and features are used or developed among various possibilities having different designs and topologies. Due this high number of options, two issues are considered to be important when determining the combination scheme.

First of all, determination of the number and the type of classifiers are critical and it has been concluded in (Kuncheva 2004), that these issues depend on the specific task at hand. Secondly, combining the results from single classifiers is essential to yield the best performance. The conclusion that different classifier designs potentially offered complementary information also means that the misclassified data by different classifiers would not necessarily overlap. These hypotheses motivate interest in combining classifiers.

In combining classifiers, a single decision making procedure is not used but instead the outputs of all classifiers are combined using the result of each. Although, various applications have experimentally demonstrated that appropriate combinations of multiple classifiers outperform a single classifier, there is still no general proof showing if such combinations perform better for all circumstances. A common conclusion on this behavior is that efficiency is increased in multilevel classifier combinations due to use of simpler classifiers and features in combination with a reject option. For function approximation, this reject option can be a linear combination of outputs at each level (Hashem and Schmeiser 1995), (Kittler et al. 1996), (Kittler et al. 1997).

For advanced problems of classification or function approximation, diverse feature sets are used sequential, pipelined (El-Shishini et al. 1989), (Pudil et al. 1992), hierarchical (Zhou and Pavlidis 1994), (Kurzynski 1989), (Ha 1998), (Ferrari et al 2004) or other studies based on gradual reduction of the set of possible classes (Fairhust and Abdel Wahab 1990), (Denisov and Dudkin 1994), (Kimura and Shridhar 1991), (Tung et al. 1994).

Another approach for generating such combinations is modular neural networks which are based on decomposing the task at hand into simpler sub-tasks, each of which are handled by a module. Then, the results of these sub-tasks are combined via a decision making strategy. Different successful models for modular neural networks have been developed in the literature some of which are, Decoupled, Other-output (de Bollivier et al. 1991), ART-BP (Tsai et al. 1994), Hierarchical (Corwin et al. 1994), Multiple-experts (Jacobs et al. 1991), Ensembles (Alpaydin 1993); (Battiti and Colla 1994)., and Merge-glue (Hackbarth and Mantel 1991).

In this chapter, a kind of hierarchical classifiers is presented as an exemplary new model in multilevel data classification using Hierarchical Neural Networks (HNN). In the proposed strategy, each feature set and each classifier extracts its own representation from the raw data which results with measurements extracted from the original data that are unique to level (i.e. each feature set – classifier combination). Proposed model of combining features-classifiers is especially important for applications that require integration of different types of features and classifiers. In Section 2, theoretical framework is presented with exemplary applications to radial basis function networks. In Section 3, advanced signal processing applications, that take advantage of the proposed approach, have been presented. These applications consist of multidimensional transfer function approximation for three dimensional medical volume visualization via interactive volume rendering, quality classification of marble slabs and classification of radar data.

8.2 Theoretical Framework

As introduced in previous section, features and classifiers should be combined in an efficient and computationally tractable manner in order to realize the hierarchical classification scheme effectively and accurately.

For classification problems, the divide-and-conquer strategy (Jordan 1993), (Chen 1998), is proposed to combine classes with similar characteristics into one class, which can be separated later at the succeeding layers. Applications of this strategy is presented in (Chen and Chi 1998) for speaker identification by combining multiple probabilistic classifiers on different feature sets. An automatic feature rank mechanism is proposed to use different feature sets in an optimal way and a linear combination scheme has been implemented in (Chen et al. 1998) and (Chen 2005). Pruning of training tokens with good interclass discrimination and then successively optimizing features and classifier topologies for the remaining tokens is another successful application of hierarchical classifiers is (Kil and Shin 1996). The main idea behind that approach is iterative optimization of classifiers as a function of a reduced training feature subset which is constructed by collecting samples that might belong to more than a single class based on a predefined criterion. In contrast to that approach, our system carries unclassified samples, which can be called as rejected samples since they do not fall inside any class at that level of hierarchy, to the next level of hierarchy where another feature space and/or another classifier set will proceed (Fig. 8.1). The overall system (i.e. all levels of hierarchy together) can consist of both simple systems (i.e. a single perceptron etc.) and complex ones (i.e. Multi Layer Perceptron (MLP), Radial Basis Function Network (RBFN) etc.) which can be combined based on the specific task at hand. Based on application, a data-dependent and automated switching mechanism, which decides to apply one of the feature/classifier couple at each step, can be implemented. In that case, the switching would be based on the rejected samples, which are then used as the input of another classifier possibly in another feature space. For very specific tasks, the switching mechanism can be fixed to perform a pre-defined order of feature extraction and classification methods.

For function approximation, the above mentioned set of diverse features and classifiers are again used in a cascaded manner. However, in function approximation, rejected data corresponds to residual data which is obtained by removing the approximated data from the original data at each level of the hierarchy. Thus, the design consist of a number of approximators, some of which are simple and therefore efficient in time and memory requirements and the others are complex providing a high approximation performance by providing better a better fit. Following subsections provides theoretical information for hierarchical function approximation and classification strategies, respectively. In both of these subsections, an exemplary application using RBFN is also given.

8.2.1 Function Approximation Using HNN

For function approximation problems, multilevel HNN performs a mapping $f(\bullet)$: $R^D \to R$, as the sum of K approximations $\{l_i(\cdot)\}_{i=1,2,\ldots,K}$:

$$f(x) = \sum_{i=1}^{K} l_i(x)$$

Herein, approximation layers, $l_i(\cdot)$'s, are sub-networks; they are indeed not structural layers but just functional layers, i.e. approximation layers, constructed in a successive manner along the training phase. The complete output of the network is the combination of all approximation layers. The first level of approximation, $l_1(\cdot)$, is performed using the original data, $f(x)$. After the first approximation, a residual is calculated point-wise, i.e. for each point x^n used in the training.

$$r_1(x^n) = f(x^n) - l_1(x^n)$$

Multilevel Data Classification and Function Approximation

The next approximation level considers the residual data found in the previous level as the new function to be approximated and the new approximation is done

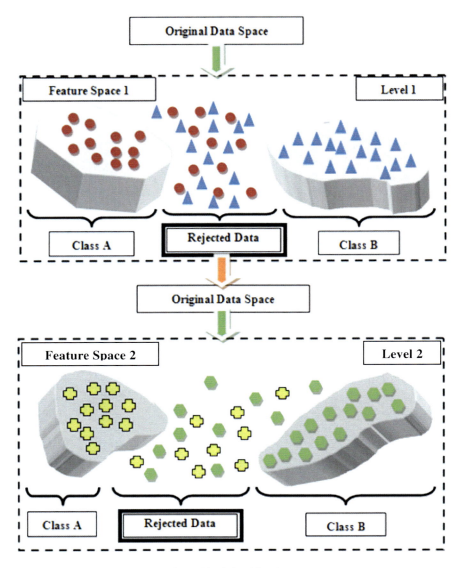

Fig. 8.1 Illustration of proposed hierarchical classification strategy.

for that residual. The general expression of a residual data to be approximated at i^{th} layer can be given as:

$$r_{i-1}(x^n) = f(x^n) - \sum_{j=1}^{i-1} l_j(x^n)$$
$$= r_{i-2}(x^n) - l_{i-2}(x^n)$$

To approximate the original data, this procedure can continue for as many layers as wanted thus it can be further iterated until a satisfactory degree of approximation is obtained. In the following paragraphs, an exemplary formulation using RBFN is presented. The Hierarchical RBFN (HRBFN) was first proposed in (Ha 1998) where some of the input data were rejected based on an error criterion at the end of each level. These rejected data become input to the next level where the number of neurons (Gaussian units) is determined as a logarithmic function of the number of rejected data. Recently, a new approach has been proposed in (Ferrari et al. 2004), in which approximation is achieved through a neural network model. It is a particular multiscale version of RBFN that self-organizes to allocate more units when the data contain higher frequencies. The quasi real time implementation of the proposed HRBFN is presented in (D'Apuzzo 2002). HRBFN is also used for classification in various applications (Chen et al. 2006), such as for recognition of facial expressions (Lin and Chen, 1999).

A single-output RBFN, which consists nonlinear neurons having a Gaussian transfer function at the hidden layer and a linear neuron performing a weighted sum at the output layer, defines the following function:

$$h(x) = \sum_{j=1}^{N} w_j \cdot g(x - c_j; \sigma_j)$$

where $\{c_j \in R^D\}$ corresponds to the center of jth Gaussian unit, $\{\sigma_j \in R\}$ to the width, and $\{w_j \in R\}$ to the jth linear weight and

$$g(x - c_j; \sigma_j) = \exp\left(-(x - c_j)^T \cdot (1/\sigma_j^2) \cdot (x - c_j)\right).$$

Then, each $l_i(\cdot)$ can be written as

$$l_i(x) = \sum_{j=1}^{M_i} w_{i,j} \cdot g(x - c_{i,j}; \sigma_{i,j})$$

which means that each layer has M_i gaussian units and the network has So the network has totally $M = \sum_{i=1}^{K} M_i$ units in total.

8.2.2 Classification Using HNN

Similar to function approximation problems, multilevel HNN performs the mapping $f(\bullet): R^D \rightarrow R$, as the union of P classifications, $\{l_i(\cdot)\}_{i=1,2,\ldots,P}$

$$y(x) = \bigcup_{i=1}^{K} l_i(x) = l_1(x) \bigcup l_2(x) \bigcup l_3(x) ... \bigcup l_P(x)$$

Here, i represents the index of each network, called level, and $\bigcup$ represents the union operation. When all of the P levels, $\{l_i(\cdot)\}_{i=1,2,...,P}$, are combined, they construct $y(x)$ which represents the overall classification result. Each $l_i(\cdot)$ is a union of a number clusters each of which represents a class.

Starting from the original data set, $y(x)$, features are calculated for the first level. After classification using the network assigned to that level, the rejected data are determined as the outliers which do not belong to any class of that level.

The next level only receives features extracted from the rejected data of the previous level as input. The general formulas for calculating the rejected data for level i is given as:

$$r_{i-1}(x^n) = y(x^n) - \sum_{j=1}^{i-1} l_j(x^n)$$
$$= r_{i-2}(x^n) - l_{i-2}(x^n)$$

Considering RBFN case again, each $l_i(\cdot)$ is a union of Q_i Gaussian units each of which represents a class and contains correctly classified samples for that level,

$$l_i(x) = \bigcup_{j=1}^{Q_i} w_{i,j} \cdot g_{ij}(x - c_{i,j}; \sigma_{i,j})$$

where $g_{ij}(\cdot)$ represents the jth Gaussian unit at the ith level and is defined by, $g_{ij}(\cdot) = \exp\left(-(x - c_{i,j})^T \cdot (1/\sigma_{i,j}^2) \cdot (x - c_{i,j})\right)$. Here, $\{c_{i,j} | c_{i,j} \in R^D\}$ represents centers and $\{\sigma_{i,j} | \sigma_{i,j} \in R^D\}$ denotes widths of Gaussian units in each level with $\{w_{i,j} | w_{i,j} \in R\}$ denoting the synaptic weights. The complete output of the HRBFN can be regarded as the combination of all levels where a class is constructed by combining all sub-classes at each level.

8.3 Applications

Advanced signal processing applications, that take advantage of the multilevel data approximation and classification approach described in previous section, have been presented in this section. These applications consist of multidimensional transfer function approximation for three dimensional medical volume visualization via interactive volume rendering, quality classification of marble slabs and classification of radar data.

8.3.1 Transfer Function Initialization for Three Dimensional Medical Volume Visualization

The goal of medical visualization is to produce clear and informative pictures of the important structures in a medical data set. Volume rendering is an important visualization technique since 3-Dimensional images are produced directly from the original data set. The advantage of this reconstruction procedure is that combinations of the selected volume appearance parameters (i.e., opacity and color) can be determined interactively as an element of rendering pipeline. This interaction is provided by adjusting a function, namely Transfer Function (TF) that maps the original data space (i.e., intensity values, gradients etc.) to appearance parameters. Therefore, it is important to design effective techniques for this mapping.

The traditional way of TF design for a physician is based on defining tissues by determining their locations in the function domain (i.e., usually intensity value of the pixels) and then assigning visual properties (i.e., opacity, color) to them. However, this design procedure is a very time consuming, tedious and expertise dependent task. Whenever an unknown dataset is used or tissues with overlapping intensity distributions are needed to be visualized, this simple manual assignment cannot produce informative images. In order to create a useful TF that provides a good basis prior to optimization, an effective domain for the TF is needed to be used and moreover an automatic tissue detection method that locates tissues in that domain is needed. Unfortunately, the tradeoff between using extensive search spaces and fulfilling the physician's expectations with interactive data exploration tools and interfaces makes the design process even harder. Thus, at the final design, it is necessary to integrate different features into the TF without losing user interaction.

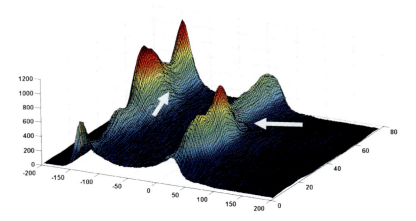

Fig. 8.2 Volume Histogram Stack (VHS) data for a CT Angiography dataset of 70 slices.

By addressing the problem of finding a suitable domain, a Volume Histogram Stack (VHS) is introduced in (Selver and Güzeliş 2009) as a new domain that is constructed by aligning the histograms of the image slices of a medical image series. VHS incorporates spatial domain knowledge with local distributions of the tissues and their intensities since a tissue/organ usually get a lobe like distribution which is a shape similar to a radially asymmetric (elliptical) Gaussian in VHS (Fig. 8.2). The construction of the VHS and its representation capabilities are discussed in detail in (Selver and Güzeliş 2009).

To be able to use VHS effectively in TF initialization, a function approximation strategy is needed to recognize the lobes of VHS. Moreover, this approximation should be able to locate suppressed lobes corresponding to suppressed tissues in VHS. Having lobes like Gaussian distributions, using an RBFN is one of the appropriate choices for finding the lobes of a VHS. However, the simplest strategy of using fixed centers and widths for units of RBFN cannot guarantee the determination of suppressed lobes. Even if we assume that appropriate number of hidden neurons is known as well as appropriate center locations and widths, simulations show that the RBFN approximation tends to fit only major lobes and skip the suppressed information carried by the minor lobes (shown with the arrows in Fig. 8.2). To overcome this drawback, the HRBFN (Ferrari et al. 2004), would be an appropriate choice since it assigns Gaussians in all scales to represent all details of the function to be approximated. It is observed that a slight modification of HRBFN is required since it produces a huge number of hidden neurons when it is applied to approximate VHS. This is not a desired property for TF design since the physician should deal with a small number of units to obtain an efficient interaction mechanism.

Considering the above reasons, HRBFN is used as the network for approximating to VHS data but with a new learning strategy, called as Self Generating HRBFN (SEG-HRBFN) (Selver and Güzeliş 2009). The developed SEG-HRBFN provides a procedure for capturing all suppressed lobes of importance in a successive manner by associating the lobes with a minimum number of Gaussian bases.

Thus, the SEG-HRBFN is designed by a hierarchical learning strategy in which all lobes are captured in a successive manner by associating the lobes with the Gaussian bases. The hierarchical strategy is due to assigning additional Gaussians bases at each level of approximation to the residual VHS, which is the remainder after removing the already obtained approximation from the original VHS. SEG-HRBFN allows an approximation with a minimum set of basis functions that can be further adjusted by the physician to optimize the TF in an interactive way.

At each layer, the SEG-HRBFN contains a number of Gaussian units that are used to approximate to the lobes in VHS. The number of these units and their structural parameters (i.e. centers, widths) and linear weights are determined automatically.

After the lobes in VHS data is characterized by a Gaussian base, then, they are merged to construct groups each of which corresponds to a tissue or organ. The selection of Gaussian bases to merge is done by a physician based on the positions of Gaussian bases. Later on, opacity and color properties are assigned to some of these groups to construct a 3-Dimensional image. Finally, this combination of the user selected units and visual properties construct an accurate initial TF.

The application of the proposed method to several medical datasets shows its effectiveness, especially in visualization of abdominal organs (Selver and Güzeliş 2009). The overall goal of this application is to improve the rendering quality for visualizing the tissues of overlapping intensities and also to shorten the physician-controlled optimization stage in TF design.

Here in this chapter, the comparison of SEG-HRBFN with HRBFN is given. As mentioned above, the main advantage of SEG-HRBFN over HRBFN is its capability of capturing all suppressed lobes and overlapping regions of importance in a successive manner as associating the lobes with a minimum number of Gaussian bases. This advantage does not result with losing any details in approximation. To prove this experimentally, SEG-HRBFN is compared with HRBFN (Ferrari et al 2004) and also with Fast-HRBFN (Ferrari et al 2005) in 3-D reconstruction of range data presented in (Ferrari et al 2005) which is known to be a successful application of HRBFN. HRBFN model of (Ferrari et al 2004) also creates an increasing number of Gaussians at each level. However, fixed widths are determined for each level without considering the signal characteristics at that level. This approach results with generation of several Gaussians most of which are not used in the reconstruction of the original data. The advantage of HRBFN is building an effective network in a very short time.

SEG-HRBFN differs from HRBFN as preventing the above mentioned redundancy by choosing small numbers of appropriate basis functions. SEG-HRBFN is able to obtain a reconstruction of the function to be approximated in same detail but with fewer units. In order to compare SEG-HRBFN with HRBFN (Ferrari et al 2004) and Fast-HRBFN (Ferrari et al 2005) the auto-scan system (Ferrari et al 2005) range data from a baby doll face is used (Fig. 8.3.a). In the reconstruction of this data, connecting the points to form a triangular mesh produces an undesirable wavy mesh and traditional linear filtering cannot be applied to clean the surface since data are not equally spaced. Moreover, the highly variable spatial frequency content of a face requires an adaptive approach.

This requirements make the HRBFN based approaches suitable for that problem since the quality of the network output increases with the number of levels by adding details mainly in the most difficult regions like the nose, the eyes, and the lips (Fig. 8.3). These details are obtained by means of Gaussian clusters at smaller scales in the higher levels. In the HRBFN and Fast HRBFN approaches, these clusters are created by the network itself at the configuration period by inserting a Gaussian only where the local reconstruction error is larger than the measurement error. However, these local operations produce a huge number of units.

The results show the effectiveness of SEG-HRBFN compared to HRBFN and Fast HRBFN in the reconstruction of a 3-D signal with a similar quality but using less number of units which is an important issue in TF initialization problem.

For the same RMS error, HRBF produces 7205 Gaussian units while Fast HRBFN produces 8087 Gaussian units. On the other hand, SEG-HRBFN produces only 2044 Gaussian units.

Multilevel data classification with HNN is used for 2-Dimensional function approximation in this application. The advantage of using HNN, in particular HRBFN or SEG-HRBFN, over traditional RBFN is the capability of representing the Gaussian-like lobes in VHS very efficiently and ability to find suppressed lobes by using residual VHS at each level of approximation. This is not possible with RBFN approach due to the tradeoff between using too many Gaussian bases to locate suppressed lobes and tending to fit major lobes without recognizing minor (suppressed) lobes.

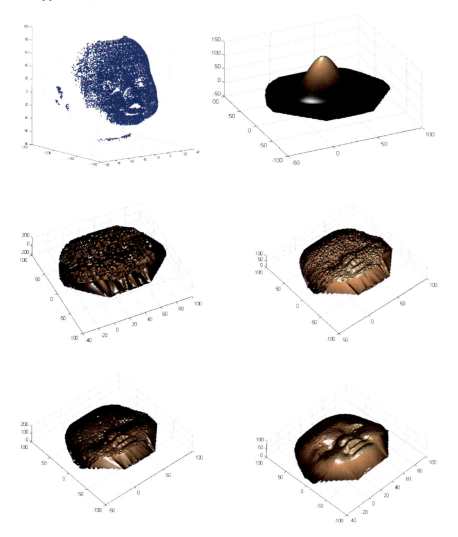

Fig. 8.3 Reconstruction of the baby doll face by using multilayer SEG-HRBFN reconstruction. (a) Autoscan range data that consist of over 61.000 data points to represent the doll face, (b) Reconstruction up to layer 1, (c) Reconstruction up to layer 8, (d) Reconstruction up to layer 14, (e) Reconstruction up to layer 22, (f) Reconstruction up to layer 29.

8.3.2 Quality Classification of Marble Slabs

For decorative applications of natural building stones, limestones should present attractive colors as well as similar pattern choices. The determination of the quality of a marble slab is based on the homogeneity, texture and color of its surface. Based on these properties, four quality groups can be defined for marbles slabs collected from the marble mines in Izmir, Turkey (Selver et al 2009):

1) Homogenous distribution of limestone (beige colored) (Fig. 8.4.a),
2) Thin joints filled by cohesive material (red-brown colored veins) (Fig. 8.4.b, Fig. 8.4.c)
3) Unified joints of cohesive materials (Fig. 8.4.d, Fig. 8.4.e)
4) Homogeneous distribution of cohesive material (Fig. 8.4.f)

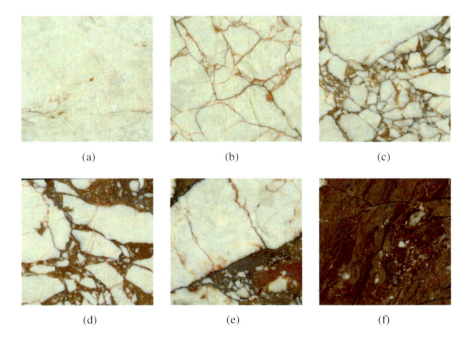

Fig. 8.4 Typical sample images from each quality group. (a) Homogenous limestone. (b), (c) Limestone with veins. (d), (e) Samples containing grains (limestone) that are separated by unified cohesive matrix regions. (f) Homogenous cohesive matrix.

False classification of marble slabs can cause major economic problems since they are exported to overseas and therefore, it is necessary to classify marble slabs correctly according to their quality and appearance.

Several methods that include different features and classifiers have been proposed for classification of marbles (Luis-Delgado et al 2003), (Martinez-Alajarin 2004),

(Boukovalas et al 1998) (Tsai and Huang 2003), (Martinez-Alajarin et al 2005) some of which produce acceptable results for industrial applications (Luis-Delgado et al 2003), (Martinez-Alajarin et al 2005). First, these strategies are tested on our marble data set which consists of 1158 marble slabs (i.e. 172 from quality group 1, 388 from quality group 2, 411 from quality group 3, and 187 from quality group 4) using different color spaces and classifiers. Our simulations showed that the application of the existing methods mentioned above to our diverse and large dataset cannot provide successful performance results comparable to ones reported in the literature.

The reason behind the poor performance of these simulations is observed to be the representation of different sub-group(s) in a quality group at each feature space (i.e. texture, spectral etc.). In other words, at each feature space, a sub-group of a quality group is represented effectively instead of the whole quality group (Selver et al 2009).

As an example, we can analyze the pattern distributions of marble slap samples from quality groups 2 (Fig. 8.4.c) and 3 (Fig. 8.4.d). These two groups cannot be separated correctly using most of the features. These challenging samples have almost same limestone area, textural properties and spectral energy. Although the above mentioned features are very useful in the classification of other marble slabs, they are not descriptive for these challenging samples. For these samples, the classification criteria of the human experts are based on the distribution of the veins and therefore, the structure of the veins should be extracted as a feature for classification of these samples. However, as presented in Table 8.1, constructing a composite feature vector using each of these useful features is not the best solution for classification of marble slabs with high performance (Selver et al 2009).

Because of the above mentioned reasons, it makes sense to apply feature extraction in a cascaded manner where a sub-group of a quality group is classified at each feature space. Thus, different features are extracted in a successive manner and the samples that can be correctly classified by that specific feature set is collected. Finally, overall classification result is obtained by merging all sub-groups of a quality group. As a realization, a Hierarchical MLP Network (HMLPN) topology is designed in which correctly classified marble samples are taken out of the dataset at each level and a different (generally more complex) feature extraction method is used for the rest of the samples at the next level.

Since there is a fixed number of quality groups, which is equal to four, proposed MLPN system consists of four classes at each level (i.e. namely sub-class). Four feature spaces are used in cascaded manner thus the network contains 16 sub-classes in total. These 16 sub-classes construct the final quality groups when they are merged at the end of the process.

As seen from Table 8.1, the proposed application of HMLPN outperforms other classification strategies. When MLP design is used, textural, spectral, and morphological features similar classification performances all of which are limited in discriminating quality groups 2 and 3. When a composite feature is constructed by lumping textural, spectral, and morphological features; only a slightly higher performance increase can be achieved. When HMLPN design is used, a significantly higher performance increase can be achieved.

Table 8.1 Results of quality classification with different feature space-neural network designs.

NN Design	Feature Set	Classification Performance (CC) G1	G2	G3	G4
MLP	Textural (Martinez-Alajarin et al 2005)	96,52	91,79	94,53	99,25
	Spectral (Luis-Delgado et al 2003)	95,77	91,04	93,78	98,01
	Morphological (Selver et al 2009)	94,11	93,24	95,56	98,13
	Textural + Spectral + Morphological (Composite feature)	97,69	93,48	94,97	99,06
HMLPN	Level 1 (Area Ratio) (Limestone / Total) Level 2 (Textural) Level 3 (Spectral) Level 4 (Morphological)	98,85	96,22	97,44	99,13

8.3.3 Classification of Radar Data

An antenna array refers to multiple antennas that are coupled to a common source or load to produce a directive radiation pattern which is also affected by the spatial relationship of the antennas. Radar data, which are obtained from a phased array of high-frequency antennas, are commonly classified as a composite feature that is formed by lumping diverse features from different antennas together. This approach has disadvantages including curse of dimensionality, formation difficulty and redundancy due to dependent components. Ionosphere data (Sigillito et al 1989) is one of the benchmarks for these kinds of datasets since its source consists of a phased array of 16 high-frequency antennas.

Received signals in Ionosphere data were processed using an autocorrelation function whose arguments are the time of a pulse and the pulse number. There were 17 pulse numbers system providing 34 features (Instances are described by 2 attributes per pulse number, corresponding to the complex values returned by the complex EM signal). These features construct the 'Complete dataset'. The targets of the Ionosphere dataset were free electrons in the ionosphere where "Good" radar returns are the ones showing evidence of some type of structure in the ionosphere while "Bad" returns show that their signals pass through the ionosphere.

In several studies, this data is classified using different methods (please see UCI Machine Learning Repository Database archive for the complete list of references). Several different network models (i.e. MLP, RBFN, and Probabilistic

Neural Networks (PNN) etc.), different topologies, alternative learning methods (i.e. Back-propagation, perceptron training algorithms etc.) are applied.

In studies such as (Wing et al 2003), (Salankar and Patre 2006), the original goal of the research is to demonstrate that neural networks could operate at a level of performance high enough to be a real aid in the automation of the classification task. In (Wing et al 2003), it is shown that that classification of radar backscattered signals (i.e. Ionosphere data) is a task for which neural networks are very well suited. It is also reported that neural networks with hidden nodes outperform those without hidden nodes in terms of accuracy, sensitivity and specificity measures. In a very recent study (Salankar and Patre 2006), a wide range of classifiers are tested by adjusting several parameters (i.e. learning method, number of neurons etc.) and their performances in classification are discussed using different evaluation measures (i.e. mean squared error, normalized mean squared error, correlation coefficients and ROC curve). The results of this study show the superiority of the RBFN over MLP in the classification of the Ionosphere data, as well as the importance of the correct determination of the parameters after extensive experimental work.

As discussed in the previous section, combining multiple classifiers with diverse feature subset vectors yields improved performance in classification problems, instead of using all features as a composite vector and giving them as the input of a single and essentially more complex classifier. Besides using Ionosphere feature vector as a composite (i.e. 34 dimensional) vector, several subsets of this feature set are used to construct the hierarchical structures. In general, (n x m + p) formula is used to create a hierarchical structure where n is the number of features for each level of the hierarchy, m is the total number of levels and p is the number of remaining features. For example, Hierarchical 16x2+2 indicates that there will be two levels of hierarchy where each has 16 features and finally remaining 2 features will be used in the final level of hierarchy (i.e. level 3). In other words, Hierarchical 16x2+2 presents the results when 16 features are fed to the network in the first level, then rejected samples are classified with another 16 features at the second level and finally remaining 2 features are used for the rest of the samples that are not classified in first and second levels.

Three different classifiers are used in the simulations. In each of the experiments, a single type of classifier (i.e. MLP, RBF, PNN) is used at all levels.

In the first group of simulations, an MLP network, which has m hidden layer neurons (m=4, 8, 20, 50), is trained using back-propagation with adaptive learning rate. Since there are two target classes, output layer have 2 neurons with linear activation functions. The network goal is chosen to be 0.001 and the maximum number of iterations is determined as 5000 epochs. The adaptive learning rate is initialized to 0.01. If performance decreases towards the goal, the learning rate increases with a ratio of 1.05, otherwise it decreases with a ratio of 0.7.

RBFN and PNN are also used to test the effect of the classifier on the performance. The RBFN used in our simulations is a two-layer network. The first layer has neurons with Gaussian activation functions. The second layer has neurons with linear activation functions. The same topology is also used for the PNN. Spread of the Gaussians is an important parameter for both networks. The important condition to meet is to make sure that spread is large enough so that the active input regions of the neurons in the first layer overlap which makes the network function smoother and results in better generalization for new input vectors. However, spread should not be so large that each neuron is effectively responding in the same large area of the input space. Spread is chosen to be 0.7 for the RBFN and 0.1 for the PNN. These parameters are chosen randomly since our aim is not to achieve the best classification performance but to observe the performance differences between the traditional and hierarchical approaches. The details and results of the simulations are given in Table 8.2.

Table 8.2 Results of classification simulations (CC: Correct Classification, SE: Sensitivity, SP: Specificity). In first row (Composite) all 34 features are fed to the network. The second row (Hierarchical 16x2+2) presents the results when 16 features are fed to the network in the first level, then rejected samples are classified with another 16 features at the second level and finally remaining 2 features are used for the rest of the samples that are not classified in first and second levels. Third, fourth and fifth rows presents results using different number of levels.

Test Results (Average of 20 simulations)	MLP			PNN			RBFN		
	CC	SE	SP	CC	SE	SP	CC	SE	SP
Composite (34)	82.86	92.12	58.33	78.44	99.25	34.23	79.57	98.17	36.54
Hierachical 16x2+2	86.74	97.50	70.28	80.26	98.02	68.14	80.26	98.22	65.16
Hierachical 8x4+2	93.27	98.50	87.10	86.63	97.98	70.56	90.63	96.94	77.08
Hierachical 4x8+2	88.81	96.48	76.13	85.14	98.26	72.77	85.14	98.11	74.71
Hierachical 2x17	89.66	96.19	56.50	83.22	98.10	65.70	83.22	98.18	77.32

8.4 Discussions and Conclusions

The problem of combining classifiers/approximators which use different representations of the data via diverse features is an open research field. In this paper, we have described a combination of classifiers and approximators based on the HNN strategy in design. As a new method for multilevel data approximation and classification with diverse features and combining multiple classifiers, our method adopts a single stage learning process to multiple levels in each of which a different input feature space is used for data representation and each NN for approximation or classification tasks. The developed strategy for classifier/approximator can be considered as special case of HNN strategies and it is experimentally shown that a higher performance than composite feature-single NN based approaches.

In this context, the first application presented a multi-dimensional function approximation problem using a modified HRBFN. The hierarchical learning (design) strategy that is carried out with SEG-HRBFN allows the recognition of suppressed lobes corresponding to suppressed tissues and the representation of overlapping regions where the intensity ranges of the tissues overlap in VHS. By automatically determining the number of necessary network parameters, this approach significantly reduces the number of Gaussian units produced at the end of approximation. Approximation to VHS with a minimum set of basis functions (i.e. Gaussian units) also provides the construction of the network in a reasonable time.

In the second application, quality based classification of marble slabs is considered. Different feature sets have been used including color properties, textural features, spectral features (multi resolution wavelets), morphological features and their combinations. The experimental results showed that there are no significant differences in the correct classification rates when using these feature sets solely or their combinations via a composite feature. When the misclassified samples are observed, it is also found that that different feature sets represent different subgroup(s) in a quality group rather than representing the complete quality group.

Therefore, the above mentioned features are used in a cascaded manner so that each feature set is used only for the sub-group(s) that can be correctly classified by that feature set. This approach was realized by MLPN topology, in which correctly classified marble samples are taken out of the dataset at each level and a different feature extraction method is used for the remaining samples at the next level.

The proposed system is shown to have better performance than the previously proposed systems for the diverse and large dataset used in this study. The MLPN approach is shown to be very useful for marble classification applications, because using different feature sets to classify different sample sub-groups is seem to be very efficient compared to using a single feature set for the whole dataset.

In the third application, proposed multi-level HNN strategy is implemented and tested for a phased array antenna radar data (i.e. Ionosphere data). The main objective was to compare the hierarchical approach, which relies on a token pruning strategy by optimizing features and classifiers at each level of hierarchy, against traditional composite feature-single classifier approach. The results show that hierarchical usage of features coming from different antennas, instead of combining them to make a composite feature, improves the classification performance not only in terms of accuracy but also selectivity and specificity.

One important point is the necessity of making careful choices in terms of utilized features which depends on the priorities of the radar application at hand. This cannot be done in this study since the Ionosphere raw data is not available and only an already processed format where an autocorrelation function, whose arguments are the time of a pulse and the pulse number, is used. Therefore, the hierarchical subsets are chosen arbitrarily from this feature set.

In all of the applications described above, a significant performance increase in classification and approximation performance is achieved with multilevel HNN strategy against using a composite feature formed by lumping diverse features together with a single and essentially very complex classifier. However, empirical or arbitrary selection of the feature sets at each level might cause sub-optimality which may be considered as a weakness of the proposed application (but not the methodology). This necessitates making careful choices in terms of order of classifiers/approximators and utilized features dependent on the priorities (i.e. success rate, time etc.) of the application at hand. For instance, performing a theoretical analysis for extending the methodology to realize the joint use of different feature sets (i.e. information coming from different antennas) and classifiers in an optimal way could be noted as a challenging future study direction for radar data classification where raw radar data and prior information about the antennas of the system should be available. On the other hand, performing a detailed theoretical analysis for extending the methodology to realize the joint use of different feature sets and classifiers in an optimal way could be noted as a challenging future study in general which also depends strongly on application at hand.

References

Alexandre, L.A., Campilho, A.C., Kamel, M.: On combining classifiers using sum and product rules. Pattern Recognition Letters 22, 1283–1289 (2001)
Alpaydin, E.: Multiple networks for function learning. In: Proc. Internat. Conf. on Neural networks, CA, vol. 1, pp. 9–14 (1993)
Battiti, R., Colla, A.: Democracy in Neural Nets: Voting schemes for classification. Neural Networks 7(4), 691–707 (1994)
Boukouvalas, C., Natale, F.D., Toni, G.D., et al.: ASSIST: Automatic system for surface inspection and sorting of tiles. J. Mater. Process. Technol. 82(1-3), 179–188 (1998)
Cao, J., Shridhar, M., Ahmadi, M.: Fusion of classifiers with fuzzy integrals. In: Proceedings of the Third International Conference on Document Analysis and Recognition, vol. 1, pp. 108–111 (1995)
Chen, K.: A connectionist method for pattern classification with diverse features. Pattern Recognit. Lett. 19(7), 545–558 (1998)
Chen, K.: On the use of different speech representations for speaker modeling. IEEE Trans. Syst., Man, Cybern. C, Appl. Rev. 35(3), 301–314 (2005)
Chen, K., Chi, H.: A method of combining multiple probabilistic classifiers through soft competition on different feature sets. Neurocomputing 20, 227–252 (1998)
Chen, K., Wang, L., Chi, H.: Methods of combining multiple classifiers with different features and their applications to text-independent speaker identification. Int. J. Pattern Recog. Artif. Int. 11(3), 417–445 (1997)
Chen, Y., Peng, L., Abraham, A.: Hierarchical radial basis function neural networks for classification problems. In: Wang, J., Yi, Z., Żurada, J.M., Lu, B.-L., Yin, H. (eds.) ISNN 2006. LNCS, vol. 3971, pp. 873–879. Springer, Heidelberg (2006)
Chen, K., Wang, L., Chi, H.: Method of combining multiple classifiers with different features and their applications to text-independent speaker recognition. Internat. J. Pattern Recogn. Artif. Intell. 11(3), 417–445 (1997)
D'Apuzzo, N.: Modeling human faces with multi-image photogrammetry. In: Proc. SPIE Three-Dim. Image Capture Appl., vol. 4661, pp. 191–197. SPIE, San Jose (2002)
Deng, D., Zhang, J.: Combining multiple precision-boosted classifiers for indoor-outdoor scene classification. Information Technology and Applications 1(4-7), 720–725 (2005)
Denisov, D.A., Dudkin, A.K.: Model-Based Chromosome Recognition Via Hypotheses Construction/Verification. Pattern Recognition Letters 15(3), 299–307 (1994)
El-Shishini, H., Abdel-Mottaleb, M.S., El-Raey, M., et al.: A Multistage Algorithm for Fast Classification of Patterns. Pattern Recognition Letters 10(4), 211–215 (1989)
Fairhurst, M.C., Abdel Wahab, H.M.S.: An Interactive Two- Level Architecture for a Memory Network Pattern Classifier. Pattern Recognition Letters 11(8), 537–540 (1990)

Ferrari, S., Maggioni, M., Borghese, N.A.: Multiscale Approximation with Hierarchical Radial Basis Function Networks. IEEE Transactions on Neural Networks 15(14), 178–188 (2004)

Ferrari, S., Frosio, I., Piuri, V., et al.: Automatic Multiscale Meshing Through HRBF Networks. IEEE Transactions on Inst. and Measurement 54(4), 1463–1470 (2005)

Ha, K.V.: Hierarchical Radial Basis Function Networks. Neural Network Proceedings, 1893–1898 (1998)

Hackbarth, H., Mantel, J.: Modular connectionist structure for 100-word recognition. In: Proc. Internat. Joint Conf. on Neural networks, Seattle, vol. 2, pp. 845–849 (1991)

Hashem, A., Schmeiser, B.: Improving Model Accuracy Using Optimal Linear Combinations of Trained Neural Networks. IEEE Trans. Neural Networks 6(3), 792–794 (1995)

Haykin, S.: Neural Networks: A comprehensive foundation, 2nd edn. Prentice-Hall, Upper Saddle River (1999)

Ho, T.K., Hull, J.J., Srihari, S.N.: Decision Combination in Multiple Classifier Systems. IEEE Trans. Pattern Analysis and Machine Intelligence 16(1), 66–75 (1994)

Ho, T., Hull, J., Srihari, S.: Decision combination in multiple classifier systems. IEEE Trans. Pattern Anal. Machine Intell. 16(1), 66–75 (1994)

Huang, Y., Suen, C.: A method of combining multiple experts for the recognition of unconstrained handwritten numerals. IEEE Trans. Pattern Anal. Machine Intell. 17(1), 90–94 (1995)

Jordan, M.I., Jacobs, R.A.: Hierarchical mixtures of experts and the EM algorithm. In: Int. Joint Conf. Neural Netw., vol. 2, pp. 1339–1344 (1993)

Kil, D., Shin, F.: Pattern Recognition and Prediction With Applications to Signal Characterization. AIP, New York (1996)

Kimura, F., Shridhar, M.: Handwritten Numerical Recognition Based on Multiple Algorithms. Pattern Recognition 24(10), 969–983 (1991)

Kittler, J., Hatef, M., Duin, R.P.W.: Combining Classifiers. In: Proc. 13th Int'l Conf. Pattern Recognition, vol. 2, pp. 897–901 (1996)

Kittler, J., Hojjatoleslami, A., Windeatt, T.: Weighting Factors in Multiple Expert Fusion. In: Proc. British Machine Vision Conf., Colchester, England, pp. 41–50 (1997)

Kittler, J., Hatef, M., Duin, R.P.W., et al.: On combining classifiers. IEEE Transactions on Pattern Analysis and Machine Intelligence 20(3), 226–239 (1998)

Kong, Z., Cai, Z.: Advances of research in fuzzy integral for classifier's fusion. In: Proceedings of the 8th ACIS International Conference on Software Engineering, Artificial Intelligence, Networking and Parallel/Distributed Computing, vol. 2, pp. 809–814 (2007)

Kumar, S., Ghosh, J., Crawford, M.M., et al.: Hierarchical fusion of multiple classifiers for hyperspectral data analysis. Pattern Analysis and Applications 5, 210–220 (2002)

Kuncheva, L.I.: Fuzzy vs non-fuzzy. combining classifiers designed by boosting. IEEE Transactions on Fuzzy Systems 11(6), 729–741 (2003)

Kuncheva, L.I.: Combining pattern classifiers: Methods and algorithms. Wiley, New York (2004)

Kurzynski, M.W.: On the Identity of Optimal Strategies for Multistage Classifiers. Pattern Recognition Letters 10(1), 39–46 (1989)

Lin, D.T., Chen, J.: Facial expressions classification with hierarchical radial basis function networks. In: Proc. Int. Conf. Neural Inf. (ICONIP 1999), vol. 3, pp. 1202–1207 (1999)

Luis-Delgado, J.D., Martinez-Alajarin, J., Tomas-Balibrea, L.M.: Classification of marble surfaces using wavelets. IEE Electronics Letters 39(9), 714–715 (2003)

Martinez-Alajarin, J.: Supervised classification of marble textures using support vector machines. IEE Electronics Letters 40, 664–666 (2004)

Martinez-Alajarin, J., Luis-Delgado, J.D., Tomas-Balibrea, L.M.: Automatic system for quality-based classification of marble textures. IEEE Trans. Syst., Man, Cyber.-Part C: Appl. and Reviews 35(4), 488–497 (2005)

Partridge, D., Griffith, N.: Multiple classifier systems: Software engineered, automatically modular leading to a taxonomic overview. Pattern Analysis and Applications 5, 180–188 (2002)

Perrone, M.: Improving regression estimation: averaging methods of variance reduction with extensions to general convex measure optimization. Ph.D. Thesis. Department of Physics, Brown University (1993)

Pudil, P., Novovicova, J., Blaha, S., et al.: Multistage Pattern Recognition With Reject Option. In: Proc. 11th IAPR Int'l Conf. Pattern Recognition, Conf. B: Pattern Recognition Methodology and Systems, vol. 2, pp. 92–95 (1992)

Selver, M.A., Güzeliş, C.: Hierarchical Neural Networks for Accurate Classification of Radar Data. In: Proceedings of International Symposium on Electromagnetic Fields in Mechatronics Electrical and Electronic Engineering (ISEF 2009), Arras, France, pp. 545–546 (2009)

Selver, M.A., Akay, O., Ardalı, E., et al.: Cascaded and Hierarchical Neural Networks for Classifying Surface Images of Marble Slabs. IEEE Transactions on Systems, Man, and Cybernetics Part C: Applications and Reviews 39(4), 426–439 (2009)

Selver, M.A., Güzeliş, C.: Semi-Automatic Transfer Function Initialization for Abdominal Visualization using Self Generating Hierarchical Radial Basis Function Networks. IEEE Transactions on Visualization and Computer Graphics 15(3), 395–409 (2009)

Suen, C.Y., Legault, R., Nadal, C., et al.: Building a new generation of handwriting recognition systems. PatternRecognition Letters 14(4), 303–315 (1993)

Tsai, D.M., Huang, T.Y.: Automated surface inspection for statistical textures. Image Vis. Comput. 21, 307–323 (2003)

Tung, C.H., Lee, H.J., Tsai, J.Y.: Multi-Stage Pre-Candidate Selection in Handwritten Chinese Character Recognition Systems. Pattern Recognition 27(8), 1.093–1.102 (1994)

Valdovinos, R.M., Sánchez, J.S., Barandela, R.: Dynamic and static weighting in classifier fusion. In: Marques, J.S., Pérez de la Blanca, N., Pina, P. (eds.) IbPRIA 2005. LNCS, vol. 3523, pp. 59–66. Springer, Heidelberg (2005)

Xu, L., Amari, S.-I.: Combining classifiers and learning mixture-of-experts. In: Rabuñal-Dopico, J.R., Dorado, J., Pazos, A. (eds.) Encyclopedia of artificial intelligence, pp. 318–326. IGI Global (IGI) publishing company (2008)

Xu, L., Krzyzak, A., Sun, C.Y.: Several methods for combining multiple classifiers and their applications in handwritten character recognition. IEEE Transactions on Systems, Man and Cybernetics 22, 418–435 (1992)

Zhou, J.Y., Pavlidis, T.: Discrimination of Characters by a Multi-Stage Recognition Process. Pattern Recognition 27(11), 1.539–1.549 (1994)

Chapter 9
Parametric Identification of a Three-Phase Machine with Genetic Algorithms

L. Simón and J.M. Monzón

Electric Engineering Department, ULPGC, Spain
`lsimon@pas.ulpgc.es, jmonzon@die.ulpgc.es`

Abstract. In this chapter a three-phase magnetic induction motor squirrel-cage is analyzed with the Finite Element Method (FEM). Five variations of the rotor geometry design are analyzed. The analysis has been made with simulations of static configurations. For each geometry an identification of the parametric model has been obtained. For the optimization of the parameters, Genetic Algorithms (GA) have been used as a robust optimization method.

9.1 Introduction

Computational tools have allowed to carry out numerical techniques that were once unattainable. The simulation of equivalent circuit models, based on their equilibrium equations and the discrete modeling methods, like FEM, are well known [5, 13, 22]. However, these numerical techniques have been applied only with the emergence of these powerful computational tools. The identification of the induction machine and its control are important in the modeling process prior to manufacture [1, 9, 18, 19]. Also, its optimization is necessary because the existing industrial processes require specific designs [10, 11, 18].

The main contribution of this paper is the parametric identification of a three-phase magnetic induction motor, with five rotor geometry designs (Fig. 9.1). The parameters of the proposed circuit model of the motor, are adjusted applying Genetic Algorithms to each geometry [3, 6, 8]. The adjustment is done, minimizing the square error of the results obtained with the continuous FEM model, for all the different geometries [4, 15, 19].

The FEM is an approximation method, to resolve continuous problems [2, 14, 20, 22]. Its approach is based on transforming a continuous object, in an approximate discrete model. This transformation is known as a discretization of the model. The knowledge of what is inside the body of this model, is obtained by interpolation of known values at the nodes. It is therefore an approximation of the values of function, from the knowledge of a specific and finite number of points [7, 17, 21].

Parallel to the technological evolution that has allowed the design of powerful electronic tools for calculations, the algorithms have evolved in complexity and abstraction, to solve many scientific problems. Genetic Algorithms are a proof of it [10, 16, 18]. Once again, a natural process, as the selection and evolution of individuals from a population, has been applied to the scientific world to observe the evolution of the systems and observe the trend, starting from a certain initial condition.

They apply the rules of the nature. The individuals of a population exchange information, through the evolutionary operators. There are three main genetic operators: selection, crossover and mutation [11, 15]. By the selection, the individuals are chosen from the population, according to the value of the adaptation function, to undergo the future action with the other operators. The crossover operator, is responsible for exchanging components of the individuals selected, to produce new solutions, taking care of the transfer by inheritance of the characteristics of the best individuals from one generation to the next. The mutation, is used as an operator which aims to explore at random, the new portions of the search space by introducing new genetic material in the search for solutions.

The major advantage of these algorithms in comparison with others optimization methods, is that they are able to locate an absolute extreme of the function, rather than a local extreme, and that the initial estimate does not have to be close to the actual values. Moreover, the Genetic Algorithms are working directly over the function and do not require the use of its derivative, or any other auxiliary function [3, 6].

Optimization is a process where multiple and often conflicting objectives need to be satisfied. To solve these problems traditionally all the objectives were converted into a single objective problem by a weighted function. The solution is an optimized result to obtain a compromise between all objectives, finding a solution that minimizes or maximizes the single function of the problems. The goal is to obtain the function that gives a value with a compromise between all objectives.

There are limitations in this approach to solve multi objective optimization problems. Some of them are: the relative importance of a priori knowledge about the objectives and the constraints; the function leads to only one solution; the difficult evaluation of the trade-offs between objectives; and the unattainable solution.

To resolve these limitations there is a set of optimal solutions that leads several trade-off solutions and can be obtained by the so called "Pareto optimal front" [4, 15]. These values are a set of all the optimal solutions to a multi objective problem. The preferred solution or the decision maker is selected from the Pareto optimal values set.

All this methodology applied to the study of induction machines, allows a new viewpoint and increases the knowledge of induction machines. The finite element program used is the Finite Element Method Magnetics (FEMM) [8]. It is a highly versatile free software tool. For the analysis of Genetic Algorithms, the *GAlib library* open source written by Matthew Wall at the Massachusetts Institute of Technology, and the *ga-library* as toolbox for *Scilab* written by Yann Collette, has been used [23, 25]. This software contains a set of Genetic Algorithms in C++ object code.

The compilation was done by the compiler *GCC++*, a C/C++ under public license. For the analysis of the torque-speed curves and the intensities we have developed a program using *Scilab*, a scientific software package for numerical computations. For all the graphic representations of data we have used the program *Gnuplot*, that is also a public license software.

The article as a whole has been produced with free software and/or open source, in the word processor for scientific documents *Latex*, and under the *openSUSE*-Linux operating system.

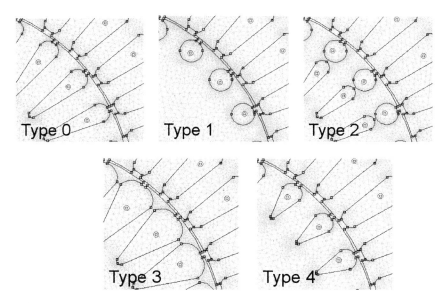

Fig. 9.1 The five rotor geometries designs analizaed.

9.2 Continuous Model FEM and Differential Formulation of Electromagnetic Field

9.2.1 *Magnetostatic*

The equations for magnetic field and windings, for two-dimensional quasi-static field in an electrical machine are described by the Maxwell's equations

$$\nabla \times \vec{H} = \frac{\partial \vec{D}}{\partial t} + \vec{J} \qquad (1)$$

$$\nabla \times \vec{E} = -\frac{\partial \vec{B}}{\partial t} \qquad (2)$$

If the analysis is referred to as quasi-static field, the polarization and displacement currents are negligible at low frequency and then the component $\frac{\partial \vec{D}}{\partial t} \ll \vec{J}$, where $\vec{D}$ [Culomb / m²] is the magnetic flux density, is ommited in (1).

The material equation using the reluctivity v where the material is not isotropic, must be replaced by a tensor that take into account the magnetizing direction (2).

If $\vec{H}$ is the magnetic field intensity at (3), the magnetic flux density $\vec{B}$ is defined by the magnetic vector potential $\vec{A}$ (4) as constitutive equation [2, 24].

$$\vec{H} = v\vec{B} \qquad (3)$$

$$\vec{B} = \nabla \times \vec{A} \qquad (4)$$

The fundamental equation of the vector potential formulation for magnetic vector is obtained by the substitution of (3) and (4) at (5)

$$\nabla \times (v\nabla \times \vec{A}) = \vec{J} \qquad (5)$$

The purpose of this chapter is to analyze various rotor geometry bars (Fig. 9.1). The FEMM program has implemented all the magnetic field equations to the study of electrical machines.

The current density $\vec{J}$ is defined so that its orientation is perpendicular to the plane, so that the vectorial field $\vec{A}$, which is a three-components vector, has the same orientation.

Then, $\vec{A}$ is defined as a function of $A_z(x, y)u_z$ and so $\vec{J}$ is defined as a function of $J_z(x, y)u_z$, where u_z denotes the unit vector in the z-axis direction.

$$-\nabla \cdot (v\nabla \vec{A}) = \vec{J} \qquad (6)$$

So a three-dimensional problem, that has a flat or plane symmetry becomes a two-dimensional problem [5, 12, 13].

9.2.2 Magnetodynamic

The harmonic form of the equation of diffusion of the magnetic potential vector for a problem in two-dimensions, where $\vec{A}_z$ is an unknown component as

$$(\vec{\nabla} \times \vec{H}) = \vec{\nabla} \times \left[\frac{1}{\mu} (\vec{\nabla} \times \vec{A}_z) \right] = \vec{J}_z \qquad (7)$$

$\vec{A}[Wb/m]$ is the magnetic vector potential. $\vec{J}_z[A/m^2]$ is the current density. $\vec{H}[A/m]$ the magnetic field intensity and $\mu[H/m]$ the magnetic permeability.

About the source of the field, the current density can be determined from the material equation by its constitutive equation as

$$\vec{J} = \sigma\vec{E} \tag{8}$$

where σ is the volumetric electrical conductivity.

Combinig (2) with (4) gives

$$\nabla \times \vec{E} = -\frac{\partial}{\partial t}(\nabla \times \vec{A}) \tag{9}$$

Then the current density is defined as (10), where ϕ is the electric scalar potential.

$$\vec{J} = -\sigma\frac{\partial \vec{A}}{\partial t} - \sigma\nabla\phi \tag{10}$$

If the excitation is in the frequency domain and the materials are linear, with a symmetry plane the expression (7) is simplified as follows

$$\vec{\nabla} v \vec{\nabla} A_z - \sigma j \omega A_z = -J_z \tag{11}$$

where v is the reluctivity, that is the inverse of the magnetic permeability μ, the angular frequency ω, electric conductivity σ, and A_z, J_z are the normal components u_z of the magnetic vector potential and the current density in complex form respectively.

For the derivation of the equations of finite elements in its application to 2D electrical machines of the existing methods the Galerkin's method is widely used. This method is a special case of the weighted residuals method where the weighting function is just the shape function of the finite element [20].

9.3 Discrete Model FEM

The application of the Galerkin's method is applied begining with the definition of a differential operator $\Im(x) = 0$ and a residual $\Im(\hat{x}) = R$ from an approximate solution $\hat{x} \neq x$ [17, 20].

The weighting function with the same form of the finite element is selected. If $\hat{A}$ is an approximation of A, the residual R is

$$R = \frac{\partial}{\partial x}\frac{1}{\mu}\frac{\partial \hat{A}}{\partial x} + \frac{\partial}{\partial y}\frac{1}{\mu}\frac{\partial \hat{A}}{\partial y} - j\omega\sigma\hat{A} + J_0 \quad (12)$$

if W is the weighted function, at the region Ω on the boundary C

$$\int_\Omega RW dxdy = 0 \quad (13)$$

Replacing at (13) the value of R at (12), and integrating by parts the result obtained is

$$\sum_M \left\{ \iint_\Omega \frac{1}{\mu^e}\left(\frac{\partial W^e}{\partial x}\frac{\partial A^e}{\partial x} + \frac{\partial W^e}{\partial y}\frac{\partial A^e}{\partial y}\right)dxdy \right\} + j\omega\sigma^e \iint_{\Omega_e} W^e A^e dxdy$$
$$-\frac{\partial A}{\partial \hat{n}}\int_C W^e dc = J_0 \iint_{\Omega_e} W^e dxdy \quad (14)$$

This is the breakdown of the surface integrals into summations over small areas where M is the numbers of the finite elements, e is the element (triangle) and $\hat{n}$ is the normal unit vector. The surface is meshes with triangles that are the finite element. The integral over the domain C is replaced by the summation of the integral over the individual triangles.

9.4 Boundary Conditions of the Continuous Model

The implementation of the FEM implies a finite domain. To problems with no border or open border, the Kelvin transformation is applied [7, 17, 21].

At the far field the material is homogeneous. But in the air the conductivity is $\sigma = 0$, the permeability is μ_0 and the current density is null (no sources).

The description of the magnetic vector potential A is a differential equation given by the Laplace equation (15).

$$\nabla^2 A = \mu_0(-J_0 + j\omega\sigma A) = 0 \quad (15)$$

Then the second member is null at the harmonic form of the equation of diffusion of A, in Cartesian notation (15) [14]. The polar notation of this equation is (16)

$$\frac{1}{r}\frac{\partial}{\partial r}\left(r\frac{\partial A}{\partial r}\right) + \frac{1}{r^2}\frac{\partial^2 A}{\partial \theta^2} = 0 \quad (16)$$

where r is the radio and θ is the angle.

The region with no border, is mapped into a limited circular region with border, where the problems are more easily resolved (Fig. 9.2).

In 2D case, the exterior can be modeled by another circular region, which represents the far field. In this region, in its center, the potential value is $A = 0$.

The expression to this external region have the same form that the internal (16).

At the limit of the edge circle that limits both regions, periodic boundary conditions are imposed. This is achieved by forcing the continuity of the potential A at the edges in both regions.

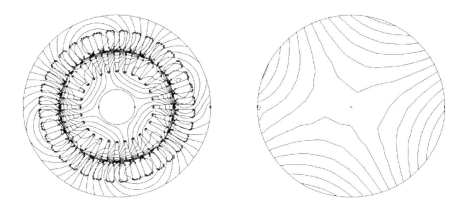

Fig. 9.2 The region of the machine and the far field.

Moreover, with the purpose of making more efficient the study, analysis is limited to the first quadrant of the machine section (Fig. 9.4).

We have proved that the values of the distribution calculated only for the first quadrant are equivalent to a quarter of the total value calculated for the complete section.

The Fig. 9.3 shows the magnetic induction at the first quarter of the complete section of the machine.

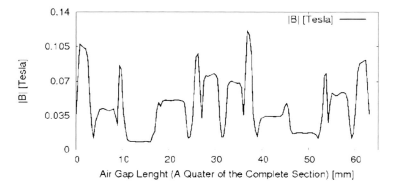

Fig. 9.3 The quarter part of the magnetic induction of the complete section of the machine.

The comparison of the Fig. 9.3 with the Fig. 9.13 that represent the same values for the results for quarter only, show the very similar values obtained between both simulations, really making the error negligible.

To obtain these results the contour conditions on both sides adjacent to the other quadrants are modeled as anti-periodic. This greatly reduced the time and computational costs. Then, all these analysis are limited to the first quadrant.

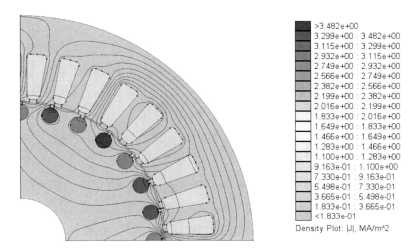

Fig. 9.4 Module of the current density from FEM continuous model type 1.

9.5 Lumped Parametric Model Identification

An induction machine with rotor movement can be modeled using a relatively simple circuital model [2, 6, 13]. Although the circuit parameters often can be approximated by expression forms in explicit terms regarding the geometry of the motor, the identification of these parameters by the FEM analysis consist of the validation of the approximations and simplifications that inevitably must be made in the derivation of the designer analytical formulas [9]. To identify parameters in a model of the induction motor, it is necessary to use a simple reasonable model (Fig. 9.5, 9.6 and 9.7).

The choice of the model depends of the numbers and the nature of parameters used.

The model shown in Fig. 9.5, is used with deep-bar rotor by use of the double-cage rotor model dividing the rotor bar into a top section and a bottom section.

The subscripts 1 and 2 are referring to the outer and inner cages, respectively. R_s is the stator winding resistance, jX_s is the stator leakage inductance, jX_m is the magnetizing inductance, jX_{r1} is the outer cage rotor leakage inductance, R_{r1} is the outer cage rotor resistance, jX_{r2} is the inner cage rotor leakage inductance, R_{r2} is the inner cage rotor resistance and V the per-phase voltage. The slip is shown as s, and all rotor parameters are referred to the stator.

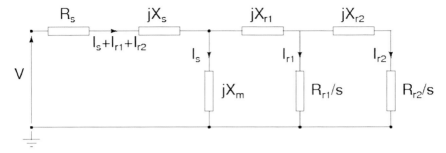

Fig. 9.5 Double-cage rotor induction motor model.

Another model is represented in Fig. 9.6, where [10, 11] proposes a slightly different model with some differences.

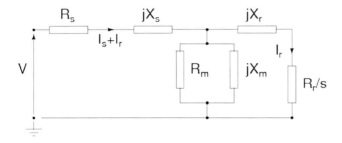

Fig. 9.6 Per-phase equivalent circuit model induction motor.

They give a per-phase equivalent circuit model where R_s is the stator resistance, jX_s the stator leakage reactance, jX_m the magnetizing reactance, R_m the magnetizing resistance, jX_r rotor leakage reactance referred to stator, R_r the rotor resistance referred to stator and V the per-phase voltage.

9.5.1 Parameter Estimation to All the Geometries

As we have seen, there are many different models that can be used as equivalent circuit. The Fig. 9.7 shows the circuit model used to this work [9, 13, 22].

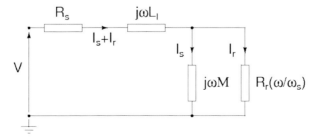

Fig. 9.7 Steady-state simple circuit model per phase.

In this model, the leakage is lumped on the stator side of the circuit in the inductance L_l. Inductance M is the inductance of the magnetic circuit between the rotor and the stator. $R_r(\omega/\omega_s)$ represents the work dissipated as heat in the rotor and the load as mechanical power [1, 9, 12]. If ω is the angular frequency in [rad/s] and p is the number of pairs of poles, the slip frequency ω_s in terms of electrical frequency and mechanical rotor speed ω_r is defined as follows

$$\omega_s = \omega - p\omega_r \qquad (17)$$

therefore, ω_s is defined as the difference between electrical frequency and rotor's mechanical frequency. Useful relations are obtained from (17). In particular we obtain the slip frequency-dependent inductance (18), showing the result in real and complex components

$$L(\omega_s) = \left(L_l + \frac{M}{1+(\tau\omega_s)^2}\right) - j\left(\frac{\tau\omega_s M}{1+(\tau\omega_s)^2}\right) \qquad (18)$$

The dependence of $L(\omega_s)$ on the slip frequency would identify some parameters of the machine as the mutual inductance M, the dispersion inductance L_l and rotor resistance R_r [9, 12]. The rotor time constant τ is defined as M/R_r. As validation of the model we have obtained the expression (19) perfectly valid for the electromagnetic torque T.

$$T = 3pMi^2\left(\frac{\tau\omega_s}{1+(\tau\omega_s)^2}\right) \qquad (19)$$

If the current i is taken as constant and the slip frequency is varied, a curve with a maximum torque in $\tau\omega_s = 1$ is obtained.

9.6 FEM Analysis

Using the FEMM, the rotor is statically analyzed. In the case of zero speed, the slip frequency simply degenerates to $\omega_s = \omega$ [22]. The logical thing would be to identify the parameters based on the results analyzed in (19), using a stator current constant over a range of frequencies, simulating the movement of the rotor. But then the calculation of the torque must be made through the Maxwell stress tensor, which is less precise in the integrated development program that incorporates the FEMM.

The Maxwell stress tensor force density is defined to a circular motor, by the tangential (20) and normal (21) components of the flux density.

$$p_t = \frac{B_n B_t}{\mu_0} \qquad (20)$$

$$p_n = \frac{B_n^2 - B_t^2}{2\mu_0} \qquad (21)$$

The module of the flux density is $|B| = \sqrt{B_n^2 + B_t^2}$ as represented in the Fig. 9.13. Then, the electromagnetic torque along the air gap is defined as

$$T = \oint r\left(\frac{B_n B_t}{\mu_0}\right) ds \qquad (22)$$

However, another option would be to adjust the results of inductance obtained by the analysis of the $\overline{A} \cdot \overline{J}$ integral block, consisting of the magnetic vector potential and current density on the volume of the interested winding, in the range of frequencies studied [9, 12, 14]. This provides high accuracy, since the integral volume (23) is closely related to the stored energy.

$$L = \int \frac{\overline{A} \cdot \overline{J}}{|i|^2} dv \qquad (23)$$

9.7 Parametric Adjustment by Genetic Algorithms

9.7.1 Optimality

The generic multi-objective optimization problem is formulated by a function which maps a set of constraint variables to a set of objective values (decision maker), and then choose among them. Additional criteria can help to refine the search [5, 8, 15].

The general formulation of multi-objective problem requiring the optimization of N objectives is obtained with the vector

$$\vec{x}^* = [x_1^*, x_2^*, \ldots, x_N^*]^T \qquad (24)$$

Which will satisfy the m inequality constraints

$$g_i(\vec{x}) \geq 0 |_{i=1,2,\ldots,m} \qquad (25)$$

The p equality constraints

$$h_i(\vec{x}) = 0 |_{i=1,2,\ldots,p} \qquad (26)$$

And optimize the vector function

$$\max or \min \vec{y} = \vec{f}(\vec{x}) = [f_1(\vec{x}), f_2(\vec{x}), \ldots, f_N(\vec{x})]^T \qquad (27)$$

Where the vector representing the decision variables is

$$\vec{x} = [x_1, x_2, ..., x_n]^T \qquad (28)$$

The space spanned by the objectives vectors is called the objective space and the subspace of the objective vectors that satisfies the constraints is called the feasible search region [4, 10].

A solution could be best, worst and also indifferent to other solutions, i.e. not dominating or dominated with respect to the objectives values. Best solution means a solution not worst in any of the objectives and at last better in one objective than the other.

An optimal solution is the solution that is not dominated by any other solution in the search space. An optimal solution is called Pareto optimal and the entire set of optimal trade-offs solutions are called Pareto optimal set [4, 15].

The Fig. 9.8 shows a Pareto optimal front solutions that fitness two objectives.

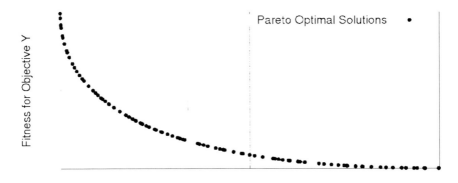

Fig. 9.8 Pareto front optimality.

9.7.2 Fitness Procedure

At this point, first we obtain the analytical results of the model for five different geometries of the rotor (Fig. 9.1), and second we obtain the results of simulations made by finite element analysis by FEMM.

The results of both tests have given a parametric profile of the induction machine for each geometry under consideration. In the frequency analysis carried out to simulate the motion of the rotor, we have obtained the evolution of some significant parameters such as per phase flux and the torque from the FEMM Maxwell stress tensor (20)(21) calculated for different sliding frequencies, so it is possible to identify the parameters M, L_l, and τ of the induction motor [13, 16, 22].

These parameters have been adjusted with GA, taking as objective function, the real part and imaginary part of inductance as a frequency-dependent sliding function, closely linked with the identifying of these parameters.

Parametric Identification of a Three-Phase Machine with Genetic Algorithms 179

The multi objective genetic algorithm used is an optimization toolbox made by Yann Collette for *Scilab* that implements several genetic algorithms [25].

A system identification consist of a structural identification of the equations in the model and a parameter identification of the model's parameters. About the fitness function, the sum of squared error (SSE) is one of the most widely used prediction error measure and was used to adjust the objective function in the GA [11, 16, 19].

$$SSE = \sum_{i=1}^{n} \hat{e}_i^2 \qquad (29)$$

Where $\hat{e}$ is the error term defined as $\hat{e} = y - \hat{y}$ and where y is the observed value and $\hat{y}$ is the predicted value.

The procedure of GA is to minimize the mean square error (29), between the values obtained from the simulations in frequency with the FEM (23) and the values obtained from the equivalent circuit (18) for each geometry [4].

$$E_{objective} = \sum_{k=0}^{10} \left(\text{Re}\left[\int \frac{\overline{A^k \cdot J^k}}{|i|^2} dv \right] - \left[L_l + \frac{M}{1 + (\tau \omega_s^k)^2} \right] \right)^2 \qquad (30)$$

The same expression (30) is applied to the imaginary part of the induction. The GA obtains an optimal for M, L_l and τ.

9.8 Results

The optimal adjustment for the types of geometry studied and for the values of the evolutionary operators of selection, crossover and mutation with optimal setting, are shown in Table 9.1. The settings are the same for all geometries.

Table 9.1 Optimal results of the adjustment by GA for all geometries.

	Type 0	Type 1	Type 2	Type 3	Type 4
τ[s]	0.1635770	0.0703136	0.0619364	0.2384990	0.0944839
M[H]	0.313115	0.164614	0.140215	0.139849	0.164065
L₁[H]	0.01556420	0.00682070	0.00938430	0.00677501	0.00888070

These settings are based on a binary-to-decimal genome with a 16 bits resolution, a population size = 50, a number of generations = 200, a probability of mutation = 0.01 and a probability of crossover = 0.6.

The figures 9.9, 9.10 and 9.11, show an example represented from the settings for the geometry type 1, which has provided values for $\tau = 0.0703136$ [s], $M = 0.164614$ [H], $L_l = 0.00682070$ [H] as an adjustment solution in the GA. These values are determined by the adjustment between the FEM simulations and the analytical results.

The objective function was raised and adjusted as a multi-objective function with two fitness functions, the real part and the imaginary part of the inductance. The results obtained by the GA have determined that the fitness functions are not contradictory. Therefore, there is not a front of solutions at the Pareto diagram, it collapses to a point.

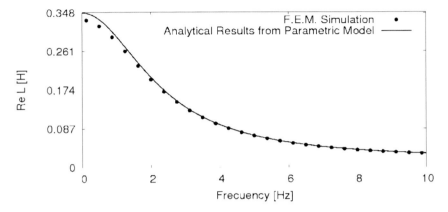

Fig. 9.9 Real part adjustment of the induction by GA.

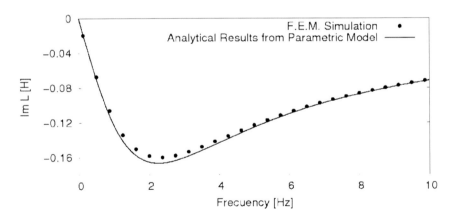

Fig. 9.10 Imaginary part adjustment of the induction by GA.

Fig. 9.9 and Fig. 9.10 show the adjustment determined by the GA of the real and imaginary parts of the inductance L [H], for the geometry type 1. The adjustment of the model has been determined with respect to the circuital model simulated by FEM continuous model, varying the slip frequency from 0 to 10 Hz. The values that enable this setting are listed in Table 9.1 together with the values adjusted to other geometries.

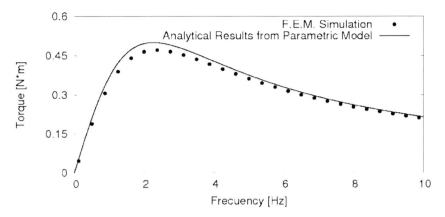

Fig. 9.11 Validation of the electromagnetic torque from parametric model and Maxwell stress tensor.

As a validation procedure we have represented in Fig. 9.11 the electromagnetic torque for the geometry type 1, and the adjustment show the reliability of the method. The FEM simulation gives the electromagnetic torque from the Maxwell stress tensor, and it is validated with the electromagnetic torque from the parametric model (22).

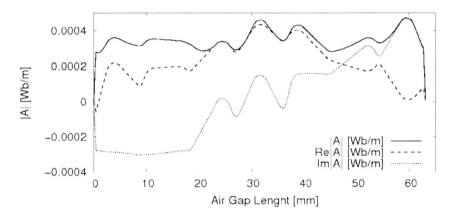

Fig. 9.12 Module of the magnetic vector potential and its components complex.

Figure 9.12 shows the module of the magnetic vector potential, with the real part and imaginary part of it. This representation is shown along the air gap in the first quadrant. The air gap length is 65 mm.

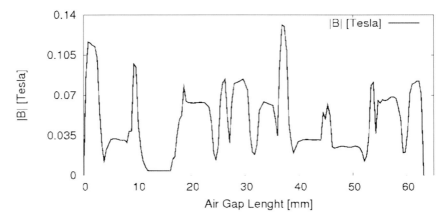

Fig. 9.13 Module of the magnetic induction.

Figure 9.13 shows the module of the magnetic induction along the air gap of the first quadrant. These values are necessary to obtain the electromagnetic torque from the Maxwell stress tensor. The relationship between Fig. 9.12 and Fig. 9.13 is given by (18).

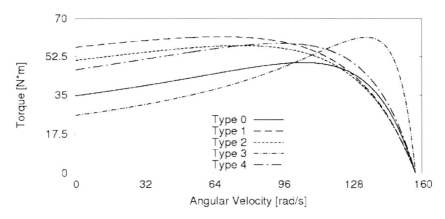

Fig. 9.14 Torque-velocity curves representations for all geometries types.

As part of the results, a representation of the characteristic curves of torque-speed and the intensity of a phase, are shown in Fig. 9.14 and Fig. 9.15 respectively, for all the geometries of the rotor bars that have been studied.

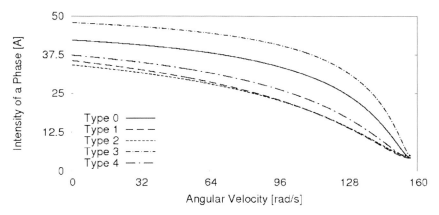

Fig. 9.15 Intensity of a phase, for all geometries types.

9.9 Conclusions

A steady-state equivalent circuit model for an electrical phase has been created.

The parameters have been identified from the inductance values, for their accuracy.

The identification method is a frequency analysis of the FEM and of the circuit model.

The parametric model of the five rotor geometries have been adjusted with the FEM continuous model.

The adjustment of the circuit model parameters of a three-phase induction machine with the continuous model FEM, has been obtained with GA for all geometries.

As genetic algorithms, the evolutionary algorithms have been used for the optimization, because they are robust and accurate.

The evolutionary algorithms toolbox libraries made to *Scilab*, have been used for its versatility, its easy procedure to implement the objective function and because we have obtained the correct solution to the problem.

The problem was studied for two objectives, but the results indicate that the problem can be reduced to a mono objective.

All this work has been done with open source software under Linux.

References

[1] Amuliu, B., Ali, K.: Identification of Variable Frequency Induction Motor Models From Operating Data. IEEE Transactions on Energy Conversion 17(1), 24–31 (2002)
[2] Belmans, R., Findlay, R.D., Geysen, W.: A Circuit Approach to Finite Element Analysis of a Double Squirrel Cage Induction Motor. IEEE Transactions on Energy Conversion 5(4), 719–724 (1990)
[3] Boonruang, M., Nittaya, M., Anant, O.: Dynamic Model Identification of Induction Motors using Intelligent Search Techniques with taking Core Loss into Account. In: 6th WSEAS International Conference on Power Systems, Lisbon, Portugal, pp. 108–115 (2006)

[4] David, A., Gary, B.: Evolutionary Computation and Convergence to a Pareto Front, pp. 221–228. Stanford University, California (1998)
[5] Haque, T., Nolan, R., Pillay, P., et al.: Parameter Determination for Induction Motors, pp. 45–49. IEEE, Los Alamitos (1994)
[6] Harry, H., Charles, W.: Estimation of Induction Motor Parameters by Genetic Algorithm, pp. 21–28. IEEE, Los Alamitos (2003)
[7] Lowther, D.A., Freeman, E.M.: Further aspects of the Kelvin transformation method for dealing with open boundaries. IEEE Transactions on Magnetics 28(2), 1667–1670
[8] von Lücken, C.D.: Algoritmos Evolutivos para Optimización Multiobjetivo: un Estudio Comparativo en un Ambiente Paralelo Asíncrono. Universidad Nacional de Asunción (2003)
[9] Meeker, D.: Induction Motor Example. IEEE, Los Alamitos (2002)
[10] Mehmet, Ç., Ramazan, A.: Design optimization of induction motor by genetic algorithm and comparison with existing motor. Mathematical and Computational Applications 11(3), 193–203 (2006)
[11] Mehmet, Ç., Ramazan, A., Osman, B.: Cost optimization of submersible motors using a genetic algorithm and a finite element method. Int. J. Adv. Manuf. Technol. 33, 223–232 (2006)
[12] Mirafzal, B., Gary, L., Rangarajan, M.: Determination of Parameters in the Universal Induction Motor Model. IEEE Transaction on Industry Applications 45(1), 1207–1216 (2007)
[13] Molinar, D., De Weerdt, R., Belmans, R., et al.: Calculation of two-axis induction motor model parameters using finite elements. IEEE, Los Alamitos (1996)
[14] Nerg, J., Pyrhonen, J., Partanen, J.: Finite element modelling of the magnetizing inductance of an induction motor as a function of torque. IEEE Transactions on M 40(4), 2047–2049 (2004)
[15] Patrick, N., Anahita, Z., El-Sharkawi, M.A.: Pareto Multi Objective Optimization, pp. 84–91. IEEE, Los Alamitos (2005)
[16] Phumiphak, T., Chat-Uthai, C.: Estimation of Induction Motor Parameters Based on Field Test Coupled with Genetic Algorithm, pp. 1199–1203. IEEE, Los Alamitos (2002)
[17] Qiushi, C., Adalbert, K.: A Review of Finite Element Open Boundary Techniques for Static and Quasi-Static Electromagnetic Field Problems. IEEE Transactions on Magnetics 33(1), 663–676 (1997)
[18] Rasmus, K.U.: Models for Evolutionary Algoritms and Their Applications in System Identification and Control Optimization. University of Aarhus (2003)
[19] Rasmus, K., Pierré, V.: Parameter identification of induction motors using stochastic optimization algorithms. Applied Soft. Computing 4, 49–64 (2004), doi:10.1016/j.asoc.2003.08.002
[20] Salon, S.J.: Finite Element Analysis of Electrical Machines. Kluwer Academic Publishers, Boston (1995)
[21] Stochniol, A.: A General Transformation for Open Boundary Finite Element Method for Electromagnetic Problems. IEEE Transactions on Magnetics 28(2), 1679–1681 (1992)
[22] Tandom, S.C.: Finite Element Analysis of Induction Machines. IEEE, Los Alamitos (1982)
[23] Wall, M.: GAlib documentation. MIT, Cambridge (2000), http://lancet.mit.edu/ga/dist/
[24] Yamazaki, K.: An efficient procedure to calculate equivalent circuit parameters of induction motor using 3-D nonlinear time-stepping finite-element method. IEEE Transactions on Magnetics 38(2), 1281–1284 (2002)
[25] Yann, C.: Personal Website (2009), http://ycollette.free.fr

Chapter 10
Ridge Polynomial Neural Network for Non-destructive Eddy Current Evaluation

Tarik Hacib[1], Yann Le Bihan[2], Mohammed Rachid Mekideche[1], and Nassira Ferkha[1]

[1] Laboratoire d'études et de modélisation en électrotechnique, Faculté des Sciences de l'Ingénieur, Univ. Jijel, BP 98, Ouled Aissa, 18000 Jijel, Algérie
[2] Laboratoire de Génie Electrique de Paris, SUPELEC, UMR 8507 CNRS, UPMC Univ. Paris 06, Univ. Paris-Sud 11, 11 Rue Joliot-Curie, Plateau de Moulon 91192 Gif-sur-Yvette Cedex, France
`tarik.hacib@gmail.com`

Abstract. Motivated by the slow learning properties of Multi-Layer Perceptrons (MLP) which utilize computationally intensive training algorithms, such as the backpropagation learning algorithm, and can get trapped in local minima, this work deals with ridge Polynomial Neural Networks (RPNN), which maintain fast learning properties and powerful mapping capabilities of single layer High Order Neural Networks (HONN). The RPNN is constructed from a number of increasing orders of Pi-Sigma units, which are used to solving inverse problems in electromagnetic Non-Destructive Evaluation (NDE). The mentioned inverse problems were solved using Artificial Neural Network (ANN) for building polynomial functions to approximate the correlation between searched parameters and field distribution over the surface. The inversion methodology combines the RPNN network and the Finite Element Method (FEM). The RPNN are used as inverse models. FEM allows the generation of the data sets required by the RPNN parameter adjustment. A data set is constituted of input (normalized impedance, frequency) and output (lift-off and conductivity) pairs. In particular, this paper investigates a method for measurement the lift-off and the electrical conductivity of conductive workpiece. The results show the applicability of RPNN to solve non-destructive eddy current problems instead of using traditional iterative inversion methods which can be very time-consuming. RPNN results clearly demonstrate that the network generate higher profit returns with fast convergence on various noisy NDE signals.

10.1 Introduction

Electromagnetic Non-Destructive Evaluations (NDE) methods consist in applying excitation field through the surface of the device under control, and observe the changes in corresponding field characteristics over the same surface. The changes depend on excitation conditions, materials properties, geometrical dimensions of the tested devices, and on the physical characteristics of the existing defects. Identification of searched parameter during the inspection is achieved by processing the information about field distribution. The development of electromagnetic NDE shows that they generalize forward and inverse problem solutions in the investigated engineering structure.

There are two approaches for the electromagnetic NDE inverse problem solution. The first one is based on the use of global optimization methods, and the second on the use of polynomial functions to determine correlation between searched parameters and field characteristics over the observed surface.

The application of global optimization methods assumes construction of an iterative procedure that uses the forward problem solution technique. The polynomial functions approach uses forward problem solution technique as well. The field distribution data of different but known values of model parameters are used to construct the desired functions. There are different methods for this construction, but usually it is done with the help of Artificial Neural Network (ANN).

The highly popularized Multi-Layer Perceptrons (MLP) models have been successfully applied in electromagnetic NDE [1] [2]. However, MLP utilize computationally intensive training algorithms such as the error back-propagation [3] and can get stuck in local minima. In addition, these networks have problems in dealing with large amounts of training data, while demonstrating poor interpolation properties, when using reduced training sets. High Order Neural Networks (HONN) are type of feedforward Neural Networks (NN) which have the ability to transform the nonlinear input space into higher dimensional space where linear separability is possible [4]. In contrast to HONN, ordinary feedforward networks cannot elude the problem of slow learning, especially when they are used to solve complex nonlinear problems [5]. On the other hand, high order terms or product units in HONN can increase the information capacity of NN. The representational power of high order terms can help solving complex problems with construction of significantly smaller network while maintaining fast learning capabilities [6]. HONN are simple in their architectures and require fewer numbers of weights to learn the underlying equation [6]. This potentially reduces the number of required training parameters. As a result, they can learn faster since each iteration of the training procedure takes less time. This makes them suitable for complex problem solving. However, they suffer from the combinatorial explosion of the higher-order terms and demonstrate slow learning, when the order of the network becomes excessively high.

In this paper, Ridge Polynomial Neural Networks (RPNN) are proposed to evaluate lift-off and electrical conductivity in conductive workpiece from the impedance measurements. The Finite Element Method (FEM) provides the data set required for the training of RPNN. A data set is constituted of input (normalized impedance, frequency) and output (h,σ) pairs. The results suggest that the lift-off and electrical conductivity of conductive workpiece can be measured with an excellent accuracy.

10.2 Higher Order Neural Network

10.2.1 Pi-Sigma Neural Network

The Pi-Sigma Neural Network (PSNN) was introduced by Shin and Ghosh [7]. It is a feedforward network with a single hidden layer and product units at the output layer [8]. The PSNN uses the product of the sums of the input components, instead of the sum of products as in functional link neural networks (FLNN) [7]. In contrast to FLNNs, the number of free parameters in the PSNN increases linearly with the order of the network. The reduction in the number of weights, as compared to FLNNs, allows the PSNN to enjoy faster training. Ghosh and Shin [8] showed that the PSNN requires fewer numbers of adjustable weights for the same order and the same number of inputs and outputs, when compared to the FLNN. The structure of the PSNN avoids the problem of the combinatorial explosion of the higher order terms. The PSNN is able to learn in a stable manner even with fairly large learning rates [9]. In addition, the use of linear summing units makes the convergence analysis of the learning rules for PSNNs more accurate and tractable.

Shin and Ghosh [9] investigated the applicability of PSNN for shift, scale and rotation invariant pattern recognition. Results for function approximation and classification were encouraging, when compared to backpropagation networks for achieving similar performance. Ghosh and Shin [7] argued that the PSNN requires less memory, and at least two orders of magnitude less number of computations, when compared to MLPs for similar performance levels, and over a broad class of problems. Fig. 10.1 shows a PSNN, whose output is determined according to the following equations

$$h_j = w_j^T x = \sum_{k=1}^{N} w_{kj} x_k + w_{0j} \tag{1}$$

$$y = \sigma\left(\prod_{j=1}^{k} h_j\right) \tag{2}$$

where w_{kj} is the adjustable weight, x_k is the input vector, K is the number of summing unit, N is number of input nodes, and σ is a suitable nonlinear transfer function. PSNN demonstrated competent ability to solve scientific and engineering problems despite not being universal approximators [7], [9].

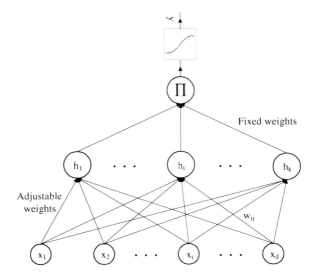

Fig. 10.1 Pi-Sigma Neural Network

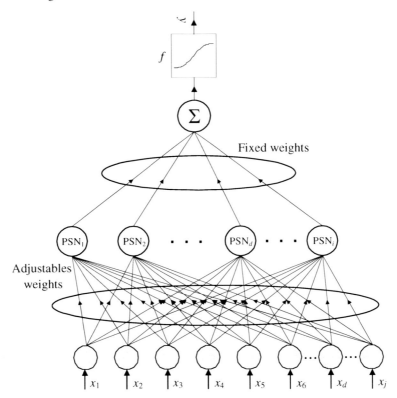

Fig. 10.2 Ridge polynomial neural network

10.2.2 Ridge Polynomial Neural Network

The RPNN was introduced by Shin and Ghosh [9]. The network is constructed by adding gradually more complex PSNNs, denoted by $P_i(x)$ in equation (3). RPNNs can approximate any multivariate continuous function defined on a compact set in multidimensional input space, with arbitrary degree of accuracy. Similar to the PSNN, the RPNN has only a single layer of adaptive weights as shown in Fig. 10.2, and hence the network preserves all the advantages of PSNN.

Any multivariate polynomial can be represented in the form of a ridge polynomial and realized by the RPNN [9], whose output is determined according to the following equations:

$$y \approx \sigma\left(\sum_{i=1}^{N} P_i(x)\right) \quad (3)$$

$$P_i(x) = \prod_{j=1}^{i}\left(\langle W_j, X\rangle + W_{j0}\right), \ i = 1, \ldots, N. \quad (4)$$

where $\langle W_j, X \rangle$ is the inner product between the trainable weights matrix W, and the input vector X, W_{j0} are the biases of the summing units in the corresponding PSNN units, N is the number of PSNN units used (or alternatively, the order of the RPNN), and σ denotes a suitable nonlinear transfer function, typically the sigmoid transfer function.

The RPNN provides a natural mechanism for incremental network growth, by which the number of free parameters is gradually increased with the addition of Pi–Sigma units of higher orders. The structure of the RPNN is highly regular, in the sense that Pi–Sigma units are added incrementally until an appropriate order of the network or a predefined error level criterion is achieved.

Shin and Ghosh [9] tested the RPNN in a surface fitting problem, the classification of high dimensional data, and the realization of a multivariate polynomial function. They highlighted the capabilities of the RPNN in comparison to MLP, cascade correlation, and optimal brain damage. Results showed that an RPNN trained with the constructive learning algorithm provided smooth and steady learning and used much less computations and memory, in terms of the number of units and weights.

10.3 Forward Problem Definition

The FEM is used as a forward problem solver to collect information for field distribution in the region under investigation. This region contains an eddy current probe with a cylindrical E-shaped ferrite-cored, disposed over a conductive specimen. The probe consists of a coil placed inside an E-shaped ferrite-cored (μ_r = 2200) which is placed over a conductive workpiece as is depicted in Fig. 10.3 which is operating in the harmonic regime at frequency depending on the problem (typically between a few Hz to a few MHz). The aim of the ferrite core is to focus the magnetic fields into the workpiece, to increase the probe sensitivity to the physical and geometrical properties of the workpiece.

The modelling of this configuration is done by using the FEM in order to generate the data sets required by the RPNN.

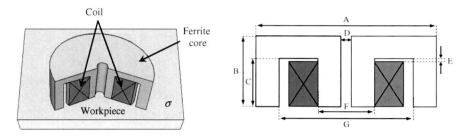

Fig. 10.3 Probe constituted by a coil placed inside an E-shaped core

The probe dimensions are given in Table 10.1. The interaction between electromagnetic field, excited from the probe, and the specimen is analyzed numerically in 2D.

Table 10.1 Coil and E-shaped ferrite-cored dimensions.

Coil (mm)	E shaped core (mm)
External diameter: 7.60	D: 3.10
Internal diameter: 14.20	F: 7.48
Height: 3.24	G: 14.98
Number of turns 2450	A: 18.00
Lift-off : 0.36	B: 5.30
	C: 3.60
	E: 0

During the analysis all used materials are accepted to be linear and homogeneous. Electromagnetic field is excited by the coil, supplied by a voltage source of 0.1 V. The resulting magnetic field distribution is governed by partial differential equation (5).

$$rot\left(\frac{1}{\mu} rot(A)\right) + \sigma\left(\frac{\partial A}{\partial t} + grad(V)\right) = 0 \qquad (5)$$

where μ is the magnetic permeability, σ the electric conductivity and V the scalar potential.

The numerical solution of equation (5) was performed at zero value boundary conditions using FEM and commercial software package COMSOL 3.2. Thanks to the symmetry of the system, only a half sector of the geometry is meshed (Fig. 10.4). Fig. 10.5 shows distribution of potential vector magnetic A.

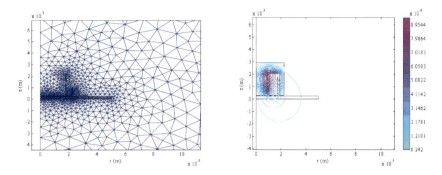

Fig. 10.4 View of the meshed measurement cell   **Fig. 10.5** Distribution of vector magnetic *A*

10.4 Implementation of RPNN for Solving the Inverse Problem

The NN approach for the described inverse problem solution requires input and target data sets for the training. The input data set is collected either by data acquisition system during physical experiments or by numerical simulations. In the examined problem, this set was formed with the variation of probe impedance above the workpiece, with frequency. This is usually considered via impedance information for probe coil. The target data set contains data for workpiece parameters.

During the training process, the values of the weights and the biases of RPNN are determined. Despite of the large number of neurons, RPNN network is trained very quickly.

For the purpose of the described inverse problem solution, two RPNN are used, one to compute the conductivity (RPNN1), and the other for the lift-off (RPNN2). They use identical input data set with information for probe normalized impedance diagram values at two different frequencies values (1 khz and 150 khz). This information was prepared using results from numerically simulated experiments. Both networks are with approximately four inputs and a single output. They differ in neurons parameters and the used target data sets. The first target data set contains information for lift-off *h* and the second one was with information for workpieces conductivity σ.

The estimation of the two quantities *h* and σ requires to follows tree steps (Figure 10.5):

1) Generation of the samples of the data sets (training, validation and test sets) and pre-processing of them (centring and normalisation).
2) Training, validation and test of the RPNN models.
3) Utilisation of the designed RPNN for non-destructive eddy current data inversion.

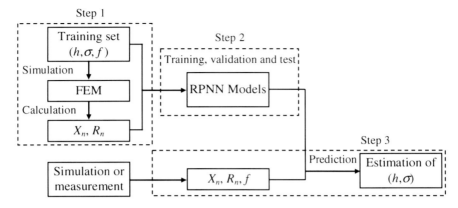

Fig. 10.5 Conductive workpieces parameter extracting procedure

10.5 Results

The physical experiment was simulated numerically using FEM. The described forward problem was solved at different values of lift-off and conductivity parameters, and the obtained resistance and reluctance values were used to reconstruct the impedances. In this work the specimen's parameters are provided by Goodfellow SARL.

Fig. 10.6 and 10.7 show respectively the influence of lift-off parameter on the resistance and reactance variation at fixed conductivity of the workpiece, and Fig. 10.8 and 10.9 represent the influence of the workpiece conductivity if the lift-off did not change.

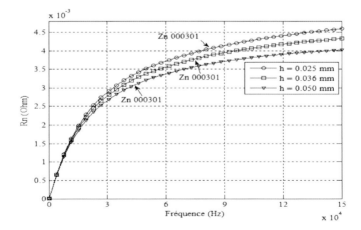

Fig. 10.6 Resistance deviation for three values of lift-off (conductivity = 3.7453 10^7 S/m)

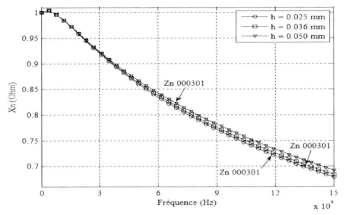

Fig. 10.7 Reactance deviation for three values of lift-off (conductivity = 3.7453 10^7 S/m)

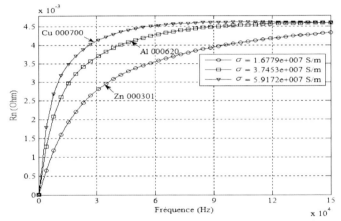

Fig. 10.8 Resistance deviation for three values of conductivity (lift-off = 0.036 mm)

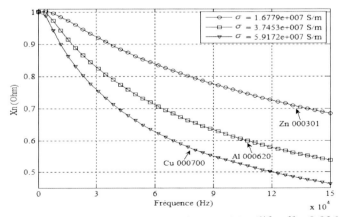

Fig. 10.9 Reactance deviation for three values of conductivity (lift-off = 0.036 mm)

The training process of the RPNN requires two data sets. The input data set contains information for the impedance. The target data set contains information about workpiece parameters. Both data sets were prepared numerically as it was described above. The information in the data sets corresponds to the results from ideal physical experiment. The network is trained with 300 learning examples.

To facilitate the identification of workpiece parameters, two RPNN were prepared, one for lift-off and another for conductivity identification. Both networks were trained with identical input data but with different target data sets. Then the RPNN is trained with the Probabilistic learning rule method [7].

10.5.1 Training without Noise

Fig. 10.10 and 10.11 show the evolution of the networks mean square error (MSE) during the learning process.

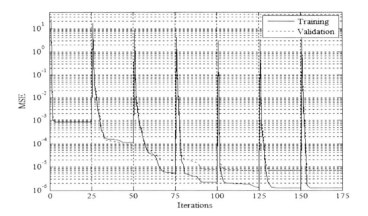

Fig. 10.10 Performance of the RPNN1 during a training session

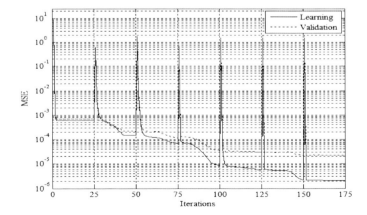

Fig. 10.11 Performance of the RPNN2 during a training session

These results are obtained where the input signals of the test data set have been corrupted with additive white Gaussian noise. The noise level is described by signal to noise ratio (SNR), which is defined as

$$SNR = 10\,Log\left(P_s / \sigma^2\right) \qquad (6)$$

where P_s and σ^2 are signal power and noise power, respectively.

After RPNN training and respective validations, new parameters were simulated by the FEM, for posteriori identification by the network. The expected and obtained values together with the relative accuracy are shown in Table 10.2. In this work the specimen's parameters are provided by Goodfellow SARL.

Table 10.2 Actual values of the parameters and those obtained by RPNN

Material	Reference	Electrical Conductivity [S/m]			Lift-off [mm]	
		expected	RPNN1	τ [%]	expected	RPNN2 τ [%]
Aluminium	Al 000620	3.7453×10^7	3.7405×10^7	0.1281	0.025	0.02491 0.3600
Zinc	Zn 000312	1.6779×10^7	1.6765×10^7	0.0834	0.036	0.03698 2.7222
Laiton	Cu 020450	1.5625×10^7	1.5618×10^7	0.0448	0.050	0.04975 0.5000

As we can see, the results obtained by the RPNN agree very well with the expected ones. Fig. 10.12 shows the evolution of relative errors, depending on the signal to noise ratio, obtained with RPNN. The RPNN show good results in simulations.

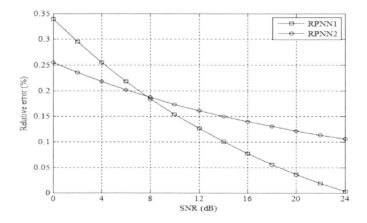

Fig. 10.12 Relative errors obtained for estimating σ and h

To illustrate the generalization capability of the two models, the plots of the predicted values against the desired one are shown for the test set in Fig. 10.13.

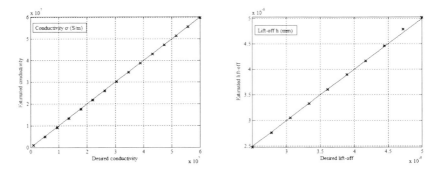

Fig. 10.13 Comparison between the conductivity and lift-off provided by the RPNN and the conductivity and lift-off contained in the test set (with 22 dB noise)

The more the points are concentrated around the diagonal line, the better the prediction of the model. The results show that most of the points locate on the diagonal line, which means the RPNN models can predict the values of lift-off and electrical conductivity accurately.

10.5.2 *Training with Noise*

In this part, the input signals of the training data set have been corrupted with additive white Gaussian noise (Learning with 22 dB noise on the input signal). Fig. 10.14 and 10.15 show the evolution of the networks mean square error (MSE) during the learning process.

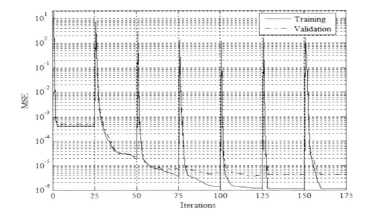

Fig. 10.14 Performance of the RPNN1 during a training session

Ridge Polynomial Neural Network for Non-destructive Eddy Current Evaluation

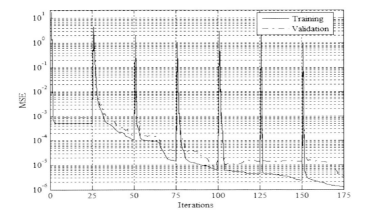

Fig. 10.15 Performance of the RPNN2 during a training session

Table 10.3 shows the expected and obtained values together with the relative accuracy. As above, the specimen's parameters are provided by Goodfellow SARL.

Table 10.3 Actual values of the parameters and those obtained by RPNN

Material	Reference	Electrical Conductivity [S/m] expected	RPNN1	τ [%]	Lift-off [mm] expected	RPNN2	τ [%]
Aluminium	Al 000620	3.7453×10^7	3.7399×10^7	0.1441	0.025	0.02509	0.3600
Zinc	Zn 000312	1.6779×10^7	1.6772×10^7	0.0417	0.036	0.03715	3.1944
Laiton	Cu 020450	1.5625×10^7	1.5614×10^7	0.0704	0.050	0.05278	5.5600

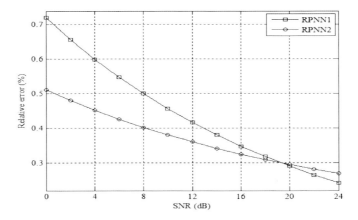

Fig. 10.16 Relative errors obtained for estimating σ and h (Learning with noise)

As we can see, the results obtained by the RPNN agree very well with the expected ones. Fig. 10.16 shows the evolution of relative errors, depending on the signal to noise ratio, obtained with RPNN (Learning with 22 dB noise on the input signal). The RPNN show good results in simulations.

Furthermore, to illustrate the generalization capability of the two models, the plots of the predicted values against the desired one are shown for the test set in Fig. 10.17 (with 22 dB noise).

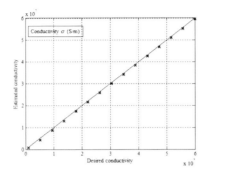 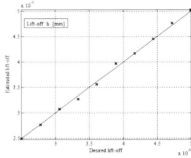

Fig. 10.17 Comparison between the conductivity and lift-off provided by the RPNN and the conductivity and lift-off contained in the test set (with 22 dB noise)

The results show that most of the points distribute along the diagonal line, which means the SVM models can predict the values of permittivity accurately.

These results show the applicability of RPNN to solve electromagnetic inverse problems instead of using traditional iterative inversion methods which can be very time-consuming.

10.6 Conclusion

This work presents an investigation of the use of RPNN for the inverse problem solution in the field of electromagnetic NDE. The solution of the forward problem was obtained using 2D FEM solver.

The results from numerical simulation of the experiment show that the correlation between the impedance and workpiece parameters is non-linear. This allows ANN to be used as a tool for inverse problem solution. The obtained two RPNN were adjusted and tested with numerically computed data from the forward problem solution.

The proposed approach was found to be highly effective in identification of parameters through solving inverse problems in electromagnetic NDE.

References

[1] Low, T.S., Chao, B.: The use of finite elements and neural networks for the solution of inverse electromagnetic problems. IEEE Trans. Magnetics 28(5), 1931–1934 (1992)
[2] De Alcantara, N.P., Alexandre, J., De Carvalho, M.: Computational investigation on the use of FEM and ANN in the non-destructive analysis of metallic tubes. In: 10th Biennial conference on electromagnetic field computation ICEFC, Italy (2002)
[3] Lawrence, S., Giles, C.L.: Overfitting and Neural Networks: Conjugate Gradient and Backpropagation. In: International Joint Conference on Neural Network, Italy, pp. 114–119 (2000)
[4] Pao, Y.: Adaptive pattern recognition and neural networks. Addison Wesley, Reading (1989)
[5] Chen, A.S., Leung, M.T.: Regression Neural Network for error correction in foreign exchange forecasting and trading. Computers & Operations Research, 1049–1068 (2004)
[6] Leerink, L.R., Giles, C.L., Horne, B.G., Jabri, M.A.: Learning with product units. Advances in Neural Information Processing Systems, 537–544 (1995)
[7] Shin, Y., Ghosh, J.: The Pi-Sigma Networks: An efficient Higherorder Neural Network for pattern classification and function approximation. In: Proceedings of International Joint Conference on Neural Networks, Washington, vol. 1, pp. 13–18 (July 1991)
[8] Ghosh, J., Shin, Y.: Efficient Higher-order Neural Networks for function approximation and classification. Int. J. Neural Systems 3(4), 323–350 (1992)
[9] Shin, Y., Ghosh, J.: Computationally efficient invariant pattern recognition with higher order Pi-Sigma Networks. The University of Texas at Austin (1992)

Chapter 11
Structural-Systematic Approach in Magnetic Separators Design

Vasiliy F. Shinkarenko[1], Mikhaylo V. Zagirnyak[2], and Irina A. Shvedchikova[3]

[1] National Technical University of Ukraine "Kyiv Polytechnic Institute", Kyiv, 03056, Ukraine
ntuukafem@ua.fm
[2] Kremenchuk Mykhaylo Ostrogradskiy State University, Kremenchuk, 39614, Ukraine
mzagirn@kdu.edu.ua
[3] East-Ukrainian Volodymyr Dal National University, Lugansk, 91034, Ukraine
snu2008@mail.ru

Abstract. A necessity of development of fundamentally new variants of unconventional embodiment devices in the modern conditions of progressive increase of structural and functional variety of magnetic separators has been grounded. It has been demonstrated that structural-systematic approach is a methodological basis able to provide completeness of magnetic separator structures synthesis at the stage of a search design and to make this synthesis directed. It has been proved that homologous series law is an important methodological instrument of structural-systematic approach having a powerful heuristic potential and able to provide a directed new structures search and synthesis at an interspecific level.

11.1 Introduction

In modern conditions a tendency of progressive increase of structural and functional variety of magnetic separators can be observed (Svoboda 1987, Unkelbach 1990, Zagirnyak 2008). This is caused by the necessity to create qualitatively new devices able to provide reliable functioning and required characteristics of the systems in conformity with new operating conditions. In its turn, it stimulates the development of principally new variants of unconventional embodiment magnetic separators.

The problems of search and synthesis of principally new structural varieties of magnetic separators refer to a complex class of indeterminated problems of search design. The role of these problems in new devices design may come up to 80%.

Ambiguity inherent in search problems and fuzziness of source data result in the fact that most of such problems are solved by means of approximate (heuristic) methods. Results of search problems solution are of a random character and do not guarantee obtaining optimal structural variants. Under these conditions search directed to development of such methodological approaches to design which are able to provide orientation and completeness of the synthesis of required magnetic separator structures is topical.

During search design information about typical design solutions, published patents and the design object analogs, about computation methods and tests results, as well as reference and catalog information, is extensively used. Usually such information is distributed among numerous information search system data bases. Existing approaches to data bases organization are not notable for rigor, as practically always there are either unaccounted data or data allocated in several data bases simultaneously (data redundancy). In this case search of the required data involves considerable time expenditure. That is why practical realization of new methods of magnetic separators design requires development of principally new concept of search design procedures dataware. This concept is to guarantee completeness of information delivery with a possibility of its redundancy limitation.

11.2 Search Methodology Substantiation

To solve the above stated problems of search design it is proposed to apply methodological instruments developed within a new scientific electromechanics trend generalized by a concept of structural-systematic search. Structural-systematic approach as an independent trend was formed after the discovery of genetic classification (GC) of primary field sources (PFS). Results of structural-systematic analysis of GC periodical structure and invariant properties of PFS as integral electromagnetic structures containing genetic information were the basis for development of the theory of structural organization and genetic evolution of electromechanical systems (Shinkarenko 1996). Direct relation between fundamental principles of preservation of electromagnetic symmetry and topology, genetic code and the principle of PFS genetic information preservation, on the one part, Species genetic nature and laws of micro- and macroevolution of real structural classes of electromechanical energy converters (EMEC), on the other part, was first scientifically substantiated within the new trend.

Genetic information presented by a universal PFS genetic code in GC structure acts as a peculiar transfer function in the hierarchy of complicated genetic levels of EMEC arbitrary Species structural organization. A genetic code structure consists of two parts – alphabetic and numerical ones. An alphabetic part denotes a contracted name of the corresponding sculpted surface geometric class to which PFS in GC structure belongs. The first big period is formed by six geometric classes (Fig. 11.1): *CL* – cylindrical; *CN* – conic; *PL* – plane; *SPH* – spherical; *TP* – toroid plane; *TCL* – toroid cylindrical field sources. A genetic code numerical part represents topologic features and kind of PFS electromagnetic symmetry, i.e. points out presence or absence of PFS surface edges (dissymetrizing factors): in the direction of field wave propagation (the first code figure) and in the perpendicular direction (the second code figure). A genetic code numerical part may assume the following numerical values: *0* – absolute electromagnetic symmetry (dissymetrizing factors or surface edges are absent); *1* – electromagnetic

Structural-Systematic Approach in Magnetic Separators Design

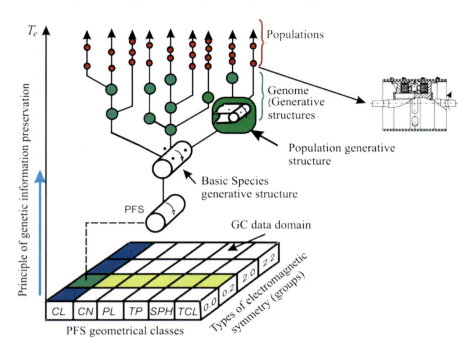

Fig. 11.1 Genetic model of magnetic separators Species internal structure (T_e - evolution time)

dissymmetry (partial dissymmetry due to presence of one surface edge); 2 – electromagnetic asymmetry (absence of symmetry due to presence of two surface edges on the way of electromagnetic wave propagation). Field sources belong to the class of orientable surfaces. Every PFS geometric surface may have two possible variants of field surface wave orientation: a longitudinal (x) and a transversal (y) one which are represented in the code structure by a corresponding index.

According to the principle of genetic information preservation, basic Species serve as system invariants in the increasing variety of EMEC functional classes. Scientific substantiation of the Species category was preceded by the development of genetic theory of speciation. The concepts of this theory formed the basis of macro-genetic analysis and macro-evolutional synthesis of EMEC species diversity.

A species presents a genetically isolated integral system with a typical interspecies structure of genetically related populations (groups) of real and implicit electromechanical objects. Genetic models (e.g. the one showed in Fig. 11.1) are intended to describe internal, genetically conditioned Species structure. A genetic model is informative and dynamic, as it models the process of complication of an arbitrary Species structural organization levels in time. This process can be presented in the following form: "Primary field source (PFS)" – "Set of generative electromagnetic structures (Species genome)" – "Populations" – "Species". PFS

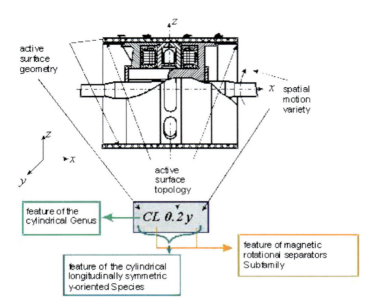

Fig. 11.2 Relation of a genetic code and magnetic separator essential features and taxonomic categories (taking a representative of cylindrical, longitudinally symmetric, y-oriented species as an example)

GC periodic structure serves as source information when a genetic model is being created (Shinkarenko, Zagirnyak, Shvedchikova 2009). Basic GC structural units are a period and a group (by analogy with Chemical elements periodic table). GC data domain (Fig. 11.1) can be regarded as an orthogonal matrix structure formed by PFS four electromagnetic classes (groups) and six geometrical classes.

Determination of belonging of a real electromechanical object to a certain Species is carried out by the procedure of identification (recognition) of the given object genetic information, the formalized basis of which consists in the principle of PFS genetic information preservation (Fig. 11.2).

A genetic code also points out that the object belongs to other superspecies systematic units, e.g., a Genus and a Subfamily. So, an electromagnetic pulley (Fig. *11*.2.), as a representative of *CL* 0.2*y* belongs to the Genus of cylindrical and the Subfamily of magnetic rotational separators.

EMEC systematics is a logical consequence of development of electromechanical systems speciation genetic theory based on GC periodical structure and genetic information preservation principle. Systematics implies determination of electromechanical objects diversity and their regulation by means of attribution to a definite systematic category. Availability of PFS genetic codes clear correlation with major taxonomic features provides the possibility of creating genetic systematics of both really-informative Species and potentially possible (implicit) Species which are absent at the present moment of EMEC functional classes evolution. The problem of determination and application of implicit Species innovation potential presents the essence of systematics anticipation function.

The problem posed in this chapter is to show practical expediency of application of genetic systematics methods to solution of electromagnetic separators design search problems.

11.3 Determination of Electromagnetic Separators Species Diversity

The problem of providing search design procedures dataware implies determination of bounds, quantity and structure of complete Species composition of the device class being researched. Availability of systemized information about species structure makes it possible to determine accurate search limits and perform directed synthesis of structural variants of magnetic separators having a given objective function.

Generality of PFS and corresponding species genetic information allows one to determine species diversity of the retrieved class using the concept of generative electromagnetic structures existence domain (Shinkarenko, Zagirnyak, Shvedchikova 2009). A primary structure whose genetic information determines evolution of a certain group (population) of genetically related electromechanical objects or the Species as a whole is called a generative electromagnetic structure.

To determine existence domain Q_{MS} of generative structures of open-type magnetic separators whose workspace with magnetic field is external in relation to the pole system, it is necessary to point out their essential features, the totality of which is generated by search objective function F_O:

$$F_O = (p_1, p_2, p_3, p_4),$$

where p_1 is availability of magnetic field inductor; p_2 is availability of movable secondary discrete structure (ferromagnetic working substances); p_3 is availability of the necessary working zone area, i.e. big enough length and height of the working zone; p_4 is possibility of placement of a non-magnetic unloading screen providing rotary (a curved trajectory of ferromagnetic bodies travel in the working zone) or forward (a straight trajectory of ferromagnetic bodies travel) motion, in the space between the primary and secondary parts.

To solve the search problem correctly the following limits are imposed on domain Q_{MS}:

1. The search is performed within the first large GC period ($P^I \subset <S_0>$, where $<S_0>$ is an ordered set of primary sources of electromagnetic field in GC periodic structure).
2. The required structure of the considered class is limited by consideration of the diversity of two subclasses Q_{MSr} and Q_{MSf}, providing rotary and forward motions, correspondingly.
3. A generative structure of a random Species is presented as an electromechanical pair created as a result of junction of solid-body primary and secondary discrete structures. A generative structure allows of the possibility of spatial combination with subsystems of another genetic nature (e.g. with non-magnetic unloading screens), as well as the possibility of formation using a module principle.

4. At this stage of the problem solution the sources-isotopes determining the diversity of twin-Species and complicated variants of joint systems with multi-element and hybrid structures are excluded from the consideration.

Taking into account the above said, existence domains of generative structures of magnetic separators of rotary and forward motions, correspondingly, may be written down in the following way:

$$Q_{MSr} = \begin{vmatrix} 0.0 & TP0.0\,ó \\ 0.2 & TP0.2\,ó, CL0.2\,ó, CN0.2\,ó, CPH0.2\,ó, TCL0.2\,ó \\ 2.2 & TP2.2\,ó, CL2.2\,ó, CN2.2\,ó, CPH2.2\,ō, SPH2.2\,ó, TCL2.2\,ó \end{vmatrix}, \quad (1)$$

$$Q_{MSf} = \begin{vmatrix} 0.0 & TP0.0\,ō \\ 0.2 & PL0.2\,ó \\ 2.0 & PL2.0\,ō, TP2.0\,ō, CL2.0\,ō, CN2.0\,ō \\ 2.2 & PL2.2\,ō, PL2.2\,ó, TP2.2\,ō, CL2.2\,ō, CN2.2\,ō \end{vmatrix}, \quad (2)$$

where *TP 0.0y, ..., TCL 2.2y, TP 0.0x, ..., CN 2.2x* – genetic codes of corresponding basic level PFS. Electromagnetically asymmetric field sources (code numerical part 2.2) on spherical (*SPH*) and plane (*PL*) surfaces with transverse (*y*) and longitudinal (*x*) field orientation are geometrically and electromagnetically equivalent.

Let us determine correspondence between essential features of open-type magnetic separators and genetic information:

1. Electromechanical objects with transverse electromagnetic field wave orientation on cylindrical (*CL*), conic (*CN*), toroid plane (*TP*), toroid cylindrical (*TCL*) and spherical (*SPH*) surfaces allow of the possibility of junction and functioning with non-magnetic unloading screens of rotary motion (with a curved trajectory of ferromagnetic bodies travel in the working zone).
2. Electromechanical objects with longitudinal electromagnetic field wave orientation on cylindrical (*CL*), conic (*CN*), toroid plane (*TP*) surfaces, as well as electromechanical objects with transverse and longitudinal electromagnetic field wave orientation on plane (*PL*) surfaces allow of the possibility of junction and functioning with non-magnetic loading screens of forward motion (with a straight trajectory of ferromagnetic bodies travel in the working zone).

Existence domain of magnetic separators generative structures includes only one electromagnetically symmetrical field source (code numerical part *0.0*), referring to geometric class – toroid plane (*TP*) which completely meets the condition of working zone openness (objective function p_3). The active zone internal surface of electromagnetically symmetrical classes: cylindrical (*CL*), conic (*CN*), toroid cylindrical (*TCL*), spherical (*SPH*) and plane (*PL*), is closed, which is contrary to search objective function p_3. Electromechanical objects with longitudinal electromagnetic field wave orientation on toroid cylindrical (*TCL*) and spherical (*SPH*)

Structural-Systematic Approach in Magnetic Separators Design   207

surfaces are excluded from the consideration due to rather complicated character of their surface, which hampers junction with nonmagnetic unloading screens (objective function p_4).

Thus, existence domain Q_{MSr} of generative structures of rotary motion magnetic separators, as it is evident from expression (1), includes twelve basic level sources put in order within three electromagnetic symmetry groups and five geometric classes. According to expression (2), existence domain Q_{MSf} of generative structures of forward motion magnetic separators includes eleven basic level sources put in order within four symmetry groups and four geometric classes.

Availability of correspondence between generative field sources and system Species category makes it possible to determine the species composition of the basic level of the functional class of rotary and forward motion magnetic separators:

$$\{HS_{MSr}\} = \begin{pmatrix} TP \\ CL \\ CN \\ SPH \\ TCL \end{pmatrix} \begin{pmatrix} TP\ 0.0y,\ TP\ 0.2y,\ TP\ 2.2y \\ CL\ 0.2y,\ CL\ 2.2y \\ CN\ 0.2y,\ CN\ 2.2y \\ SPH\ 0.2y,\ SPH\ 2.2xy \\ TCL\ 0.2y,\ TCL\ 2.2y \end{pmatrix}, \quad (3)$$

$$\{HS_{MSf}\} = \begin{pmatrix} TP \\ PL \\ CL \\ CN \end{pmatrix} \begin{pmatrix} TP\ 0.0x,\ TP\ 2.0x,\ TP\ 2.2x \\ PL\ 0.2y,\ PL\ 2.0x,\ PL\ 2.2xy \\ CL\ 0.2x,\ CL\ 2.2x \\ CN\ 2.0x,\ CN\ 2.2x \end{pmatrix}, \quad (4)$$

where $\{HS_{MSr}\}$ and $\{HS_{MSf}\}$ are species composition of the basic level (in genetic codes) of magnetic separators of rotary and forward motions, correspondingly. Thus, species composition of rotary and forward motion separators is presented by 23 (12+11) basic level Species.

Analysis of structural diversity of open-type magnetic separators (Fig. 11.3.) shows that all the known engineering solutions at the current evolution stage are realized by structural specimens of only ten real-informative basic species (43.5% of the potential of all the basic Species of the class). Generative structures of really-informative species were determined according to the results of patent information retrieval (the period from 1930 till 2000). According to Fig. 11.3, five really-informative Species (*CL 0.2y, CL 2.2y, TP 0.2y, TP 0.0y, CN 2.2y*) refer to Subfamily of rotary motion magnetic separators and five (*PL 2.2x, PL 2.2y, TP 2.0x, PL 0.2y, CN 2.0x*) – to Subfamily of forward motion magnetic separators. Species *CL 2.2y, CL 0.2y, PL 2.2x(y)* are dominant really-informative basic Species for magnetic separators class.

Information about the number $N_{really-informative}$ of really-informative Species and information about species diversity (number N) of "ideal" class provides the possibility to determine innovation potential of the class representing the

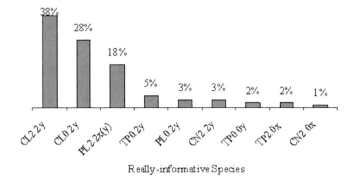

Fig. 11.3 Distribution of really-informative magnetic separators species according to their number

information about number $N_{implicit}$ of implicit Species, i.e. about the class structural representatives which are absent at the present evolution stage (were not found in the course of patent information retrieval)

$$N_{implicit} = N - N_{really\text{-}informative}. \qquad (5)$$

For the devices class being researched the innovation potential is represented by thirteen ($N_{implicit} = 23 - 10$) implicit Species, which makes 56.5% of the total species number. Thus, first of all, implicit species and not numerous really-informative Species CN 2.0x, TP 2.0x, TP 0.0y (Fig. 11.3) are peculiar reference points showing the direction of the search for new patentable engineering solutions.

At the present moment there are a number of patent pending engineering solutions obtained with application of systematics anticipation function.

11.4 Systematics Rank Structure as the Basis for a New Concept of Design Procedures Dataware

Availability of system information about Species number and genetic structure makes it possible to determine the class major systematic units rank structure in which the Species performs the function of the main systematic category for superspecies level taxons: "Species" → "Genus" → "Subfamily" → "Family" (Fig. 11.4.). Rank sequence of the major systematic units is universal for electromechanical systems arbitrary functional classes, which provides methodological unity and invariance of systematics structure.

Species structural diversity is determined by the totality of genetically related species united by the common PFS spatial geometry. Historically formed classes of rotary and forward motion separators have the status of a Subfamily as they unite the corresponding separators Genus taxons according to the character of their spatial motion. The analysis of magnetic separators systematics ranking

Structural-Systematic Approach in Magnetic Separators Design

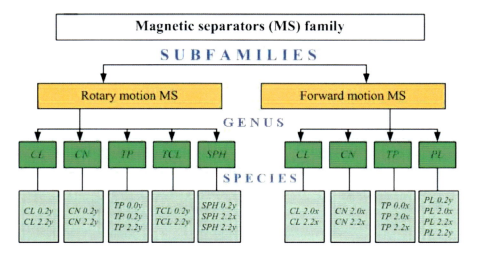

Fig. 11.4 Rank structure of magnetic separators systematics

structure shows that structural diversity of rotary and forward motion magnetic separators is presented by five and four Genera, correspondingly. The Subfamily of rotary motion separators is determined by twelve basic Species with axisymmetric magnetic field sources. Species diversity of the subfamily of forward motion separators is put in order by eleven Species with traveling magnetic field sources.

Genetic systematics rank structure is the basis for development of new system concept of search design dataware. This concept implies structured information supply in the form of catalogs and electronic databases about both existing (really-informative) and potentially possible (implicit) magnetic separators species.

A catalog of magnetic separators species has been developed within the adopted concept. The catalog contains information about species genetic nature with indication of their complete names and genetic codes, as well as information about species objects system properties genetically conditioned: the character of movable elements motion, active zone geometry and topology (Table 11.1). Information about Species genetic nature and their system specific features is invariable.

Systemized information from the catalog was used for development of electronic database "Systematics of magnetic separators family species" with description of both really-informative and implicit magnetic separators species in Russian and English. In this case genetic systematics rank structure is a guarantor of information supply completeness and solves the problem of data redundancy.

The electronic database was developed in MySQL medium using PHP programming language. MySQL and PHP systems are cross-platform and free, which was the basic reason for their choice. MySQL system provides high-speed data access, makes it possible to carry out storage, completion, regulation, retrieval and data exchange functions most efficiently and also provides the possibility for several users to work with the data in the network simultaneously.

Table 11.1 Example of describing genetic features of electromagnetic separators Species *CL 0.2y* in the catalog

CL 0.2y		Magnetic separators (MS) family
Basic Species	Cylindrical, longitudinally symmetrical, y-oriented	Subfamily of rotary motion MS
		The Cylindrical Genus
Active surface geometry – cylindrical;		
Active surface topological features – unilateral active zone;		
Character of nonmagnetic unloading screen motion – rotary;		
Character of ferromagnetic bodies motion – curvilinear.		

The proposed electronic database contains information from the catalog about species genetic nature, system and additional information useful during search design. System information is presented by the data about species status (really-informative or implicit) and about their evolution level, about species generative structures, fields of practical application and geography of the species objects manufacturers. Additional information is presented by description of Species structural representatives contained in patents.

The offered form of information supply in the mentioned electronic database includes all the necessary information characterizing species of the Family of rotary and forward motion magnetic separators. The database can be constantly augmented with new information without disturbance of its structure.

Information about really-informative and implicit species of the Family of magnetic separators, which is included into the electronic database, may be used as informational-innovation basis for magnetic separators search design.

11.5 Interspecific Synthesis of New Magnetic Separator Structures Using the Law of Electromechanical Systems Homologous Series

As stated above, genetic systematics possessing a prognostic function sets peculiar guiding lines and magnetic separators new Species search and synthesis should be performed in accordance with their direction. First of all, these guiding lines include implicit Species and, to a smaller extend, rare really-informative Species (Fig. 11.5). The homologous series law (HSL) is an important methodological genetic synthesis instrument having a powerful heuristic potential and able to provide a directed new structures search and synthesis at an interspecific level.

Homology is a close alliance (similarity) of objects according to particular structural features conditioned by their electromagnetic symmetry group generality. The HSL basis is formed by PFS GC. The essence of HSL consists in the fact that features parallelism is typical of allied basic Species synthesized from the sources of one topologically equivalent series.

Structural-Systematic Approach in Magnetic Separators Design

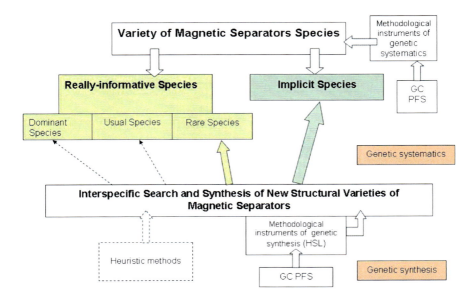

Fig. 11.5 Interconnection of two main directions of genetic electromechanics: genosystematics and new structures genetic synthesis

The problem of directed search and synthesis of magnetic separator structures using HSL implies acquisition of information about structural filling of homologous series of both existing and implicit magnetic separator Species structures. Completeness of such series elements is provided by the principle of topological invariance within an arbitrary GC group in accordance with the rules of information transfer and corresponding methods of topological transformations.

Thus, HSL presents the basis for a fundamentally new approach to the methods of search problems solution, which is grounded on the generation of homological structures by topological transformations methods. By means of HSL, using genosystematics prognostic function results, in particular, macrogenetic analysis results, it is possible to realize a directed synthesis of Species composition of magnetic separators, information about which is inaccessible or unavailable at the present moment of the class evolution.

Let us consider the basic structure-prototype S_p of a magnetic separator for bulk material separation (Fig.11. 6) (Nevzlin, Zagirnyak 1984), presented in the form of an informational (the claim of the patent specification) and geometric (figures explaining the device operation) models, to be output information for the statement of the problem of directed search and synthesis of magnetic separators new structures.

Analysis of genetic information of the device for bulk material magnetic separation (Fig. 11.6) shows that the device of this type belongs to the subfamily of translational motion magnetic separators, the Species of plain (*PL*) ones and is characterized by electromagnetic asymmetry both in longitudinal (x) and in transverse (y) directions of electromagnetic field wave propagation. The peculiarity of the structure-prototype consists in the presence of two electromagnetic systems (magnetic field inductors) mounted above the transporting device astride its symmetry axis at an angle of α = 60-90° with the transporter basis (*2 PL 2.2y*).

To determine structures homologous series Species composition Q_T it is necessary to single out structure-prototype S_p essential features meeting the synthesis objective function:

$$F_S = (p_{S1}, p_{S2}, p_{S3}, p_{S4}). \qquad (6)$$

Structure-prototype essential features determining invariant properties of homologous series structures include: p_{S1} – presence of two magnetic field static inductors which can be mounted above the transporter astride its symmetry axis; p_{S2} – possibility of mounting a movable nonmagnetic unloading screen in the space between the inductor and the transporting device; p_{S3} – y-orientation of the field traveling wave; p_{S4} – field source electromagnetic asymmetry (electromagnetic symmetry group 2.2).

To solve the synthesis problem correctly the following limitations $L=f([x)$ are imposed on domain Q_T:

1. Synthesis is performed within the GC first big period ($P^I \subset <S_0>$, where $<S_0>$ – electromagnetic field primary sources ordered set in GC periodic structure).
2. Electromagnetic systems with movable inductor and complicated variants of combined systems containing multi-element and hybrid structures are excluded from consideration at the present stage of synthesis problem solution.

Taking the above stated into account, the problem of directed search and synthesis of magnetic separators new structural varieties can be formulated in the following way: to determine the domain of existence Q_T according to known basic

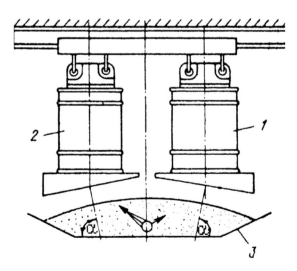

Fig. 11.6 Device for bulk materials magnetic separation (*2 PL 2.2y*): 1,2 – electromagnetic systems; 3 – transporting device (α - angle of electromagnetic systems installation)

structure (structure-prototype) S_p belonging to basic Species *PL2.2y* with known synthesis objective function $F_S = (p_{S1}, p_{S2}, p_{S3}, p_{S4})$ and assigned limitations totality $L=f(x_1, x_2)$, and to synthesize a finite set of structures belonging to other Species of homologous series *T* and meeting function F_S.

Domain Q_T of directed search and synthesis of homologous series structures can be presented in the following form (Table 11.2).

Table 11.2 Domain Q_T of directed search and synthesis of homologous series structures

Basic level generative structures	Domain Q_T	
	Generative structures built on resources-isotopes.	
PL 2.2y	PL 2.2y$_1$	PL 2.2y$_2$
CL 2.2y	CL 2.2y$_1$	CL 2.2y$_2$
CN 2.2y	CN 2.2y$_1$	CN 2.2y$_2$
TP 2.2y	TP 2.2x$_2$	TP 2.2y$_2$
SPH 2.2y	SPH 2.2y$_1$	SPH 2.2y$_2$
CL 2.2y	TCL 2.2x$_2$	TCL 2.2y$_2$

It follows from Table 11.2 that to obtain an allied structures directed synthesis domain it is necessary to single out the series whose structure is allied to the structure-prototype and which can meet synthesis objective function F_S, from the domain Q_{MC} of magnetic separators class existence (see formulas (3) and (4)).

Obtained domain Q_T contains 18 generative structures (including six basic level generative structures and 12 structures-isotopes). In the obtained directed synthesis domain Q_T there is a structure of one symmetry group and six spatial forms.

Applying the information transfer method in relation to the assigned structure-prototype, we get a Species composition of structures homologically similar to the chosen structure-prototype and meeting the synthesis objective function F_{S1}. The results of structures synthesis are presented in the form of an information database (Table 11.3). The following designations are adopted in Table 11.3: **2PL 2.2y ... 2SPH 2.2y** – basic level structures; 2 *PL2.2y$_1$...2CPH2.2y$_2$* – twin structures.

Table 11.3 Information database of homologous series *T* structures

Electromagnetic symmetry group	Structures geometric classes					
	PL	CL	CN	TP	TCL	CPH
2.2	2PL2.2y	2CL 2.2y	2CN 2.2y	2TP 2.2y	2TCL2.2y	2CPH2.2y
	2PL2.2y$_1$	2CL2.2y$_1$	2CN2.2y$_1$	2TP2.2y$_1$	2TCL2.2y$_1$	2CPH2.2y$_1$
	2PL2.2y$_2$	2CL2.2y$_2$	2CN2.2y$_2$	2TP2.2y$_2$	2CL2.2y$_2$	2CPH2.2y$_2$

Analysis of the presented information (Table 11.3) shows that the homologous series includes 18 structures, three, or 16.7% of which belong to really-informative Species (in Table 11.3 the boxes with generative structures whose novelty has been certified by inventions are marked with a color). 15 structures (83.3%) define implicit (potentially possible) Species existence domain.

Graphics provide perfectly clear and concise presentation of information about synthesized structures. In this case genetic particularities of the synthesis objects comprise the core of their graphic interpretation. Visualization of synthesis results for basic level homologous structures is given in Table 11.4.

Table 11.4 Visualization of genetic synthesis results (homologous series T)

Genetic code	Synthesis results graphic interpretation	Genetic code	Synthesis results graphic interpretation
2 PL 2.2y		2 TP 2.2y	
2 CL 2.2y		2 TCL 2.2y	
2 CN 2.2y		2 CPH 2.2y	

It should be mentioned that out of three structures belonging to really-informative Species (Table 11.3) two structures (*2CN2.2* и *2TP2.2y*) were synthesized using HSL. In this case structure *2CN2.2* refers to rare Species domain (Fig. 11.7) (Shvedchikova, Sukharevskaya, Martinenko 2009), and structure *2TP2.2y* was obtained from implicit Species domain (Fig. 11.8) (Shvedchikova, Golubeva, Sukharevskaya 2009).

The main advantage of synthesized structure *2CN2.2y* as compared to structure-prototype *(2PL 2.2y)* lies in the fact that the continuity of the process of ferromagnetic objects removal and subsequent unloading is provided in a new structure. In this case, there is no necessity of periodic transference of electromagnetic systems

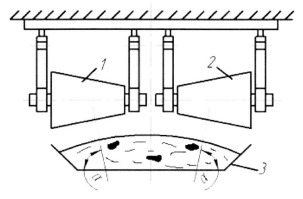

Fig. 11.7 Device for bulk material magnetic separation (*2 CN2.2y*), synthesized using HSL: 1, 2– electromagnetic systems; 3 – transporting device (α – angle of electromagnetic systems installation)

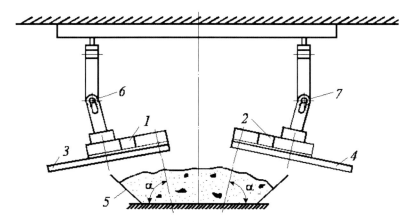

Fig. 11.8 Device for bulk material magnetic separation (*2 TP2.2y*), synthesized using HSL: 1, 2– electromagnetic systems; 3, 4 – nonmagnetic unloading disks; 5 – transporting device; 6, 7 – rotators (α – angle of electromagnetic systems installation)

into unloading area and their disconnection from electric mains. Besides, electromagnetic systems conic form contributes to decrease of contamination of extracted ferromagnetic inclusions with nonmagnetic material.

In its turn, synthesized structure *2TP2.2y* has certain advantages in comparison to both structure-prototype (*2PL2.2y*) and homologically similar structure *2CN2.2y*. In the device for bulk material magnetic separation (Fig. 11.8) electromagnetic systems 1 and 2 are made in the form of static half-disks under which rotating nonmagnetic disks 3 and 4 are located; they provide the continuity of the process of ferromagnetic objects removal and subsequent unloading. Angle α of installation of magnetic systems 1 and 2 in relation to transporter 5 basis can be changed by means of rotators 6 and 7 depending on material natural slope angle and assume an intermediate position in the range of 60-90°. Rotators 6 and 7 can also be used to adjust distance between electromagnetic systems 1 and 2 and transported material surface. Thus, the main advantage of structure *2TP2.2y* as compared to previously synthesized and homologically similar structures (*2PL 2.2y, 2CN 2.2y*) consists in improvement of ferromagnetic inclusions extraction conditions as the distance between the electromagnetic systems surfaces and the surface of material transported by the conveyor is approximately equal when the angle of material natural slope changes.

The reliability of the performed synthesis procedure is confirmed by comparative analysis of patent retrieval data and synthesis results. Belonging of a number of homologous series structures (Table 11.3) to really-informative Species as well as the possibility of directed generation of new structures, in particular, structures *2CN2.2y*, *2TP2.2y*, whose novelty has been confirmed by patents, testify to the adequacy of the performed synthesis procedure. In this case, meeting the following condition can be considered the condition of genetic synthesis adequacy:

$$(2PL\ 2.2y,\ 2CN\ 2.2y,\ 2TP\ 2.2y) \subset \grave{O}\ , \qquad (7)$$

where *2PL 2.2y, 2CN 2.2y, 2TP 2.2y* – subset of really-informative Species of structures *T* homologous series.

11.6 Conclusion

Using genetic systematics results, a system concept of innovation dataware of magnetic separators search design has been developed. It has been ascertained that species diversity of the family of electromagnetic separators is determined by 23 basic level Species.

Quantitative and qualitative composition of magnetic separators implicit species has been determined on the basis of application of genetic systematics anticipation function, which provided the possibility of carrying out a directed synthesis of patentable engineering solutions and realizing innovational potential of magnetic separators class.

Applying the results of the research, a systemized catalog of magnetic separators Species diversity has been developed for the first time. It includes information about both known and potentially possible species. Practical application of the catalog and the electronic database made it possible to improve search procedures efficiency significantly and thereby to decrease unavoidable time and resources expenditure.

Presence of structural features parallelism in allied Species of EM systems and existence of corresponding series of magnetic separators structural varieties, invariant to the operating principles and their structures complexity, provide the possibility of realization of directed search and synthesis of new structures. Structures, synthesized as a result of research, are operable and recommended for industrial application.

References

Zagirnyak, M.V.: Magnetic separators. In: Proceedings of the Seventeenth International Electrotechnical and Computer Science Conference ERK, Portorož, pp. 7–8 (2008)

Nevzlin, B.I., Zagirnyak, M.V.: Device for bulk material magnetic separation. Patent of USSR No 3445203/22-03 (1984)

Shinkarenko, V.: Synthesis of the periodical structures in the problems of unconventional electromechanical systems design. In: Proc. of the Second International Scientific and Technical Conference on Unconventional Electromechanical and Electrotechnical Systems, Szczecin, Poland, pp. 367–372 (1996)

Shinkarenko, V., Zagirnyak, M., Shvedchikova, I.: Macrogenetic analysis and rank structure of systematics of magnetic separators. Electrical engineering and electromecanics 5, 33–39 (2009) (in Russia)

Shvedchikova, I.A., Sukharevskaya, N.A., Martinenko, N.V.: Device for bulk material magnetic separation. Patent of Ukraine u, 12508 (2009)

Shvedchikova, I.A., Golubeva, S.M., Sukharevskaya, N.A.: Device for bulk material magnetic separation. Patent of Ukraine u, 08746 (2009)

Svoboda, J.: Magnetic Methods for the Treatment of minerals. Elsevier, Amsterdam (1987)

Unkelbach, K.H.: Magnetic separators mode of operation and applicability for the separation of materials. Köln, KHD Humboldt Wedag AG (1990)

Chapter 12
Weight Reduction of Electromagnet in Magnetic Levitation System for Contactless Delivery Application

Do-Kwan Hong[1], Byung-Chul Woo[1], Dae-Hyun Koo[1], and Ki-Chang Lee[1,2]

[1] Korea Electrotechnology Research Institute/Electric Motor Research Center,
70 Boolmosangil, Changwon, 641-120, Republic of Korea
dkhong@keri.re.kr, leekc@bcwoo@keri.re.kr,
dhk371@keri.re.kr
[2] Pusan National University/School of Mechanical Engineering, Geumjeong-gu,
Busan, 609-735, Republic of Korea
leekc@pusan.ac.kr

Abstract. This paper presents lightweight optimum design of electromagnet in magnetic levitation system. This paper deals with the possibility of using the response surface methodology (RSM) for optimization of an electromagnet with a higher number of the design variables. 2D and 3D magnetostatic analysis of electromagnet is performed by using ANSYS. The most effective design variables were extracted by pareto chart. The most desired set is determined and the influence of each design variables on the objective function can be obtained. This paper procedure is validated by the comparison between experimental and calculation result.

12.1 Introduction

Electromagnetically levitated and guided systems are commonly used in the field of people transport vehicles, tool machines and conveyor system because of its silent and non-contacted motion [1]-[3]. Its main drawback is complexity and make-up cost. So passive guidance controls are normally used in real implementation. The Japanese HSST and the English BAMS (Birmingham Airport Maglev System) are well-known working transport system. In both magnetically levitated trains, the guidance force needed to keep the vehicles on the track is obtained with the levitation electromagnets, thanks to particular shapes of the rails and to a clever placement of the electromagnets with respect to rails. The design of electromagnet is the most important in magnetic levitation system design. Electromagnet is an essential factor, it is being charged considerable (10%) parts in whole weight, which can largely affect influence the lightweight and stability of the magnetic levitation system. Consequently it accomplished the optimum plan for the lightweight of the magnetic electromagnet [4]-[7]. They desire to develop magnetic levitation system for contactless delivery application in Fig. 12.1. At first step, the design goal is to reduce the weight of electromagnet of magnetic levitation system with constraint of normal force considering the initial model. At second step, the most effective design variables and their levels should be determined and be arranged in a table of orthogonal array.

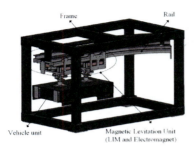
(a) 3D modeling

(b) Prototype

Fig. 12.1 Magnetic levitation system prototype for contactless delivery application

Generally Response Surface Methodology (RSM) is used with 2 or 3 design variables, however we have 7 design variables using table of mixed orthogonal array are utilized. For each design variable combination the response value is determined by 2D and 3D Finite Element Method (FEM). In this paper we use the reduced gradient algorithm, which can lead to the selection of the most desired set of variables. Fig. 12.2 shows flow chart of RSM optimization procedure. Fig. 12.3 shows the application prototype model for applying the contactless delivery application. The response is determined and we evaluate the influence of each design variable on the objective function. Based on this method the weight of the optimized electromagnet can be reduced by 11.412 % and normal force improved by 7.754 % of initially designed electromagnet.

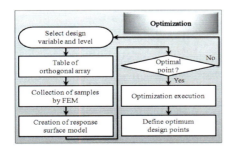

Fig. 12.2 Flow chart of optimization   **Fig. 12.3** Application prototype

12.2 Passive Guidance Control and Optimization of Electromagnet

Let's assume that the guidance forces are generated by closed-loop control of levitation electromagnets. The lateral response of the electromagnet due to the airgap control is a force increasing with the lateral offset X. This force is almost a linear function of lateral offset of the electromagnets. Then we can imagine that the electromagnet will move laterally as a mass bound by a spring. As we know, a spring working on a mass is a mechanical resonant system, whose resonance frequency is given by:

$$f_{resonance} = \frac{1}{2\pi}\sqrt{\frac{k_{guide}}{M}} \qquad (1)$$

where the M is the sum of the masses of the electromagnet and of part of the carried vehicle. We can easily imagine that any external action will cause a nondamped oscillating response of the lateral position. The value of k_{guide} can be designed to keep disturbance not to make excessive displacement. An example of design problem is shown in Fig. 12.4.

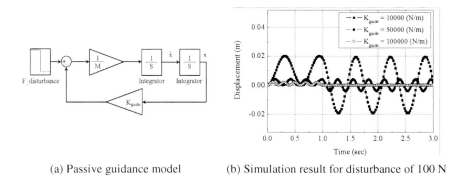

(a) Passive guidance model   (b) Simulation result for disturbance of 100 N

Fig. 12.4 Lateral position model for electromagnets under constant levitation control

Total mass of 100 kg is assumed to be levitated by constant gap. And the sum of lateral position stiffness of levitation magnets is thought as 10,000 N/m, 50,000 N/m and 100,000 N/m respectively. The lateral disturbance forces are assumed to be constant value of 100 N for about 1 second. The simulation result shows that if the lateral position stiffness remains higher than 50,000 N/m, the lateral position deviation is smaller than 5 mm under constant 100 N disturbance. In the experiment, total mass of 200 kg including 4 levitation electromagnets are levitated under small deviation of electromagnet placement as shown in Fig. 12.5. So mass of 50 kg is levitated by one electromagnet respectively. In the experiment, levitation magnets are controlled to keep constant gap length of 5 mm. Position disturbance of 3.2 mm is applied at first and removed. The lateral response of the experiment is shown in Fig. 12.5 (b). We can infer that if two electromagnets are positioned with small deviation from the original rail position, guidance forces generated each electromagnet can be larger, but resultant force becomes smaller because of differentially actuated scheme. So the position stiffness become smaller, the natural frequency becomes smaller and the lateral deviation becomes larger according to deviation of magnets becomes larger. Therefore, no guidance model is better.

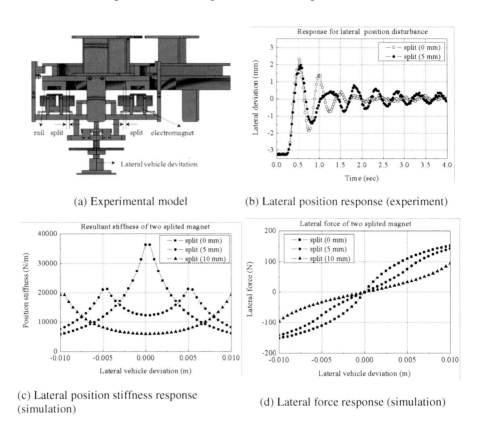

Fig. 12.5 Lateral position response of magnetically sprung mass under position disturbance

12.3 Optimum Design

Fig. 12.6 shows the electromagnet prototype of reference model which is analyzed. The comparison of the static force obtained from FEM simulation and experimental test is shown in Fig. 12.7. From the results, the use of FEM is validated as it can be observed. Since the comparison of normal force by simulation and experiment test is shown in Fig. 12.7 with good agreement. Selection of the design variables is very important setup in optimization procedure. Seven dimensions are selected as the design variables as shown in Fig. 12.8. The B-H characteristic of S20C,

Weight Reduction of Electromagnet in Magnetic Levitation System

SM490A material is as shown in Fig. 12.9. Table 12.1 shows the design variables and levels. Fig. 12.8 shows flux density vector in electromagnet of 2D and 3D FEM.

Table 12.1 Design variable and level

Design variable Level	dv_1	dv_2	dv_3	dv_4	dv_5	dv_6	dv_7
-1	16	45	16	40	7	16	144
0	20	50	20	45	11	20	180
1	24	55	24	50	15	24	216

Table 12.2 Table of mixed orthogonal array $L_{18}(2^1 \times 3^7)$

Exp.	dv_1	dv_2	dv_3	dv_4	dv_5	dv_6	dv_7	Normal force (N)	Weight (kg)
1	16	45	16	40	7	16	144	382.1	8.706
2	16	50	20	45	11	20	180	473.67	12.153
3	16	55	24	50	15	24	216	564.15	16.366
4	20	45	16	45	11	24	216	682.56	15.433
5	20	50	20	50	15	16	144	450.22	10.979
6	20	55	24	40	7	20	180	570.56	13.699
7	24	45	20	40	15	20	216	795.4	16.803
8	24	50	24	45	7	24	144	531.6	12.66
9	24	55	16	50	11	16	180	651.24	14.064
10	16	45	24	50	11	20	144	379.41	10.277
11	16	50	16	40	15	24	180	473.76	12.094
12	16	55	20	45	7	16	216	568.3	13.874
13	20	45	20	50	7	24	180	570.38	13.637
14	20	50	24	40	11	16	216	682.15	15.453
15	20	55	16	45	15	20	144	450.16	10.946
16	24	45	24	45	15	16	180	659.92	14.604
17	24	50	16	50	7	20	216	789.93	16.646
18	24	55	20	40	11	24	144	528.87	12.4

Fig. 12.6 Prototype of electromagnet

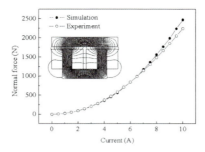

Fig. 12.7 Comparison of FEM and experiment (reference model)

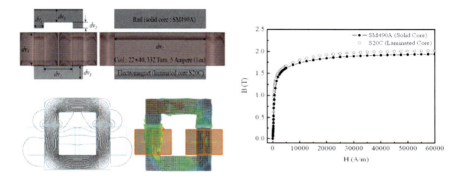

Fig. 12.8 Design variables and flux pattern (2D and 3D FEM) **Fig. 12.9** B-H Curve of used core material

12.4 Response Surface Methodology

Table 12.2 shows the table of mixed orthogonal array and simulation result by 2D FEM. Table 12.2 represents the table of mixed orthogonal array, which is determined by considering the number of the design variables and each level of them. After getting experimental data by FEM, the function to draw response surface is extracted. In order to determine equations of the response surface for response value (weight, normal force), several experimental designs have been developed to establish the approximate equation using the smallest number of experiments. The purpose of this paper is to minimize the objective function ($Weight_{total}$) with constraints of normal force (F_{normal}). The two fitted second order polynomial of objective functions for the seven design variables are as follows.

$$F_{normal} = 148.75 + 26.1 dv_1 - 2.41 dv_2 - 10.02 dv_3 - 32.18 dv_4 - 1.63 dv_5 + 35.16 dv_6 + 3.01 dv_7 \\ - 7.14E\text{-}2 dv_1^2 + 1.3E\text{-}3 dv_2^2 + 0.23 dv_3^2 + 0.35 dv_4^2 + 5.56E\text{-}2 dv_5^2 - 0.9 dv_6^2 - 3.72E\text{-}4 dv_7^2 \quad (2)$$

$$Weight_{total} = 8.765 + 0.212 dv_1 - 0.248 dv_2 - 0.154 dv_3 - 0.488 dv_4 - 0.113 dv_5 + 0.264 dv_6 + 0.065 dv_7 \\ + 1.84E\text{-}3 dv_1^2 + 2.8E\text{-}3 dv_2^2 + 6.54E\text{-}3 dv_3^2 + 5.95E\text{-}3 dv_4^2 + 7.58E\text{-}3 dv_5^2 - 4.05E\text{-}3 dv_6^2 + 2.64E\text{-}6 dv_7^2$$

Weight Reduction of Electromagnet in Magnetic Levitation System 225

The adjusted coefficients of multiple determination R^2_{adj} for normal force and weight are weight total (99 %) and Fnormal (100 %). Normal force and weight of experiment result according to change of the design variables are shown in Table 12.2. As many parameters are defined as design variables, the large simulation time is required due to a large number of the required experiments. Therefore, it is necessary for significant parameters to investigate the influence on the design result. The pareto chart of normal force and weight shows the magnitude and importance of an effect. This chart displays the absolute value of the effects. The geometries of electromagnet can be defined by 7 parameters as shown in Table 12.1 and Fig. 12.8. The most effective design variables of normal force and weight are dv_7, dv_1 by pareto chart in Fig. 12.10. The values of ineffective design variables are determined by RSM.

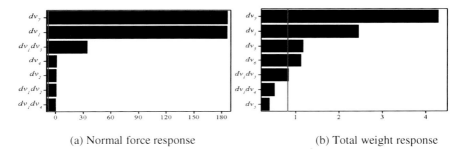

(a) Normal force response (b) Total weight response

Fig. 12.10 Pareto chart of the standardized effects (alpha=0.05) The reference line corresponds to alpha=0.05 ; 95% confidence interval

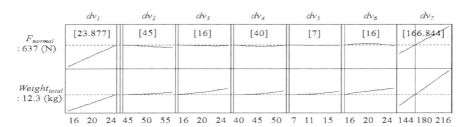

Fig. 12.11 Response optimization

Table 12.3 Optimum level and size

Design variable Level	dv_1	dv_2	dv_3	dv_4	dv_5	dv_6	dv_7
Initial size	20	50	20	40	15	20	180
Optimum size	23.877	45	16	40	7	16	166.844

In table 12.3 and 12.4, the optimal point is searched to find the point of less than 11.412 % of the weight and greater than 7.754 % of the normal force of initially designed electromagnet in magnetic levitation system. The simulation result of predicted optimum set is shown Table 12.4 with good agreement. Fig. 12.11 shows response optimization to find optimal solution according to response curves. Slope of response function shows sensitivity of design variable in Fig. 12.11. The optimum design procedure is introduced to design of electromagnet in magnetic levitation system to reduce its weight and to improve normal force of the initially designed electromagnet in magnetic levitation system with many design variables. The most effective design variables were extracted by pareto chart. The most desired set is determined and the influence of each design variables on the objective function can be obtained. Therefore, when this proposed approach is applied, it can efficiency raise the precision of the optimization and reduce the number of experiments in the optimization design.

Table 12.4 Comparison of initial and optimum model

Model		Weight (kg)	Normal force (N)
Initial	2D FEM	13.319	578.5
	3D FEM	13.319	573.48
	Error(2D vs 3D) %	0	0.86
Optimum	RSM(predicted)	12.3	637
	FEM(verification)	11.799	611.7
	Error(RSM vs FEM) %	-4.073	-3.972
Variation between initial and optimum FEM %		-11.412	7.754

References

[1] D'Arrigo, A., Rufer, A.: Integrated electromagnetic levitation and guidance system for the swissmetro project. In: MAGLEV 2000, pp. 263–268. Rio de Janeiro, Brazil (2000)
[2] Tzeng, Y.K., Wang, T.C.: Optimal design of the electromagnetic levitation with permanent and electro magnets. IEEE Trans. Magn. 30(6), 4731–4733 (1994)
[3] Kim, Y.J., Shin, P.S., Kang, D.H., Cho, Y.H.: Design and analysis of electromagnetic system in a magnetically levitated vehicle, KOMAG-01. IEEE Trans. Magn. 28(5), 3321–3323 (1992)
[4] Onuki, T., Toda, Y.P.: Optimal design of hybrid magnet in Maglev system with both permanent and electromagnets. IEEE Trans. Magn. 29(2), 1783–1786 (1993)

[5] Hong, D.K., Woo, B.C., Chang, J.H., Kang, D.H.: Optimum design of TFLM with constraints for weight reduction using characteristic function. IEEE Trans. Magn. 43(4), 1613–1616 (2007)
[6] Hong, D.K., Choi, S.C., Ahn, C.W.: Robust optimization design of overhead crane with constraint using the characteristic functions. International Journal of Precision Engineering and Manufacturing 7(2), 12–17 (2006)
[7] Kim, J.M., Lee, S.H., Choi, Y.K.: Decentralized H∞ control of Maglev systems. In: Proc. of IECON Conf., Paris, France, pp. 418–423 (2006)

Chapter 13
Genetic Algorithm Applied in Optimal Design of PM Disc Motor Using Specific Power as Objective

Goga Cvetkovski[1], Lidija Petkovska[1], and Sinclair Gair[2]

[1] Ss. Cyril & Methodius University, Faculty of Electrical Engineering and Information Technologies, Karpos II bb 1000 Skopje, Macedonia
`gogacvet@feit.ukim.edu.mk, lidijap@feit.ukim.edu.mk`
[2] University of Strathclyde, 204 George street, G1 1XW Glasgow, Scotland, UK
`s.gair@eee.strath.ac.uk`

Abstract. In general the optimal design of electrical machines is a constrained maximization/minimization problem with a big number of optimization parameters and variety of constraints. This makes it a difficult problem to solve for the deterministic methods, but on the other hand quite an easy task for the stochastic methods, especially for the Genetic Algorithms (GAs). When optimizing an electric motor, there are multiple choices of the objective function available. The objective function is the specific property of the machine to be optimized, for example efficiency, torque, volume or cost. The application of the permanent magnet disc motor (PMDM) is in electric vehicle and therefore there are several objectives that should be tackled in the design procedure, such as an increased efficiency, reduced total weight of the motor or increased power/weight ratio (specific power). In this work an optimal design of a PMDM using specific power as objective function is performed. In the design procedure performed on the PM disc motor, genetic algorithm, as an optimization tool is used. Comparative analysis of the optimal motor solution and its parameters in relation to the initial model is presented.

13.1 Introduction

In the process of optimizing an electric motor, there are multiple choices of the objective function available. The objective function is the specific property of the machine to be optimized, for example efficiency, torque, volume, cost, etc. Of course a 'good' motor is desired, but what is a 'good' motor and what to do when the different goals are contradicting? A good motor has low price, high torque density and high torque quality, high efficiency, but unfortunately, these goals are contradicting. One way to meet at least most of these goals is to perform a multi objective optimization. But, with such an optimization there is always a problem of defining the different objective functions and how to combine them into a single composite objective function. The determination of a single objective is possible with methods such as utility theory, weighted sum method, etc., but the problem lies in the correct selection of the weights or utility functions to characterize the decision-makers preferences. The second approach, which the authors of this paper have adopted, is to define an objective function which will combine some of the individual objective functions, others to be defined as variables and some to be in relation with the previous ones. In such way the objective function defined for the optimal design of the PMDM is the power/weight ratio or the specific power of the motor. The optimal design variables are some of the geometrical dimensions, as well as the electromagnetic torque of the motor.

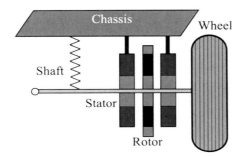 

Fig. 13.1 PMDM direct wheel drive   **Fig. 13.2** PM disc motor test rig

13.2   Permanent Magnet Disc Motor Description

The optimized motor is a brushless three phase synchronous permanent magnet disc motor, with rated torque 54 Nm and speed 750 rpm@50 Hz, fed by a pulse width modulated (PWM) inverter and rechargeable batteries or fuel cell. The PMDM is a double sided axial field motor with two laminated stators having 36 slots and a centered rotor with 8 skewed neodymium-iron-boron permanent magnets with B_r=1.17 T and H_c=-883 kA/m. The graphical presentation of the prototype motor mounted on one wheel of the vehicle is given in Fig. 13.1; its real side view test rig is shown in Fig. 13.2. In this solution the two stators are attached to the chassis of the vehicle, where the rotor is fixed to the shaft of the wheel and free to move in a vertical direction. The rotor of this machine has to be carefully constructed so that it has adequate mechanical integrity and with the least possible weight. The weight of the rotor in such a solution is increasing the unsprung mass of the vehicle, but it is far less than the mass of the whole motor as in some other suggested solutions. This axial type of motor topology beside this application can be applied in many other applications [1-3] where high power density is needed and installation space is the main constraint.

13.3   GA Optimization Method Description

Since John Holland [4] presented the GA as a computer algorithm, a wide range of applications of GA has appeared in various scientific areas, and GA has been proved powerful enough to solve the complicated problems, especially the optimal design problems. Genetic algorithms are evolutionary search algorithms based on the mechanics of natural selection and natural genetics. They implement, in the most simplistic way, the concept of survival of the fittest. The reproductive success of a solution is directly tied to the fitness value it is assigned during evaluation. In this stochastic process, the least-fit solution has a small chance at reproduction while the most-fit solution has a greater chance of reproduction. The search starts from a randomly created population representing the chromosomes and obtains optimum after a certain number of generations of genetic operations. The optimisation is based on the survival of the string structures from one generation to the next, where a new improved generation is created by using the bits of information-genes of the survivors of the previous generation.

The created optimal design program GA-ODEM (Genetic Algorithm for Optimal Design of Electrical Machines) is using the Genetic Algorithm as an optimization tool. The design variables are presented as vectors of floating-point numbers [5]. The search starts from a randomly created population of strings representing the chromosomes and obtains optimum after a certain number of generations by applying genetic operations. The search can continue indefinitely. Therefore, a stopping rule is necessary to tell the algorithm when it is time to stop. This is achieved in many different ways and is also a user's and a problem dependent. Some of the possible methods are to fix the number of generations and to use the best individual of all generations as the optimum result; to fix the time elapsed and to select the optimum similarly; or to let the entire population converge in to an average fitness with some error margin. The stopping rule applied in this GA optimal design program is the number of generations. The parameters of the GA shape the way the algorithm runs. They could be grouped in two groups such as: primary and secondary parameters. There is one primary parameter:

➢ N the population size which is the number of chromosomes in the population;

and two secondary parameters that define the occurrence probabilities of the GA operators:

➢ p_c crossover probability;
➢ p_m mutation probability.

The values that are assigned to all of them are user and problem dependent. In order to make a proper selection of the p_c and p_m value, a very complex and detailed analysis of these parameters and their influence on the quality of the GA search, has been performed [6]. The considered values of the GA parameters for this optimal design problem are: population size $N=20$, crossover probability $p_c=0.85$, mutation probability $p_m=0.07$ and number of generations $G=15000$.

The main genetic operators of the genetic algorithm in general are reproduction, crossover and mutation.

13.3.1 Reproduction

Working on the entire population, the reproduction operator creates a new generation from the old generation. Based on the fitness measure of an individual and the average fitness of the population, the reproduction operator, determines the number of copies that particular individual will have in the next generation. The underlying idea in designing the reproduction operator is to give the individual with higher fitness a better chance to be represented in the next generation but leaving the decision to a random variable. The reproduction procedure that is implemented in the GA-ODEM programme is performed by a linear search through a roulette wheel with slots weighted in proportion to string fitness values.

13.3.2 Crossover

A central feature of genetic algorithms that creates a new chromosome from two "parents" is crossover. Corresponding to biological crossover, the software version combines a pair of parents by randomly selecting a point at which pieces of the parents' vectors of numbers are swapped. Instead of using the simple crossover the swapping is done with the so called arithmetical crossover which is defined as a linear combination of two vectors x_1 and x_2, after which the resulting offspring is

$$\mathbf{x}_1' = c \cdot \mathbf{x}_1 + (1-c) \cdot \mathbf{x}_2 \qquad (1)$$

$$\mathbf{x}_2' = c \cdot \mathbf{x}_2 + (1-c) \cdot \mathbf{x}_1 \qquad (2)$$

In the previous equations c could be any number between 0 and 1 or it can be taken as a fixed number; in this case it was adopted to be equal to 0.5. This type of crossover is called uniform arithmetical crossover and with its usage it is guaranteed that the values of the new parents will always be in the domain.

13.3.3 Mutation

Another step in reproduction is mutation, which involves the random real number generation of a selected variable in its upper and lower bound domain, of the new population. The primary purpose of mutation is to introduce variation into a population. This process is carried out randomly and it is done at a randomly selected place.

Another procedure that is implemented in the optimal design programme is the fitness scaling which improves the overall performance and leads towards better reliability of the GA search.

13.3.4 Fitness Scaling

In proportional selection procedure, the selection probability of the chromosome is proportional to its fitness. This simple scheme exhibits some undesirable properties. For example, in early generations, there is a tendency for a few super chromosomes to determine the selection process; while in later generations, when the population is largely converged, competition among chromosomes is less strong and random search behaviour will emerge.

Therefore a fitness scaling is proposed to mitigate these problems. Scaling method maps raw objective function values to some positive real values, based on which the survival probability for each chromosome is determined. By implementing fitness scaling two things can be achieved:

> ➤ To maintain a reasonable differential between relative fitness ratings of chromosomes,
> ➤ To prevent a very rapid domination by some super chromosomes in order to meet the requirement to limit competition early on, but to simulate it later.

For most scaling methods, scaling parameters are problem-dependent. Since De Jong's works, scaling of objective function values has become a widely accepted practise and several scaling mechanisms have been proposed. In general, the scaled fitness f_k' from the raw fitness f_k for chromosome k can be expressed as follows:

$$f_k' = h(f_k) \qquad (3)$$

where function h transforms the raw fitness into scaled fitness. The function h may take different forms to yield different scaling methods, such as linear scaling, sigma truncation, power law scaling, logarithmic scaling, and so on. These methods can be roughly classified into two categories:

> ➤ Static scaling
> ➤ Dynamic scaling

The mapping relation between the scaled fitness and raw fitness can be constant to yield static scaling methods, or it can vary according to some factors to yield dynamic scaling methods. The dynamic scaling methods are further divided into two cases:

> ➤ Scaling parameters are adaptively adjusted according to the scatter situation of fitness values in each generation in order to keep constant selective pressure,
> ➤ Scaling parameters are dynamically changed along with the increase of the number of generations in order to increase the selective pressure accordingly.

In a previous work of the authors [7] some of the above mentioned scaling methods were implemented in the GA optimal design programme GA-ODEM. A brief description of the implemented fitness scaling methods in the following text is presented.

13.3.4.1 Linear Scaling

Linear scaling (LS) adjusts the fitness values of all chromosomes such that the best chromosome gets a fixed number of expected offspring and thus prevent it from reproducing too many. The linear scaling method can be presented with the following equation:

$$f_k' = a \cdot f_k + b \qquad (4)$$

The coefficients a and b may be chosen in a numerous ways or can be defined as:

$$a = \frac{(C-1)f_{avg}}{f_{max} - f_{avg}} \qquad b = \frac{f_{avg}(f_{max} - Cf_{avg})}{f_{max} - f_{avg}} \qquad (5)$$

where C is a constant that can be C=1.2-2.0 and the values f_{max} and f_{avg} are the maximum and average value of the fitness for each generation.

13.3.4.2 Power Law Scaling

This method takes the form of an arbitrary power of the raw fitness as it presented in the following equation:

$$f_k' = f_k^\alpha \qquad (6)$$

In general the value of α is problem-dependent. The gap of scaled fitness between the best and the worst chromosomes increases with the value of α. When α approaches to zero, the gap approaches also zero and sampling becomes random search; when $\alpha > 1$, the gap is enlarged and sampling will be allocated to fitter chromosomes.

Another possible definition of power law scaling is a combination of linear scaling and power law scaling as it is presented in the following equation:

$$f_k' = (a \cdot f_k + b)^\alpha \qquad (7)$$

13.3.4.3 Boltzmann Selection

Boltzmann selection (BS) is a misleading name for yet another scaling method for proportional selection, using the following scaling function:

$$f_k' = e^{\frac{f_k}{T}} \qquad (8)$$

Selection pressure is low when the control parameter T is high.

In the previous authors' work [7] the presented fitness scaling methods were implemented and their influence on the GA performance analysed. From the performed analysis it was concluded that the linear scaling is the best fitness scaling method for the optimal design of the PMDM. Therefore in this work the linear scaling is implemented in the GA-ODEM program.

Finlay to summarize, the main genetic operators of the GA-ODEM implemented in the GA optimization program, which were previously presented, are reproduction, crossover and mutation. The reproduction procedure that is implemented in the optimization program is performed by a linear search through a roulette wheel with slots weighted in proportion to string fitness values. The type of crossover that is used is called uniform arithmetical crossover and with its usage it is guaranteed that the values of the new parents will always be in the domain. Another step in reproduction is mutation, which involves the random real number generation of a selected variable in its upper and lower bound domain, of the new

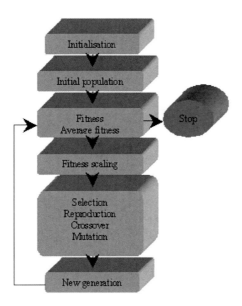

Fig. 13.3 Main steps of the GA-ODEM program

Table 13.1 Optimisation Constraints

Description	Parameters	Value
Phase voltage	U_{ph} [V]	181
Number of phases	N_{ph}	3
PM residual flux density	B_r [T]	1.17
Number of permanent magnets	N_m	8
Number of stator slots	Z	36
Stator back iron flux density	B_{mbi} [T]	≤ 1.7
Stator teeth flux density	B_{mst} [T]	≤ 1.7
Stator steel mass density	ρ_{st} [kg/m^3]	7300
Permanent magnets mass density	ρ_{PM} [kg/m^3]	7400
Rotor steel mass density	ρ_{rot} [kg/m^3]	7850
Copper mass density	ρ_{Cu} [kg/m^3]	8930

population. The primary purpose of mutation is to introduce variation into a population. This process is carried out randomly and it is done at a randomly selected place. Another procedure that is implemented in the optimal design program is the so called linear fitness scaling which improves the overall performance and leads towards better reliability of the GA search. Linear scaling adjusts the fitness values of all chromosomes such that the best chromosome gets a fixed number of expected offspring and thus prevent it from reproducing too many. After the operators perform their functions, the new generation is produced of members, which have gained new information through the exchange between pairs. The better traits of the "parent" chromosomes are carried along to the future generations. The optimal solution of the PMDM is selected as the best solution of the GA search. A block diagram presentation of the GA-ODEM program for the optimal design of the PM disc motor is presented in Fig. 13.3.

13.4 GA Optimal Design of Permanent Magnet Disc Motor

According to the design characteristics of PMDM, some of the parameters are chosen to be constant and some variable, such as: inside radius of the stator cores and PMs R_i, outside radius of the stator cores and PMs R_o, permanent magnet fraction α_m, permanent magnet axial length l_m, air-gap g, single wire diameter d_w, stator sloth width b_s and output torque T. The specific power of the motor is taken to be the objective function of the optimization:

$$specific\ power = P_{sp} = \frac{T \cdot \omega_m}{m_{Cu} + m_{Fe\ stator} + m_{PM} + m_{Fe\ rot}} \qquad (9)$$

where: ω_m-synchronous speed, m_{Cu}-total weight of the stator copper windings, $m_{Fe\ stator}$-total weight of the iron stator $m_{Fe\ rot}$-total weight of the iron rotor and m_{PM}-total weight of the permanent magnets.

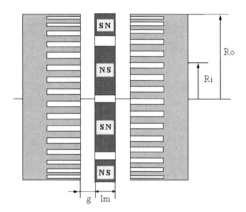

Fig. 13.4 Side view and PMDM parameter presentation

The optimal design process is a maximization problem of the objective function. Some of the optimisation variables are presented in Fig. 13.4 and Fig. 13.5. Based on some previous investigations it is decided the inside and outside radii of both stators and rotor to be the same.

Some of the optimization constraints used in the optimal design of the PMDM are of geometrical nature, and other are constraints concerning the motor performance and material characteristics, which are presented in Table 13.1.

Genetic Algorithm Applied in Optimal Design of PM Disc Motor

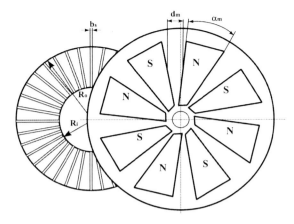

Fig. 13.5 Partial presentation of the PMDM and optimization parameters

Table 13.2 Upper and lower GA parameter optimization bounds

Parameters	Lower bound	Upper bound	Prototype
R_i [m]	0.070	0.074	0.072
R_o [m]	0.128	0.138	0.133
α_m [/]	0.6	0.730	0.6646
l_m [m]	0.009	0.0110	0.010
g [m]	0.0018	0.0022	0.002
d_w [m]	0.0006	0.0014	0.001
b_s [m]	0.0070	0.0090	0.008
T [Nm]	44	64	54

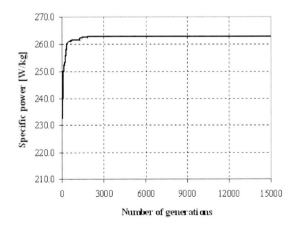

Fig. 13.6 Specific power change during GA search during generations

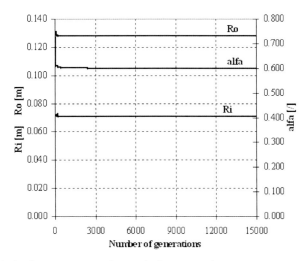

Fig. 13.7 Optimization parameters change during generations

Both types of constraints have been used to additionally reduce the number of independent design variables. This is obtained by a steady-state analysis of the motor that allows the main electrical, magnetic and mechanical quantities, including the set of motor specifications, to be expressed as functions of its dimensions and working conditions [8]. Some of the analytically calculated parameter values are in good agreement with the measured ones, such as the back emf, which proves that the mathematical model of the motor is quite realistic. In the mathematical model of the motor for the optimal design the eddy currents have not been taken into account. The stopping rule while the genetic algorithm works is selected to be the number of generations.

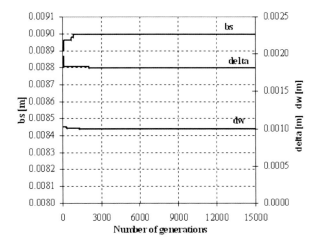

Fig. 13.8 Optimization parameters change during generations

Genetic Algorithm Applied in Optimal Design of PM Disc Motor

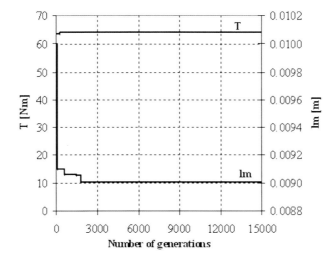

Fig. 13.9 Optimization parameters change during generations

The lower and upper bound as well as the optimization parameters values of the optimization procedure, in relation to the prototype model, are presented in Table 13.2. The comparative optimization parameters data of the initial and optimized model are presented in Table 13.3.

The convergence of the specific power of the motor as an objective function during the GA optimization search for 15000 generations is shown in Fig. 13.6. The value change for each optimization parameter during the generations is presented in Fig. 13.7, Fig. 13.8 and Fig. 13.9.

Table 13.4 Initial motor and GA solution data comparison

Parameters	Description	Initial Motor	GA Solution
$m_{Festator}$ (kg)	Total weight of the stator iron	12.687	10.565
m_{PM} (kg)	Total weight of the PM	3.714	1.427
m_{Fer} (kg)	Total weight of the rotor iron	2.993	2.714
m_{Cu} (kg)	Total weight of the winding copper	4.702	4.854
I_{ph} (A)	Phase current	8.716	10.880
R_{ph} (ohm)	Phase resistance	1.513	1.562
P_{Cu} (W)	Ohmic losses	344.95	554.778
P_{Fe} (W)	Iron losses	11.69	10.196
η (/)	Efficiency	0.833	0.826

13.5 GA Optimal Design Results of PM Disc Motor

Some specific parameters values for the GA optimal solution and for the prototype model, for the comparison analysis, are shown in Table 13.4. These values show evident improvement of the presented parameters and characteristics of the optimized PMDM model in relation to the prototype. It is evident that the GA optimized solution in relation to the prototype has less total weight, and therefore improved specific power, and a bit smaller efficiency, which is due to the increase of the total ohmic power losses of the winding. The decrease of the PM overall weight of the optimized model in relation to the prototype could lead to a reduction of the prize of the motor. This improvement in the PM weight, as well as the weight of the rotor iron could also lead to an improvement of the performance of the EV since the rotor is directly mounted on the shaft of the vehicle.

13.6 PM Disc Motor FEM Modeling and Magnetic Field Analysis

In order to be able to get the necessary data for the PM disc motor, a calculation of the magnetic field has to be performed. The 2D analysis is very suitable for this type of geometry and has a lot of advantages over the 3D calculation, such as lower memory storage and reduced time computation, which for one segment is done in several minutes. The quasi-3D method [9, 10] which is adopted for this analysis consists of a 2D FEM calculation of the magnetic field in a three dimensional radial domain of the axial field motor. For this purpose, the two stators and one rotor of the disc motor in the radial direction are divided in five segments. After the division a notional radial cut through the motor is performed, as shown in Fig. 13.10. Then one quarter of the motor is opened out into linear form, as presented in Fig. 13.11.

By using this linear quasi three-dimensional model of the disc motor, which is divided into five segments, it is possible to model the skewing of the magnets and also to simulate the vertical displacement and rotation of the rotor. Due to the symmetry of the machine the calculation of the motor is performed only for one quarter of the permanent magnet disc motor or for one pair of permanent magnets. The boundary condition on the top and bottom of each segment is defined as a Dirichlet's condition, and on the left and right side of each segment as a mixed boundary condition. The segments are not connected magnetically, but are electrically connected via the current density defined for each slot separately.

Genetic Algorithm Applied in Optimal Design of PM Disc Motor 241

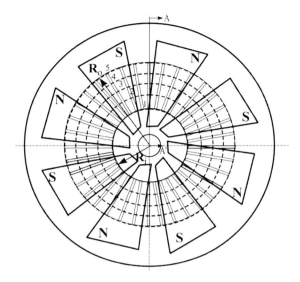

Fig. 13.10 Radial division of the motor into 5 segments

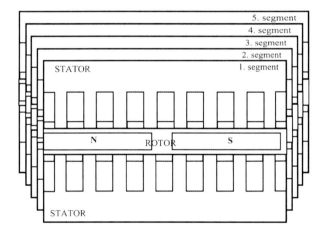

Fig. 13.11 Presentation of the 5 linear segments of the motor

After the proper modeling of the permanent magnet disc motor and the adequate mesh size refinement, especially in the air gap a magnetic field calculation is performed for each segment separately, for different current loads and different rotor displacements [11]. As an example the magnetic field distribution of the motor at no load and one rotor position for the 3rd middle segment, for the initial motor and the GA solution, is presented in Fig. 13.12 and Fig. 13.13, respectively.

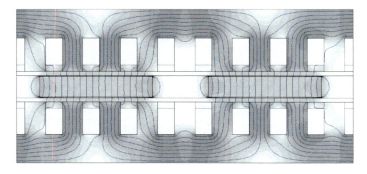

Fig. 13.12 Magnetic field distribution for the initial model at no load

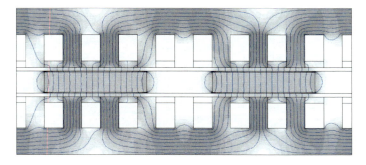

Fig. 13.13 Magnetic field distribution for the GA solution at no load

Also a presentation of the magnetic field distribution at rated load for the 3rd middle segment, for the initial motor and the GA solution, is presented in Fig. 13.14 and Fig. 13.15, respectively.

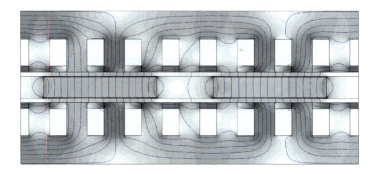

Fig. 13.14 Magnetic field distribution for the initial model at rated load

Genetic Algorithm Applied in Optimal Design of PM Disc Motor

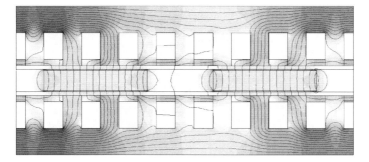

Fig. 13.15. Magnetic field distribution for the GA solution at rated load

In the postprocessor mode of the program using the data from the magnetic field calculation, the value of the air gap flux density and air gap flux in the middle of the air gap can be calculated by using equation (10) and equation (11) and solving it numerically.

$$\mathbf{B} = \text{curl } \mathbf{A} \tag{10}$$

$$\Phi_g = \int_S \vec{B} \cdot \vec{S} \tag{11}$$

The value of the air gap flux density for the axial field permanent magnet synchronous motor is calculated for different current loads and for all five segments. Due to lack of space the distribution of the air gap flux density is only presented for the middle segment of the permanent magnet disc motor models at no load and at rated current load as shown in Fig. 13.16, Fig. 13.17, Fig. 13.18, and Fig. 13.19, respectively.

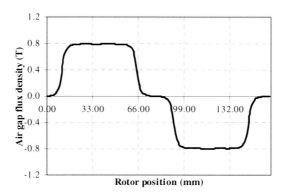

Fig. 13.16 Air gap flux density distribution at no load for the initial model

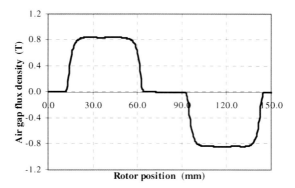

Fig. 13.17 Air gap flux density distribution at no load for the GA solution

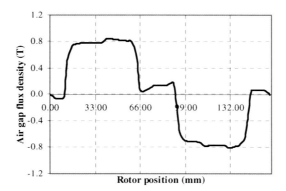

Fig. 13.18 Air gap flux density distribution at rated load for the initial model

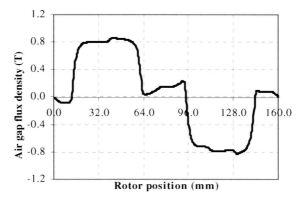

Fig. 13.19 Air gap flux density distribution at rated load for the GA solution

Table 13.5 Initial and GA solution FEM data comparison

Parameters	Initial model	GA solution
Air gap flux Φ_g (Wb)	0.00254	0.00234
Average air gap flux density B_g (T)	0.533	0.491

From the presented values in Table 13.5 it is evident that there is a change in the value of the air gap flux, air gap flux density of the GA solution in relation to the prototype, as a result of the change of the dimensions of the stator and therefore the change in the magnetic field distribution. The value of the air gap flux density under the PM for the GA solution is bigger than the value for the initial solution, but the average value of the flux density for the GA solution is smaller than the value of the initial solution due to the decrease of the PM's overlap angle. That is why the value of the air gap flux for the GA solution is smaller than the value for the initial solution. The change in the values of the flux density in different sections of the two stators also results in difference in the iron losses for the two solutions, as it is shown in Table 13.5.

13.7 Conclusion

An optimization technique based on GAs has been developed and applied to the design of a permanent magnet disc motor. A brief introduction of the genetic algorithms and their structure, performance and operators is presented. According to the results investigated above, it can be concluded that the GA is a very suitable tool for design optimization of PMDM and electromagnetic devices in general. By using GA the risk of trapping in a local maximum or minimum is reduced, especially by using some search improvements, which is very difficult to eliminate in deterministic methods. The quality of the GA optimized model has been proved through the data analysis of the prototype and optimized solution. This improvement resulted in improvement of the specific power as well as an overall weight reduction of the motor, which is very important for the improvement of the EV performance. The decrease of the PM total weight of the optimized model in relation to the prototype also leads to a reduction of the motor total cost, as well as an improvement of the electric vehicle overall performance. The reduction of the total weigh of the motor, as well as the increase of the torque leads toward increase of the specific power that was the main objective of the optimization. At the end the quality of the GA solution has been again proved by comparative analysis of the two motor models using FEM as a performance analysis tool. The proper modeling of the disc motor is presented and partial comparative results of the magnetic field for no load and rated load is presented.

As a future work a proper electromagnetic and magneto-thermal analysis of the two models is going to be performed in order to take into account all aspects of the motor performance analysis.

References

[1] Hakala, H.: Integration of Motor and Housing Machine Changes the Elevator Business. In: Proceedings of ICEM, vol. 3(4), pp. 1242–1245 (2000)
[2] Jang, G.H., Chang, J.H.: Development of an Axial-gap Spindle Motor for Computer Hard Disk Drives Using PCB Winding and Dual Air Gaps. IEEE Transactions on MAG 38(5), 3297–3299 (2002)
[3] Mongeau, P.: High torque/high power Density Permanent Magnet Motors. Naval Symposium on Electrical Machines, 9–16 (1997)
[4] Holland, J.H.: Adaptation in Natural and Artificial Systems. University of Michigan Press, Ann Arbor (1995)
[5] Janikow, Z., Michalewicz, Z.: An Experimental Comparison of Binary and Floating Point Representations in Genetic Algorithms. In: Proceedings of International Conference on GA, pp. 31–35 (1991)
[6] Cvetkovski, G.: Investigation of Methods and Contribution to the Development of Genetic Algorithms for Optimal Design of Permanent Magnet Synchronous Disc Motor. Faculty of Electrical Engineering and Information Technologies-Skopje, PhD thesis (2000)
[7] Cvetkovski, G., Petkovska, L.: Fitness Scaling Selection for Improved GA Based PMDM Optimal Design. In: Book of summaries of the 7th International Symposium on Electric and Magnetic Fields, pp. 53–54 (2006)
[8] Cvetkovski, G., Petkovska, L., et al.: Mathematical model of a permanent magnet axial field synchronous motor for GA optimization. Proceedings of ICEM 2(3), 1172–1177 (1998)
[9] Cvetkovski, G., Petkovska, L., et al.: Quasi 3D FEM in Function of an Optimization Analysis of a PM Disk Motor. Proceedings of ICEM 4(4), 1871–1875 (2000)
[10] Cho, C.P., Fussel, B.K., Hung, J.Y.: Detent Torque and Axial Force Effects in a Dual Air-Gap Axial Field Brushless Motor. IEEE Transactions on MAG 26(6), 2416–2418 (1993)
[11] Cvetkovski, G., Petkovska, L.: Performance Evaluation of an Axial Flux PM Motor Based on Finite Element Analysis. In: Proceedings of ISEF 2005 on CD, pp. 1–6 (2005)

Chapter 14
Magnetically Nonlinear Iron Core Characteristics of Transformers Determined by Differential Evolution

Gorazd Štumberger[1,*], Damir Žarko[2], Amir Tokic[3], and Drago Dolinar[1]

[1] University of Maribor, Faculty of Electrical Engineering and Computer Science, Slovenia, Smetanova 17, 2000 Maribor, Slovenia
gorazd.stumberger@uni-mb.si
[2] University of Zagreb, Faculty of Electrical Engineering and Computing, Croatia
[3] University of Tuzla, Faculty of Electrical Engineering, Bosnia and Herzegovina

Abstract. An optimization based method for determining magnetically nonlinear iron core characteristics of transformers is proposed. The method requires a magnetically nonlinear dynamic model of the transformer as well as voltages and currents measured during the switch-on of unloaded transformer. The magnetically nonlinear iron core characteristic is in the model accounted for in the form of three different approximation functions. Their parameters are determined by the stochastic search algorithm called differential evolution. The optimization goal is to find those values of approximation functions parameters where the root mean square differences between measured and calculated currents are minimal. The impact of individual approximation functions on calculated dynamic responses of the transformer are evaluated by the comparison of measured and calculated results.

14.1 Introduction

The dynamic models of transformers used for computation of power system transients are often completed by the magnetically nonlinear iron core characteristics in order to achieve better agreement between the measured and calculated responses. The iron core characteristics are normally determined either experimentally [1]-[3] or by the finite element (FE) computations [4], [5]. However, in the cases when the characteristics can be determined neither experimentally due to the limited access when the device is in operation, nor by the FE due to the missing design data, the optimization methods can be applied.

Various experimental methods for determining magnetically nonlinear characteristics of electromagnetic devices presented in [2] require voltage supply of a specific waveform. However, tests with specific voltage waveforms cannot be

---
[*] Corresponding author.

performed on all devices. Some power system devices are simply too important to be disconnected just to perform tests needed to determine their magnetically nonlinear characteristics. Thus, currents and voltages can be measured on the terminals of such a device exclusively during normal operating conditions. These currents and voltages can be applied, combining an optimization method together with the magnetically nonlinear dynamic model of the tested device, to determine an approximation of magnetically nonlinear iron core characteristic. Some of appropriate approximation functions are presented in [6]-[9].

In this work, the magnetically nonlinear iron core characteristics are approximated by three functions. Their parameters are determined in the optimization procedure using transformer's dynamic model with approximated magnetically nonlinear iron core characteristic, currents and voltages measured during the switch-on of the transformer, and a stochastic search algorithm called Differential Evolution (DE) [10]. The objective function is the root mean square difference between the measured and calculated current responses. It is minimized during the optimization. In the case study, the impact of the three obtained approximation functions on responses calculated by the transformer's dynamic model with included approximation functions is evaluated.

14.2 Dynamic Model of a Single-Phase Transformer

The dynamic model of a single phase transformer is given by Eq. (1) describing the voltage balances in the primary winding and by Eq. (2) describing the voltage balances in the secondary winding:

$$u_1 = i_1 R_1 + L_{\sigma 1} \frac{di_1}{dt} + N_1^2 \frac{\partial \phi}{\partial \theta} \frac{di_1}{dt} + N_1 N_2 \frac{\partial \phi}{\partial \theta} \frac{di_2}{dt} \tag{1}$$

$$u_2 = i_2 R_2 + L_{\sigma 2} \frac{di_2}{dt} + N_1 N_2 \frac{\partial \phi}{\partial \theta} \frac{di_1}{dt} + N_2^2 \frac{\partial \phi}{\partial \theta} \frac{di_2}{dt} \tag{2}$$

where u_1, u_2 and i_1, i_2 are the primary and the secondary voltages and currents, R_1 and R_2 are the primary and the secondary resistances, $L_{\sigma 1}$ and $L_{\sigma 2}$ are the primary and the secondary leakage inductances, N_1 and N_2 are the number of the primary and secondary turns, ϕ is the magnetic flux, while $\theta = (N_1 i_1 + N_2 i_2)$ is the magnetomotive force (mmf). The magnetically nonlinear behaviour of the tested transformer is accounted for by the characteristic $\phi(\theta)$, more precisely by its partial derivative $\partial \phi / \partial \theta$. This function is approximated by three different approximation functions whose parameters are determined in the optimization procedure.

14.3 Approximation Functions

In order to be appropriate for inclusion into the dynamic mode given by Eq. (1) and Eq. (2), the approximation function of magnetically nonlinear characteristic $\phi(\theta)$ should be given in an analytic form, while its partial derivative $\partial \phi / \partial \theta$ should be a continuous and monotonous function. Magnetically nonlinear behaviour of magnetic material is normally described by the flux density B versus magnetic field strength H characteristic. However, if the mean length of magnetic paths in the iron core l and the cross-section of the iron core A are considered as constants, the equivalent characteristic $B(H)$, which in this case represents average magnetic conditions in the entire iron core, can be transformed into the characteristic $\phi(\theta) \approx AB(lH)$ required in the dynamic model given by Eq. (1) and Eq. (2).

From many different approximation functions presented in [6]-[9] only three of them are discussed in the following subsections. They are applied in the dynamic model, given by Eq. (1) and Eq. (2), to take into account magnetically nonlinear behaviour of the entire transformer. The approximation function parameters are determined by the DE using dynamic model along with measured currents and voltages. Thus, the equivalent characteristic $B(H)$, obtained in this manner, represents magnetically nonlinear behaviour of the entire transformer, which in this case justifies the use of relations given by Eq. (3):

$$\phi = AB; \qquad \theta = Hl; \qquad \frac{\partial \phi}{\partial \theta} = \frac{A}{l}\frac{\partial B}{\partial H}. \qquad (3)$$

14.3.1 Simple Analytic Saturation Curve

A simple analytic approximation function is given by Eq. (4) [6]:

$$B(H) = \mu_0 H + \frac{2J_s}{\pi}\operatorname{arctg}\left(\frac{\pi(\mu_r-1)\mu_0 H}{2J_s}\right) \qquad (4)$$

where μ_0 is the permeability of vacuum, μ_r is the initial relative permeability, while J_s is the saturated magnetization. The approximation function parameters to be determined by the DE are μ_r and J_s. Fig. 14.1 shows an example of analytic saturation curve described by Eq. (4).

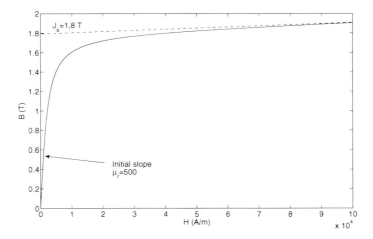

Fig. 14.1 Example of the analytic saturation curve

The partial derivative of Eq. (4) needed in the dynamic model in the form of Eqs. (1), (2) is given by Eq. (5).

$$\frac{\partial B}{\partial H} = \mu_0 + \frac{(\mu_r - 1)\mu_0}{1 + \left(\frac{\pi(\mu_r - 1)\mu_0 H}{2J_s}\right)^2} \quad (5)$$

14.3.2 Analytic Saturation Curve with Bend Adjustment

An analytic approximation function with bend adjustment is given by Eq. (6) [6]:

$$B(H) = \mu_0 H + J_s \frac{H_a + 1 - \sqrt{(H_a + 1)^2 - 4H_a(1-a)}}{2(1-a)} \quad (6)$$

$$H_a = \mu_0 H \frac{\mu_r - 1}{J_s} \quad (7)$$

where the approximation function parameters to be determined by the DE are the saturated magnetization J_s, the initial relative permeability μ_r and the bend adjustment coefficient $a \in [0, 0.5]$. The example curve is shown in Figure 14.2.

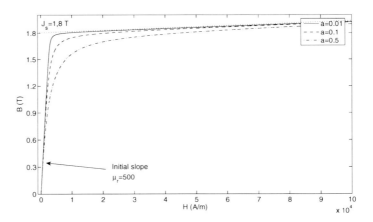

Fig. 14.2 Example of the analytic saturation curve with bend adjustment

The partial derivative of Eq. (6) is given by Eq. (8).

$$\frac{\partial B}{\partial H} = \mu_0 + \frac{\mu_0(\mu_r - 1)}{2(1-a)}\left(1 - \frac{(H_a + 1) - 2(1-a)}{\sqrt{(H_a + 1)^2 - 4H_a(1-a)}}\right) \quad (8)$$

14.3.3 Approximation with Exponential Functions

The approximation function in the form of two exponential functions is given by Eq. (9):

$$B(H) = C_0 + C_1 e^{D_1 H} + C_2 e^{D_2 H} \tag{9}$$

where C_0, C_1, C_2, D_1 and D_2 are the approximation function parameters, which have to the determined by the DE. The Partial derivative of Eq. (9) can be easily calculated by Eq. (10).

$$\frac{\partial B}{\partial H} = C_1 D_1 e^{D_1 H} + C_2 D_2 e^{D_2 H} \tag{10}$$

14.4 Determining Approximation Function Parameters by Differential Evolution

When measured characteristic $B(H)$ is available, the approximation function parameters can be determined by different methods, assuring the best possible agreement between measured characteristic and its approximation. However, in some cases magnetically nonlinear characteristic of the device is not available because it cannot be determined by the FE computation due to the missing design data, while experimental methods cannot be applied due to the operating conditions in the power system. In such cases the approximation function parameters can be determined by the DE applying dynamic model along with the currents and voltages measured on the device terminals during normal operating conditions. One of the suitable operating conditions where the currents and voltages could be measured is the switch-on of the unloaded transformer. On the one hand is this normal operating condition while on the other hand the iron core normally becomes highly saturated during the switch-on. The effects of saturation are normally visible in the measured inrush current and can be used to determine magnetically nonlinear iron core characteristic in the form of an approximation function. The DE searches for approximation function parameters with respect to the given objective function. Thus, before the optimization procedure starts an objective function has to be determined. In order to assure acceptable agreement between the measured and calculated currents in the time and in the frequency domain, the objective function has to be determined in both domains. The rest of this section describes the DE algorithm and definitions of the objective function.

14.4.1 Description of Differential Evolution Algorithm

Differential Evolution [9] is a direct search stochastic algorithm capable of solving global optimization problems subject to nonlinear constraints. It operates on a population of candidate solutions and does not require a specific starting point. The population is of constant size *NP*. In each iteration a new generation of solutions is created and compared to the population members of the previous generation. The process is repeated until the maximum number of generations G_{max} is reached.

A nonlinear global optimization problem can be defined as follows: Find the vector of parameters $\mathbf{x} = [\, x_1, x_2, \ldots, x_D \,]$, $\mathbf{x} \in R^D$ which will minimize the function $f(\mathbf{x})$. The vector $\mathbf{x}$ is subject to m inequality constraints $g_j(\mathbf{x}) \leq 0$, $j = 1, \ldots, D$ and D boundary constraints $x_i^{(L)} \leq x_i \leq x_i^{(U)}$, $i = 1, \ldots, D$, where $x_i^{(L)}$ and $x_i^{(U)}$ are the lower and upper limits.

The population of the Gth generation can be written in the form $P_G = [\, \mathbf{x}_{1,G}, \mathbf{x}_{2,G}, \ldots, \mathbf{x}_{NP,G} \,]$, $G = 0, \ldots, G_{max}$. Each vector in P_G contains D real parameters $\mathbf{x}_{i,G} = [\, x^i_{1,G}, x^i_{2,G}, \ldots, x^i_{D,G} \,]$, $i = 1, \ldots, NP$, $G = 0, \ldots, G_{max}$.

The initial population $P_{G=0}$ is generated using random values within the given boundaries which can be written in the form of Eq. (11):

$$x^i_{j,0} = \text{rand}_j[0,1]\left(x_j^{(U)} - x_j^{(L)}\right) + x_j^{(L)}, \quad i = 1,\ldots,NP,\ j = 1,\ldots,D \quad (11)$$

where rand$_j$ [0,1] is the uniformly distributed random number on the interval [0,1] which is chosen anew for each j, while (U) and (L) denote the upper and lower boundaries of the vector parameters. In every generation, new candidate vectors are created by randomly sampling and combining the vectors from the previous generation in the following manner described by Eq. (12):

$$i = 1,\ldots,NP,\ j = 1,\ldots,D,\ G = 1,\ldots G_{max}$$

$$u^i_{j,G} = \begin{cases} x^{r3}_{j,G-1} + F\left(x^{r1}_{j,G-1} - x^{r2}_{j,G-1}\right) & \text{if rand}_j[0,1] \leq CR \text{ or } j = k \\ x^i_{j,G-1} & \text{otherwise} \end{cases} \quad (12)$$

where $F \in [0,2]$ and $CR \in [0,1]$ are DE control parameters which are kept constant during optimization, $r1, r2, r3 \in \{1, \ldots, NP\}$, $r1 \neq r2 \neq r3 \neq i$ are randomly selected vectors from the previous generation, different from each other and different from the current vector with index i, and $k \in \{1, \ldots, D\}$ is a randomly chosen index which assures that at least one $u^i_{j,G}$ is different from $x^i_{j,G-1}$.

The population for the new generation P_G will be assembled from the vectors of the previous generation P_{G-1} and the candidate vectors $\mathbf{u}^i_G$ according to the following selection scheme described by Eq. (13):

$$i = 1,\ldots,NP,\ G = 1,\ldots G_{max}$$

$$\mathbf{x}^i_G = \begin{cases} \mathbf{u}^i_G & \text{if } f\left(\mathbf{u}^i_G\right) \leq f\left(\mathbf{x}^i_{G-1}\right) \\ \mathbf{x}^i_{G-1} & \text{otherwise} \end{cases} \quad (13)$$

14.4.2 Objective Function

The objective function in the time domain q_t is defined by Eq. (14) as the mean square difference between the measured $i_{1m}(t)$ and the calculated switch-on current $i_1(t)$ in the time interval of observation $t \in [t_1, t_2]$.

$$q_t = \frac{1}{t_2 - t_1} \int_{t_1}^{t_2} e_t^2(\tau) d\tau \tag{14}$$

$$e_t(t) = i_{1m}(t) - i_1(t) \tag{15}$$

Before the objective function in the frequency domain q_f is defined, the measured current $i_{1m}(t)$ and calculated current $i_1(t)$ have to be decomposed into individual harmonic components. The switch-on (inrush) current of a transformer is normally an aperiodic function. Therefore, the harmonic decomposition of currents in the entire interval of observation $t \in [t_1, t_2]$ normally does not give useful results. For optimization purposes it is better to define a moving window in the length of one cycle of the applied supply voltage T. In this manner the currents can be decomposed into individual harmonic components inside the window $[t\text{-}T, t]$, while the window moves with the time t. The same approach is applied in differential protection algorithms for power transformers.

For M being the highest order harmonic component present in the calculated current $i_1(t)$ in the interval $[t - T, t]$, the current $i_1(t)$ can be expressed in the form of Fourier series given by Eq. (16):

$$i_1(t) = A_0(t) + \sum_{h=1}^{M} \left(a_h(t) \cos(h\omega t) + b_h(t) \sin(h\omega t) \right) \tag{16}$$

where h is the order of the harmonic component, $\omega = 2\pi T$ while $A_0(t)$, $a_h(t)$ and $b_h(t)$ are the Fourier coefficients. If $A_{0m}(t)$, $a_{hm}(t)$ and $b_{hm}(t)$ are the Fourier coefficients of the measured current $i_{1m}(t)$, then the sum of squared differences between individual Fourier coefficients of $i_1(t)$ and $i_{1m}(t)$ is given by Eq. (17).

$$e_f^2(t) = \left(A_{0m}(t) - A_0(t) \right)^2 + \sum_{h=1}^{M} \left(a_{hm}(t) - a_h(t) \right)^2 + \left(b_{hm}(t) - b_h(t) \right)^2 \tag{17}$$

The frequency domain objective function q_f for the interval $t \in [t_1, t_2]$, in which the window $[t\text{-}T, t]$ moves with the increasing time t along the interval $[t_1, t_2]$, is defined by Eq. (18).

$$q_f = \frac{1}{t_2 - t_1} \int_{t_1}^{t_2} e_f^2(\tau) d\tau \tag{18}$$

The objective functions q_t given by Eq. (14) and q_f given by Eq. (18) can be combined into a single objective function q defined by Eq. (19).

$$q = q_t + q_f \tag{19}$$

During the optimization procedure the DE generates a certain number of approximation function parameter sets called population. These sets describe magnetically nonlinear behaviour of the tested single-phase transformer. They are included into the dynamic model given by Eqs. (1) and (2) in the form of $B(H)$ characteristic. The model is supplied with the voltages measured on transformer terminals during the switch-on of the unloaded transformer. In this way the current time response is calculated. The quality of the approximation function parameter set is evaluated by the objective function (19). By simulating the principle of evolution in the natural world, the DE preserves only those sets of approximation function parameters with the best (lowest) values of the objective function. According to the previously described DE algorithm, they are used to generate a new population – new sets of approximation function parameters. The set with the best value of the objective function at the end of the optimization procedure represents magnetically nonlinear characteristic of the tested transformer.

The DE is searching for approximation function parameters by minimizing the value of the objective function q. It gives acceptable results at moderate computational effort. However, when higher computational effort is acceptable, multiobjective Pareto optimization for the set of objective functions q_t, q_f can be performed.

14.5 Results

The approximation functions, given by Eqs. (4), (6) and (9), contain two, three and five parameters, respectively. These parameters were determined by the DE. The increasing number of parameters means that the computational effort required for their determination increases as well. However, the increasing number of parameters means also more possibilities for approximation function shape adjustments.

The data of the tested small laboratory single-phase transformer are given in Table 14.1. Table 14.2 shows the DE settings applied in the optimization procedure for determining the values of approximation function parameters given in Table 14.3.

Table 14.1 Data of tested single-phase transformer

Primary resistance R_1	11.0 Ω
Secondary resistance R_2	141.8 Ω
Primary leakage inductance $L_{\sigma 1}$	0.033 H
Secondary leakage inductance $L_{\sigma 2}$	0.033 H
Number of primary turns N_1	452
Number of secondary turns N_2	1722
Mean length of magnetic path in the iron core l	0.168 m
Cross-section of the iron core A	6.02 cm^2

Table 14.2 The DE control parameters

Approximation function	Stepsize F	Crossover probability CR	Population size NP
Eq. (4)	0.7	0.5	20
Eq. (6)	0.7	0.5	30
Eq. (9)	0.7	0.5	20

Table 14.3 Approximation function parameters

Approximation function Eq. (4)	$J_s = 3.2750$		$\mu_r = 9306.9$	
Approximation function Eq. (6)	$J_s = 3.3074$	$\mu_r = 10429,0$		$a = 0,5$
Approximation function Eq. (9)	$C_0 = 0.8536$ $C_1 = 0.5923$ $C_2 = 0.2614$		$D_1 = 5.3134$	$D_2 = -100.79$

In order to evaluate different approximation functions and their parameters determined by the DE, a new test independent of the one used in the optimization procedure, was performed. Again, the applied primary voltage u_{1m} and the primary current i_{1m} were measured during the switch-on of the unloaded transformer. The transformer dynamic model with included magnetically nonlinear iron core characteristic, given by Eqs. (1) and (2), was supplied by the measured applied voltage $u_1 = u_{1m}$. The magnetically nonlinear iron core characteristic was approximated by the approximation functions given by Eqs. (4), (6) and (9), whose parameters, determined by the DE, are given in Table 14.3. The impact of individual approximation functions and their parameters on calculated currents was evaluated by the comparison of measured and calculated responses.

Fig. 14.3 shows voltage $u_1 = u_{1m}$ and current i_{1m} measured during the switch-on of the unloaded single phase test transformer. The iron core characteristic of the tested transformer is determined in the form of characteristic $\phi(\theta)$ [2]. It is shown in Fig. 14.4 together with the characteristics $\phi(\theta)$ determined by the DE, where approximation functions given with Eqs. (4), (6) and (9) are applied together with Eq. (3). The results presented in Fig. 14.4 clearly show that the agreement between the measured and DE determined characteristic $\phi(\theta)$ in the interval between 30 and 250 A turns is the best when approximation function given by Eq. (9) is applied. It seems that in the given case this interval has substantial impact on the time behaviour of currents calculated by the dynamic model. The comparison between measured i_{1m} and dynamic model calculated currents i_1 is given in Fig. 14.5 together with the differences between the measured and calculated currents ($i_{1m} - i_1$).

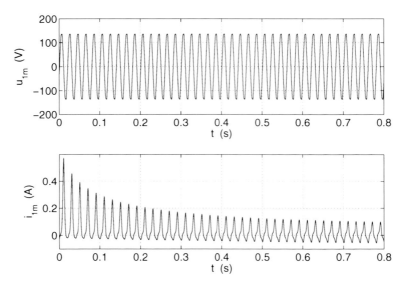

Fig. 14.3 Applied voltage u_{1m} and responding current i_{1m} measured during the switch-on of unloaded single-phase transformer

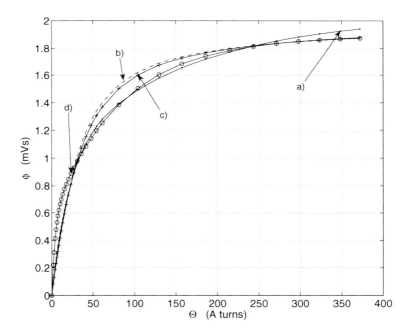

Fig. 14.4 Characteristics $\phi(\Theta)$: a) experimentally determined; b) DE determined using approximation function Eq. (4); c) DE determined using approximation function Eq. (6); d) DE determined using approximation function Eq. (9)

Magnetically Nonlinear Iron Core Characteristics

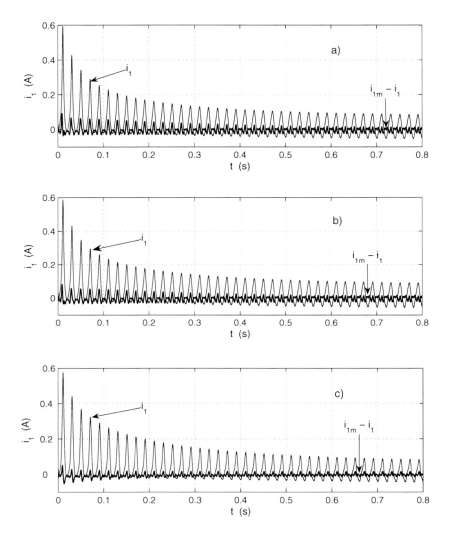

Fig. 14.5 Calculated current i_1 and difference between the measured and calculated current $i_{1m} - i_1$ for: a) approximation function Eq. (4); b) approximation function Eq. (6); c) approximation function Eq. (9)

The quality (objective function) q, given by Eq. (19), is used to evaluate the agreement between the measured and the calculated currents shown in Figure 14.2. The differences among individual calculated currents appear solely due to the different approximation functions. Thus, the objective function values given in Table 14.4 show the quality of different approximation functions and the quality of approximation function parameters determined by the DE whose values are shown in Table 14.3.

Table 14.4 Evaluation of approximation functions and their parameters

Approximation function	Eq. (4)	Eq. (6)	Eq. (9)
Quality q	8.2576	7.8230	2.9922

The results presented in Figure 14.2 and Table 14.4 show the best agreement between the measured and calculated results in the case when the approximation function given by Eq. (9) is applied. The expression for its partial derivative given by Eq. (10) is simple and can be easily determined. Thus, in opinion of the authors the approximation function given by Eq. (9) is the most suitable for the use in the proposed method for determining magnetically nonlinear characteristics of electromagnetic devices by optimization methods.

14.6 Conclusion

The optimization based method for determining magnetically nonlinear characteristics of electromagnetic devices is presented in the paper. This method requires an optimization tool like DE, a magnetically nonlinear dynamic model of the tested device, an approximation function for magnetically nonlinear characteristic of the device and currents and voltages measured on the device terminals during normal operating conditions. The approximation function parameters are determined during the optimization procedure. The proposed method can be applied for determining magnetically nonlinear characteristics of electromagnetic devices in the cases when they cannot be determined experimentally or by the finite-element computations.

Acknowledgments

This work was partly supported by the Slovenian Research Agency (ARRS), project no. L2-2060 and P2-0115.

References

[1] Calabro, S., Coppadoro, F., Crepaz, S.: The measurement of the magnetization characteristics of large power transformers and reactors through d.c. excitation. IEEE Transactions on Power Delivery 1(4), 224–232 (1986)
[2] Štumberger, G., Polajžer, B., Štumberger, B., Toman, M., Dolinar, D.: Evaluation of experimental methods for determining the magnetically nonlinear characteristics of electromagnetic devices. IEEE Transactions on Magnetics 41(10), 4030–4032 (2005)
[3] Dolinar, M., Dolinar, D., Štumberger, G., Polajžer, B., Ritonja, J.: A Three-Phase Core-Type Transformer Iron Core Model With Included Magnetic Cross Saturation. IEEE Transactions on Magnetics 42(10), 2849–2851 (2006)

[4] Ren, Z., Razek, A.: A strong coupled model for analysing dynamic behaviours of non-linear electromechanical systems. IEEE Transactions on Magnetics 30(5), 3252–3255 (1994)
[5] Dolinar, D., Štumberger, G., Grčar, B.: Calculation of the linear induction motor model parameters using finite elements. IEEE Transactions on Magnetics 34(5), 3640–3643 (1998)
[6] Flux2d user's guide, Cedrat (2000)
[7] Pedra, J., Sainz, L., Corcoles, F., Lopez, R., Salichs, M.: PSPICE computer model of a nonlinear three phase Three-legged transformer. IEEE Transactions on Power Delivery 19(1), 200–207 (2004)
[8] Perez-Rojas, C.: Fitting saturation and hysteresis via arctangent functions. IEEE Power Engineering Review 20(11), 55–57 (2000)
[9] Štumberger, G., Štumberger, B., Dolinar, D., Težak, O.: Nonlinear model of linear synchronous reluctance motor for real time applications. Compel 23(1), 316–327 (2004)
[10] Price, K.V., Storn, R.V., Lampinen, J.A.: Differential evolution: a practical approach to global optimization. Springer, Heidelberg (2005)

Chapter 15
Different Methods for Computational Electromagnetics: Their Characteristics and Typical Pratical Applications

Arnulf Kost

TU Berlin – Institut für Elektrische Energietechnik –Sekr.EM4- Einsteinufer 11
10587 Berlin, Germany
`kost@iee.tu-berlin.de`

Abstract. There is no method of electromagnetic modelling, which is well suited for any practical application, but requirements on different methods for Computational Electromagnetics, created by practical applications, are leading to certain trends of method selection and hybrid combinations of them. The comparison of the methods is accompanied by practical applications and an outlook on futural demands.

15.1 Introduction

Regarding their mode of operation, it can be observed, that not all, but a lot of electrical devices and their production is nowadays based on and/or influenced by electromagnetic fields. For instance it is a highly important question, if contemporary cellular phones meet the need for protection of the human head against electromagnetic radiation, as demanded by the limiting values of e.g. the International Radiation Protection Association (IRPA). Also the designer of a technical system will have to meet the desired functional performance regarding Electromagnetic Compatibility (EMC), but the device must also meet legal requirements in virtually all countries of the world.

To do his job the designer may use classical prototyping assisted by measurements, but more and more is using simulation tools, the need for which is evident for all the obvious reasons, such as design cycles are becoming shorter and shorter, prototyping is too expensive or even impossible. Especially and of course in the case of the system cellular phones/human head prototyping has to be ruled out. Inside devices interferences may occur by propagation within circuits but mostly via electromagnetic fields.

To overcome or reduce the latter ones, there is a growing need of Electromagnetic Modelling. The same holds for many products in communication engineering and electrical energy engineering like rotational and linear motors and actuators, loudspeaker systems, recording heads, medical scanning systems, optical pick-up units for DVD and transformers.

The related systems are of an enormous variety. Their size may be that of a high-speed train engine, of a magnet system for Magnetic Resonance Imaging (MRI), of a Printed Circuit Board (PCB) or of an Integrated Circuit (IC), just to name a few, and accordingly the geometric complexity may strongly vary. It may be an open boundary system (very often) or a closed one. Materials may have nonlinear behaviour (ferromagnetic ones) and/or frequency dependent characteristics (absorbers). The whole frequency scale from 0Hz to about 50 GHz is covered, and often system parameters have to be optimised.

Due to this variety of systems and parameters it is understandable, that not only one method of Electromagnetic Modelling will be the best choice for any system.

Therefore, the paper will give an overview of different methods, compare them and show their advantages and disadvantages, and moreover will illustrate them by typical practical applications.

15.2 The Finite Element Method (FEM)

As with further methods of Computational Electromagetics the starting point of FEM is the strategy of weighted residuals

$$\int_\Omega R.wd\Omega \tag{1}$$

applied here on the residual of e.g. the skin effect equation in Ω_{FE} (volume of FEM-region of surface Γ_1). Applying Green's first theorem, (1) is transformed into its weak form and selecting the local Galerkin procedure (the shape functions are also chosen as weighting functions) the discretisation of Ω_{FE} into e.g. triangles transforms (1) into a linear equation system, the unknowns of which are the nodal values of the vector potential inside Ω_{FE} and on Γ_1 as well as the normal derivatives $\partial A/dn$, being constant on the surface elements of Γ_1 in the case of linear shape functions:

$$(K + S)A - T\frac{\partial A}{\partial n}\Big|_{\Gamma_1} = F \tag{2}$$

K, the stiffness matrix and **S**, the influence matrix of the induced currents, form together a sparse and symmetrical matrix. **T** is a sparse matrix, **F** is a vector due to the sources.

In a similar way in high-frequency problems Maxwell's equations are transformed to an Electrical Field Integral Equation (EFIE) and Magnetic Field Integral Equation (MFIE), and applying Stratton's first theorem instead of Green's first theorem again a linear equation systems is established, now for edge variables **E** inside Ω_{FE} and on Γ_1, as well as tangential components of **H** on Γ_1(**E**: electrical, **H** magnetic field strength), details see e.g. [1].

Edge variables and the related edge elements turned out to be of striking success, by satisfying physical conditions "automatically". It has to be said, however, that FEM had not the same impact for high frequency problems as for low frequency ones.

The FEM advantages are: sparse matrices, so that iterative equation solvers like ICCG (Incomplete Cholesky Conjugate gradient method) with a solution time of order n·logn can handle very high numbers n of nodes and unknowns, which nowadays are surpassed by Multigrid Methods. FEM is moreover well suited for nonlinear and/or anisotropic materials and complicated geometrical boundaries and interfaces. Especially with triangular and tetrahedral elements adaptive mesh generation turned out to be very effective .

The FEM disadvantage is that the entire problem space has to be discretized. Problems with open boundaries are therefore not best suited for FEM.

15.3 The Boundary Element Method

Like for FEM the starting point is the strategy of weighted residuals (1). But the objective for BEM is completely different from that for FEM: with BEM a field quantity (e.g. the vector potential $\mathbf{A}_i$) in a point of position i in Ω_{BE} shall be expressed only by boundary quantities. To do this Green's second theorem is applied on the weighted residual strategy, and by the Dirac-Delta-function a field quantity in i can be filtered out of its integration over Ω_{BE}. Hereby e.g. the Poisson equation for the vector potential $\mathbf{A}$ is transformed into a boundary integral equation and discretizing Γ_1 into boundary elements, transforms (1) into an equation system, the unknowns of which are the nodal values of the vector potential and its normal derivatives on Γ_1:

$$HA|_{\Gamma_2} - G\frac{\partial A}{\partial n}\bigg|_{\Gamma_1} = A_e \qquad (3)$$

In contrast to FEM also constant approximations of $\mathbf{A}$ in a boundary element are possible and widely used.

The BEM advantages are: In 3D problems the set of unknowns has to be calculated only on surfaces, in 2D problems only on contour lines leading to much less unknowns than with FEM. Unbounded regions are especially well to handle by BEM.

The BEM disadvantages were: Fully or dense blockwise populated matrices (solution time of order n^3). The point of position i and the source point on the boundary can fall together or be very close during integration, so that singular or nearly singular integrals arise, which have to be very carefully treated. Both disadvantages nowadays have been overcome to a large extent due to the treatment of the matrices by the Fast Multipole Method, and by strong improvements to treat the singular integrals. Nonlinear or inhomogeneous materials however remain difficult to treat by BEM.

15.4 The Moments Method (MoM)

This method, as introduced by Harrington [2] is in principle only another term for the strategy of weighted residuals (1). In contrast to FEM and BEM (1) is not transformed by a Green's or Stratton's theorem, but directly exploited and therefore states a very simple method, compared to FEM or BEM. However, if differential equations of second order would be tackled by this method, their second derivatives would cause severe problems when discretizing and using first or even zero order approximations for the solution. This is the explanation, why the method is applied mostly to problems, which are naturally described by integral equations, e.g. scattering and antenna problems. Given such a problem in operator style

$$Lu = f \qquad (4)$$

the choice of suitable basis and weighting functions turned out to be more important for MoM than for FEM, where adaptive mesh generation is often used. Regarding current distributions on antennas, not only local but also global basis functions like trigonometrical functions, powers and Legendre polynomials are used.

In the case of antennas the integral equation for the current distribution on the antenna is stating the operator L. In the case of a conductive scatterer, its unknown surface currents are the solution of the MFIE integral equation, stating the L operator in that case.

Advantages of the moment method turned out to be a robust behaviour in antenna design and scattering problems by conductive objects, if the discretisation to capture the radiating surfaces does not need too many elements.

Disadvantages are the fully populated system matrix and its hereby caused failing for problems, where the unknowns have to be calculated not only on contours (linear antennas) or surfaces (patch antennas or conductive scatterers) but in volumes (lossy dielectric scatterers or absorbers), causing no more manageable large equation systems. The method is not effective for complex geometries and inhomogeneous materials.

15.5 The Finite Difference Time Domain Method (FDTD)

The FDTD is a mathematically very simple method, solving Maxwell's equations in integral form on a regular grid and by central difference approximation of the time derivatives. The volume to be modelled is represented by two interleaved grids, one for the magnetic field, the other one for the electric field to be evaluated.

The application of Faraday's law on the facing element of area A is expressed by

$$\frac{1}{A}\left[E_{z1}(t) + E_{y2}(t) - E_{z3}(t) - E_{y4}(t)\right]$$
$$= -\frac{\mu}{2\Delta t}\left[H_{x0}(t + \Delta t) - H_{x0}(t - \Delta t)\right] \qquad (5)$$

In this equation the only unknown is $H_{x0}(t+\Delta t)$, provided the other quantities have been found in the time steps t and t-Δt before. Then Ampere's law is applied to find the electric field at the time step t+2Δt and so forth. Note, that the simple equations used to update the field are completely explicit, so that no equation system has to be solved. The time stepping process is continued until a steady state solution or desired response has been obtained. The Finite Integration Theory method (FIT) follows a very similar way as FDTD.

The advantages of FDTD are the simplicity of the method, and due to the volume cell structure nonlinear and/or anisotropic materials can be easily modelled. In the meantime surface conforming techniques with non-cubic elements are available to avoid the otherwise necessary staircasing of nonrectangular problem structures. Hereby FDTD is easier to adapt to complex boundaries than TLM. The time domain results can be converted to frequency domain ones by applying Fourier transforms.

The disadvantage of FDTD is generally the fact, that the grid fineness is determined by the smallest objects of the problem that need to be modelled. Recently however subgridding techniques were developed to compensate this disadvantage to a certain extent. Furthermore the grid volume has to be large enough to encompass the problem structure and the near field. The resulting grids may be so large and dense in respective applications, that their size gets out of hand for computation. Attention has to be paid, too, to the outer boundaries of the problem to prevent unwanted reflections, like with FEM. Perfectly matched layers [3] can be used for that and are one of the most important innovations of the last decade.

15.6 The Transmission Line Matrix Method (TLM)

This method is similar to FDTD: analysis is performed in the time domain and the whole problem region is discretized by a grid. This one is however not a double one like with FDTD but a single one. Its nodes are interconnected by virtual transmission lines, so that excitations at source nodes propagate to adjacent nodes through the lines at each time step.

The connection of two neighbouring nodes is realized by a pair of orthogonally polarized transmission lines. Dielectric loads are possible by the introduction of loading nodes with reactive stubs. Lossy media can be realized by losses in the transmission lines or by loading the nodes with lossy stubs. Absorbing boundaries are easily introduced by terminating respective boundary node transmission lines by its characteristic impedance.

Advantages of the TLM method are comparable to those of the FDTD method. Disadvantages of TLM are as well comparable to those of FDTD.

15.7 Other Single Method and Need for Hybrid and Even More Effective

Among other methods, not so often used for EMC and antenna design, the Geometrical Theory of Diffraction (GTD) and the Multiple Multipole Technique

(MMP) should be mentioned here, which are both applied only in high frequency problems.

Extremely important and sometimes the only possibility to solve EMC and antenna design problems are hybrid methods, as shown in the following chapter.

In spite of the mentioned methods above and the possibility of combining them, it should be noted, that they are sometimes too slow to solve the class of complex multiphysical problems needed by research and industry, particularly together with optimisation. Therefore the trend to find more effective methods can be generally observed.

It is an open question if they will be detected or if a progress will be achieved only by improvements in computer architecture and speed.

15.8 Hybrid FEM/BEM – Method for Quasistationary Fields

The shielding of underground power cables can be provided by plates of ferromagnetic material, the nonlinearity of which has to be taken into account. This open problem, containing nonlinear material, is best tackled by a hybrid method with FEM for the plates and BEM for the rest of the infinite space, see Fig. 15.1. Again there is a trend of electromagnetic modelling to optimise an important parameter, in this case the distance h between cable and shielding.

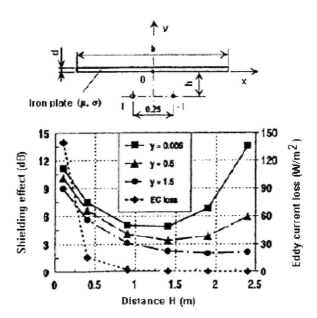

Fig. 15.1 Shielding of a high voltage cable by an iron plate and optimising the distance h .FEM/BEM Method [4].

Different formulations are possible. Among them for general 3D problems an $\vec{A} \times \vec{H}$ - combination proved to be very appropriate ($\vec{A}$: source free vector potential in the eddy current FEM region, $\vec{H}$: magnetic field strength outside in the non-conductive BEM region). An effective coupling of the FEM- and BEM-region is enabled by edge elements in the FEM-region, leading to linear approximations of $\vec{A}$ and tangential $\vec{n} \times \vec{H}$ in the interface elements and to a natural choice of a constant approximation of normal $\vec{n}(\vec{n}\vec{H})$, a behaviour also called "mixed elements".

15.9 Hybrid FEM/BEM Method for Waves

Selected here is the case of a microstrip patch antenna array, shown by Fig. 15.2.

The open problem requires generally the introduction of artificial boundary conditions like Perfectly Matched Layers (PML) [3] or other absorbing boundary conditions., leading to high costs of discretization. This disadvantage can be overcome by applying FEM in the cavity domain only and BEM for the whole rest of the space, so that finally only the cavity needs to be discretized. Weighted residuals applied to the wave equation lead to a system of linear equations for the electric field strength, which typically contains a fully populated BEM submatrix. New strategies to solve this system by the Fast Multipole Method (FMM) enable a fast solution. Hereby typical antenna characteristics like the radiation pattern can be calculated. A good agreement with measurement data is shown in Fig. 15.3. [7]

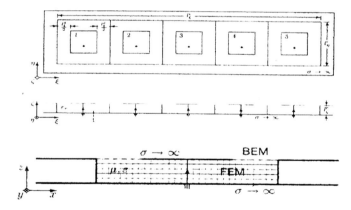

Fig. 15.2 Microstrip patch antenna . Above: Array geometry. Below: Discretization by FEM/BEM

15.10 Hybrid MoM/GTD Method

Like in a typical situation, the operating frequency of a mobile phone transmitting antenna on a roof of a building is 925 MHz, and the building size is 93 λ x 62 λ x 71 λ with the wavelength λ, resulting in a building surface of 32863 λ^2. Under

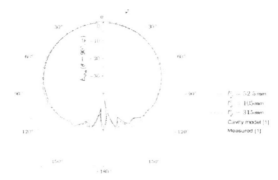

Fig. 15.3 E-plane radiation pattern of a circumferentially polarized single patch antenna

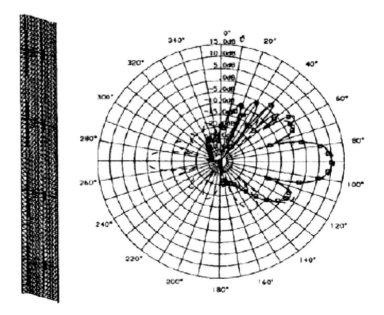

Fig. 15.4 Left: Basis station antenna, discretized into triangular patches. Right: Vertical radiation pattern (dashed; antenna without building influence, line: antenna with building influence). MoM/GTD [5].

these circumstances the application of MoM seems impossible (2.3 millions of basis functions). However, as the dimensions of the building are large compared with the wavelength, GTD (the Geometric Theory of Diffraction) can be applied to the influence of the building's reflections, to be superimposed to the field of the antenna of eight patches in front of a reflector, calculated by MoM alone and discretized by triangular patches, see Fig. 15.4. The resulting hybrid MoM/GTD was successfully tested by different cases in [5].

15.11 Hybrid BEM/IBC and FEM/IBC Method

Eddy current regions are nowadays often thin conductive layers, serving as electromagnetic shieldings in EMC problems.. If FEM would be employed inside such a layer, the elements would become very flat or would exist in a huge quantity resulting in ill-conditioned matrices. However in the thin layer the field can be well approximated by the analytically known solution of a 1D differential equation, leading to Impedance Boundary Conditions (IBC), which connect the fields on both sides outside of the layer, finally computed by BEM. If due to special reasons the region outside the thin conductive layer should be modelled by FEM instead of BEM, the introduction of the IBC into FEM is straightforward and replaces the layer by special sheet elements due to the IBC [8].

Fig. 15.5 shows an example of a cylindrical aluminum shielding enclosure with a gap. As a lot of parameters of influence exist, like size of gap, number and position of gaps, enclosure thickness and material, frequency etc., Computational Electromagnetics is nowadays much superior to study them than building prototypes and / or doing measurement, if possible at all.

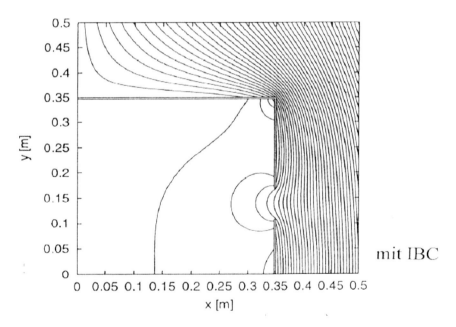

Fig. 15.5 Shielding of a homogeneous field (f=50 Hz) by a cylindrical aluminum enclosure with a gap, flux lines. d=3mm, $\delta/d=4,45$. FEM/IBC [8]

15.12 Non Linear FEM and FEM/IBC with Complex Effective Reluctivity

As the shielding of sources of electromagnetic fields at power frequency becomes more and more important, the need of an appropriate method to calculate the shielding effectiveness is growing, also in cases of strong fields driving the magnetization curves into the nonlinear behaviour. Primarily the average or RMS value of field quantities is of interest with shielding problems and only seldomly its time dependence. Therefore a complex valued effective reluctivity model is chosen for such problems, which allows a sinusoidal calculation in spite of the material´s nonlinearity. The absolute value and the phase of the complex reluctivity are determined from the hysteresis curves of the material under the conditions that the RMS value of $\vec{B}$ and the energy departed during one time period are preserved. The Newton-Raphson method is then adapted to solve the arising equation system [9].

In the case of nonlinear thin shielding materials the linear IBC conditions are extended to nonlinear ones and introduced into FEM. Fortunately hereby the system matrix keeps symmetric also in the nonlinear case. The nonlinear system is now solved by successive approximation [8].

Fig. 15.7 shows the influence of different magnetization curves on the field in the shadow zone of the classical shielding arrangement of Fig. 15.1.

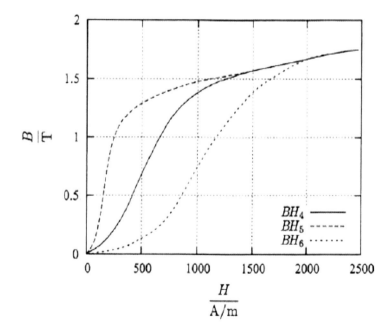

Fig. 15.6 Magnetization curves for strongly differing materials

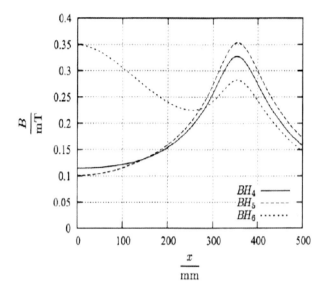

Fig. 15.7 Total flux density (RMS value) above the plate at y= 0.05m. Computed curves correspond to the magnetization curves in Fig. 15.6

15.13 Laser Trimming of IC Resistors Industrial Production Using

High precision resistors play an important role in the modern IC production. They are used to compensate any production process variation and ensure circuits functionality and reliability. Further, it is difficult to manufacture a precise resistor and therefore some of them need to be trimmed. This is aspired by laser trimmings afterwards which has become the most effective and popular method. Due to the high trim costs it is highly desirable to optimize resistor size, shape and trim figure. Approximation methods are insufficient for most situations and thus, robust methods of computational electromagnetics can overcome this problem.

BEM turned out to be the best suited method [6]. Fig. 15.8 shows the most important quantities of the process, the trim characteristics r(β) and trim sensitivity s(β), with β: trim pathway length:

$$r(\beta) = \frac{R(\beta)}{R_0} \Longrightarrow s(\beta) = \frac{dr(\beta)}{d\beta} \qquad (6)$$

15.14 Equation Solvers

Solving the equation systems in the different methods of Computational Electromagnetics is finally a mathematical problem and often the bottleneck of the whole

272 A. Kost

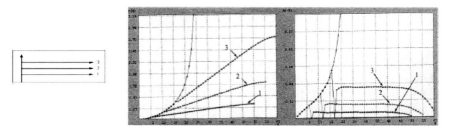

Fig. 15.8 Trim pathways, trim characteristics r(β) and trim sensitivities s(β) for examples of a bar resistor [6]

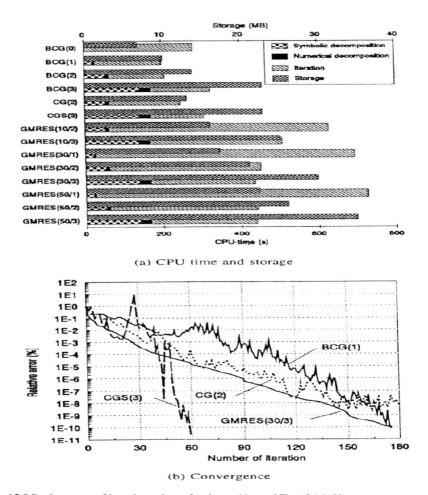

Fig. 15.9 Performance of iterative solvers for the problem of Fig.15.1 [10]

method. Fig. 15.9 shows characteristics of preconditioned iterative solvers for complex and unsymmetrical equation systems, as they arise e.g. for shielding structures in Fig. 15.1, when tackled by FEM [10]. GMRES, the General Method of Residuals, turned out to be one of the most stable solvers for such problems. Conjugate gradient solvers can be significantly faster, but bear the danger of inconvergence.

The introduction of multigrid methods can speed up the convergence enormously compared to the already fast ICCG solvers, as Fig. 15.10 shows [11].

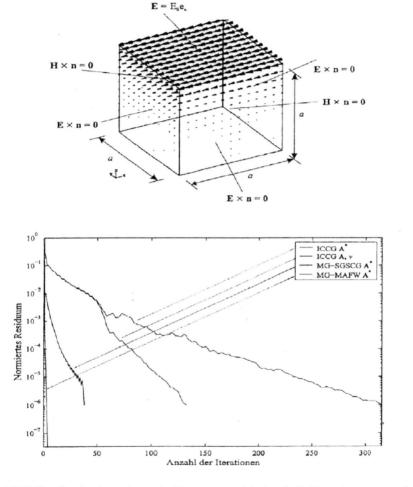

Fig. 15.10 Up: Conductive cube excited by a tangential electric field on the upper surface (sinusoidal time dependence). Bottom: Convergence of the residual for two multigrid solvers and an ICCG solver for different formulations. FEM [11].

15.15 Improvements in Adaptative Mesh Generation

Problems with several regions of very high and low field gradients can be advantageously solved by adaptive mesh generation, hereby saving computational resources (less elements, less unknowns) and improving accuracy. The advantage of adaptive mesh generation is the optimum fitting of the mesh to the problem. A suitable choice of the refinement parameters leads to a mesh generation, which is memory and time efficient. The parameters have to be chosen problem dependent and dynamically, based on local error estimation following the method of Bank/Weiser [12]. An important question is the amount of the chosen refinement percentage in dependence of the refinement step. A respective method was developed in [12], showing results for a switched reluctance motor (SRM) in Fig. 15.11. Surprisingly only the worth 2% elements should be refined during the first steps. This holds for geometrically complex structures containing a lot of edges like SRM. For smoother structures the refinement begins typically with about 10% of the elements.

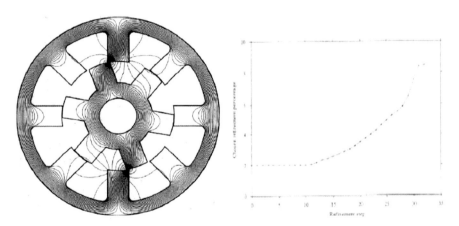

Fig. 15.11 Left: Switched reluctance motor, calculated by FEM and adaptive mesh generation. Right: Chosen percentage in dependence of the refinement step. FEM [12]

15.16 Conclusions

There is no method of electromagnetic modelling which is well suited for any practical application. Depending on the specific problem, however a powerful, "best" method can be found, which often turns out to be a hybrid one. Electromagnetic modelling has become an important tool to solve technical and industrial problems, where electromagnetic fields play a role. Electromagnetic modelling penetrates the design phase of equipment more and more and also begins to accompany the production process online. However, for the class of complex multiphysical problems, particularly in conjunction with optimisation, significant improvement of Computational Electromagnetics is desired.

References

[1] Ali, M.W., Hubing, T.H., Drewniak, J.L.: IEEE Trans. on EMC 39, 304–313 (1997)
[2] Harrington, R.F.: Field Computation by Moment Methods. Macmillan, N.Y. (1968)
[3] Berenger, J.P.: Comput. Phys. 114, 185–200 (1994)
[4] Shen, J., Kost, A.: IEEE Trans. on Mag. 32, 1493–1496 (1996)
[5] Jakobus, U.: Intelligente Kombination verschiedener numerischer Berechnungsverfahren, habilitation thesis, Univ. Stuttgart, Germany (1999)
[6] Schimmanz, K., Kost, A.: Int. Journal of Appl. Electromag. and Mechanics 19, 253–256 (2004) ISSN 1383-5416
[7] Jacobs, R.T., Kost, A., Igarashi, H., Sangster, A.J.: Analysis of planar and curved microstrip antennas. Journ. of Microwaves and Optoelectronics 6(1), 96–110 (2007)
[8] Kost, A., Bastos, J.P.A., Miethner, K., Jänicke, L.: Improvements of nonlinear impedance boundary conditions. IEEE Trans. on Magn. 38(2), 573–576 (2002)
[9] Lederer, D., Kost, A.: Modelling of nonlinear magnetic material using a complex effective reluctivity. IEEE Trans. on Magn. 34(5), 3060–3063 (1996)
[10] Shen, J., Hybler, T., Kost, A.: Comparative Study of Iterative Solvers for the Complex and Unsymmetric System of Equations. In: 7th Biennal IEEE Conference on Electromagnetic Field Computation (CEFC 1996), Okayama, Japan (1996)
[11] Weiß, B.: Multigrid-Verfahren für Kantenelemente, Dr thesis, TU Graz (1992)
[12] Jänicke, L., Kost, A.: Efficiency Properties of Adaptive Mesh Generation. Internat. Journ. of Applied Electromagnetics and Mechanics 13(1-4), 387–392 (2001/2002)

Chapter 16
Methods of Homogenization of Laminated Structures

Nabil Hihat[1,2], Piotr Napieralski[3], Jean-Philippe Lecointe[1,2], and Ewa Napieralska-Juszczak[1,2]

[1] Univ Lille Nord de France, F-59000 Lille, France
[2] UA, LSEE, F-62400 Béthune, France
[3] Technical University of Lodz, Stefanowskiego 18/22 Lodz, Poland

Abstract. This chapter covers the modeling of laminated structure made with grain-oriented electrical steel sheets. It introduced a homogenization technique which takes into account the anisotropy due to laminations rolling direction overlapping and the one due to the crystals orientation. The homogenization method is based on the minimization of energy. That is why, at the beginning of the chapter, some optimization methods are presented. The homogenization method is explained by describing its various stages. Finally, the proposed method is applied to a step lap joint of transformer and results are given.

16.1 Introduction

It is nowadays possible to produce stacks of thin ferromagnetic material layers. The behavior of these multilayer structure differs from that of massive ferromagnetic circuits. Multilayers present numerous applications, particularly in several domains such as the magnetic or magneto-optical recording, the electrical machines and the transformers. Moreover, magnetic cores of power transformers are almost always build with anisotropic laminations. Full 3D numerical modeling of such structures is computationally very demanding, both in terms of long execution times and large memory requirements.

Ferromagnetic cores are often studied simply assuming the structure as homogeneous and neglecting the effects on the field distribution produced by the eddy currents induced in the lamination depth. The reasons of the wide use of this simplifying approach is that a whole study would require a complete three-dimensional analysis, accounting for the phenomena occurring in the lamination plane and in the transversal direction. Taking into account the high number of sheets, their small thickness with respect to the other dimensions and the significant field gradient along the lamination depth produced by the eddy currents (skin effect), this analysis would require a very fine discretization. These methods are generally based on the substitution of the inhomogeneous structure, constituted by lamination and interposed dielectric, by a homogeneous equivalent material obtained by suitably averaging their characteristics. Some of these approaches introduce a permeability tensor in order to deal with the anisotropic magnetic behavior of the laminated structure [1, 2, 3, 4].

Next step consists in extending the homogenization technique to the electrical behavior of magnetic material in order to take into account the effect of eddy currents induced in the laminations [5]. The problem of eddy currents in laminated iron cores has been usually treated in terms of a single-component electric vector potential to represent the laminar flow of eddy currents density. Other possibility is to calculate these eddy currents and associated losses. It requiers to use the magnetic vector potential in the conducting media and a homogeneous material with anisotropic characteristics instead the many thin layers of the iron core.

Some attempts to treat, in a rigorous way, rapidly varying periodic structures have been proposed by authors of [6], but their application to three-dimensional field analysis is not straightforward.

In any case, all the proposed approaches cannot be used in two dimensional (2-D) field analysis, which is often conveniently applied to the study of electrical devices, because the 2-D formulations, which assume the magnetic flux to lie in the rolling plane and the currents to flow in the normal direction, intrinsically neglect the prevalent eddy current components in the considered plane. Following these formulations, the induced currents in the lamination depth and the consequent losses can be at least included a posteriori, starting from the computed field distribution obtained by the 2D analysis [7], but their effects on the field distribution in the rolling plane cannot be evaluated at all. Taking into account the intrinsic limit of the two-dimensional field formulations, the eddy current effects can be included only through the relationship linking the local values of magnetic flux density and field strength. A method based on this idea was recently proposed in [8].

The main idea of the approach presented in this chapter is to replace the complex three-dimensional structure by a homogenized "equivalent" material. A nonconducting homogeneous material is commonly considered for the stacked core and the eddy current losses are estimated in a post-processing step. The iron loss components can also be included directly in the field calculation using the constitutive law [9]. In several methods the eddy currents due to the flux perpendicular to the laminations are neglected. Some possibilities to take into account the effect of eddy currents produced by perpendicular flux is the application of conductivity tensor for the homogenized core [10].

In this chapter, different methods of homogenization are prsented, especially in the domain of electrical machines and transformers. The actual joints of power transformers are made of thin anisotropic laminations (0.18-0.35 mm), of varied rolling directions (RD) and of air gaps. The idea is to represent such a structure by an equivalent material homogenized in two or three dimensions. In the presented method the effects of eddy currents are neglected. This method permits to replace a real 3D structure by an equivalent 2D structure by establishing equivalent characteristics (and effectively "artificial"), especially to make them suitable for use in combination with commercial field modelling softwares. This model neglects the effects of hysteresis and the the normal flux influence (i.e. crossing the laminations) on the characteristics, although it is accepted that its presence and that of the associated induced currents may have an effect on the magnetic field distributions. Thus in the following derivations of the equivalent characteristics, it assumed that the normal flux density B_z is negligible compared to the components in the sheet plane : B_x and B_y. Consequently, the effects of eddy currents are also neglected. Such an approximation will inevitably causes some inaccuracies. Even if the component perpendicular to the sheets is typically only a few percent of the tangential components, because it travels through large areas. However, comparison with 3D simulations and measurements provided later that overall errors associated with our 2D treatment not exceed 3 to 5% (except in some local positions close to internal air gaps). Thus this initial simplifying assumption may be considered as acceptable for practical purposes [11]. The advantage of the proposed approach using an equivalent material and its magnetic characteristic is that it allows the assessment of resultant overlapping structures at the design stage, without the need to undertake very expensive 3D calculations. Moreover, the material homogenization and introduction of equivalent anisotropic characteristics make it possible to use the method in combination with commercial software where modifications to the source code are normally not available.

16.2 Optimization Methods

Most practical problems have several solutions and the optimal solution is, most of the time, researched. Sometimes an infinite number of solutions may be reached. The optimal solution consists in choosing the best element among some set of available alternatives. In mathematics, the term "optimization" refers to the problem of finding the minimum specified to a function [13]. The basic example of optimization amounts to adjust variables of a function to minimize scalar quantity of performance criterion:

$$minimize\ F = f(x_1, x_2, ..., x_n) \qquad (1)$$

where F is a scalar quantity of performance criterion. Variables x_1, x_2, x_n are the parameters of criterion (also called optimization variables). An optimization problem can be represented in the following way:

$$f : \mapsto R^n \qquad (2)$$

f is a function from some set A where $A \subset R^n$. This function is called objective function. The optimization task is to find feasible solution for maximization:

$$x^* \in A, \forall x \in A - \{x^*\} \Rightarrow f(x) < f(x^*) \qquad (3)$$

When the negative of that function is minimized:

$$x^* \in A, \forall x \in A - \{x^*\} \Rightarrow f(x) > f(x^*) \qquad (4)$$

The satisfiability problem, also called the feasibility problem, is to find any attainable solution without regard to objective value. This optimal point or a solution is a point that satisfies these inequalities and it has the smallest objective value among all vectors fulfilling the constraints.

16.2.1 Solving Optimization Problem

General optimization problem is very difficult to resolve ; many of even easy optimization problems cannot be solved analytically and they has a non-deterministic polynomial-time. Methods used to solve a general optimization problem have two main categories: global optimization (mostly based on convex optimization) and optimization with local solution (with only approximate results). Linear problems where the number of independent parameters does not exceed two or three, can be solved analytically [15]. These methods do not need the use of a digital computer and often give global solution. A similar method with comparable limitation is the Graphical method, where the function can be plotted to be maximized or minimized. However, most of optimization problems cannot be analytically solved. Therefore, to solve this problem in computer science, it is common practice to evaluate the performance of optimization algorithms on the basis of the experimental approach or by iterative numerical procedures to generate a series of progressively improved solutions to the optimization problem. The system is set up and the optimization variables are adjusted in the sequence and the performance criterion is measured in each case. The process is completed when some convergence criterion is satisfied. This technique may lead to optimum or near optimum operating conditions. The most important general approach to optimization in computer science is based on numerical methods. Numerical methods can be used to solve highly complex optimization problems. The discipline encompassing the theory and the practice of numerical optimization methods is known as mathematical programming.

16.2.2 Linear Programming

The first optimization technique is the Linear programming [14]. It is a technique which consists in minimizing a linear function subject to a multitude of linear inequalities. There is no analytical formula for this solution. Linear programs can be expressed in the following canonical form:

minimal $c^T x$
subject to $a_i^T x \leq b_i, i = 1, 2, ..., m$

where x represents the vector of optimization variables. $c \in R^{n \times 1}$ and $b \in R^{p \times 1}$ are vectors of coefficients and $a \in R^{p \times n}$ is a matrix of coefficients. Value of a, b and c are given.

The first method for the solution of this kind of problems is known as **the simplex method** (developed by George Dantzig in 1947). The name of this method comes from the simplex, which is a generalization of convex shapes of the triangle for more dimensions. The simplex algorithm requires the linear programming problem to be in augmented form, so that the inequalities are replaced by equalities:

$$f(x,y,t) = c^T x + o^T y + (-M) j^T t \tag{5}$$

Where $x = [x_1, x_2, , x_n] \in \chi$ are the introduced slack variables from the augmentation process.

- We consider a point polyhedron $\chi = \{x \in R^n : Ax = b, x \geq 0\}$.
- If $\chi \neq 0$ then χ has, at least, one extreme point.
- If $min\{cx : Ax = b, x \geq 0\}$ has an optimal solution then it has an optimal extreme point solution.

The algorithm steps:
1. **Initialize:**
   - $x_0 :=$ extreme point of χ
   - $k := 0$
2. **Iterative step:**
   do
   if for all edge directions D_k at x_k, the objective function is non-decreasing
   then exit and return optimal x_k **else** pick some d_k in D_k such that $cd_k < 0$
   if $d_k \geq 0$ then declare the linear program unbounded in ob-jective value and exit
   else $x_{k+1} := x_k + \theta_k d_k$, where $\theta_k = max\{\theta : x_k + \theta d_k \geq 0\}$
   $k := k + 1$
3. **End**

It is a quite efficient algorithm and it can guarantee to find the global optimum if certain precautions against cycling are taken. The condition is that, in sequence, each new vertex in the objective function improves or is unchanged. In all vertex value of objective function is verified. When it get worse value, this vertex are discarded. The next iteration step consists in the next value, located on one edge of the vertex which has already rediscovered. Iteration ends when the next vertex is viewed as the best in terms of the respective values of the objective function. The first worst-case polynomial-time algorithm for linear programming is the ellipsoid algorithm (developed by Leonid Khachiyan in 1979). The classical ellipsoid algorithm solves nonlinear programming problems for convex real-valued functions. The ellipsoid algorithm generates a sequence of ellipsoids E_k, each guaranteed to contain x^*, with the property that their volumes shrink to zero as the terms of a geometric progression. The starting ellipsoid $E_0 = \left\{x \in R^n, (x - x^0)^T Q_0^{-1} (x - x^0) \leq 1\right\}$ is the smallest ellipsoid containing U and L, with x_0 the midpoint of the bounds and Q_0 positive definite and symmetric. At each iteration, the algorithm finds the normalized gradient g of the objective function $f_0(x)$: if x_k is achievable or of a violated constraint $f_0(x)$ if x_k is infeasible. To calculate a direction:

$$\mathbf{d} = -\frac{Q_k g}{\sqrt{g^T Q_k g}} \tag{6}$$

Using **d**, we find the next ellipsoid using the updates. The algorithm goes on to generate a sequence of progressively smaller ellipsoids, each of which containing the minimizer. After a sufficiently large number of iterations, the volume of the ellipsoid shrinks to zero and the minimizer is localized. The algorithm steps are:

1. **Initialize:**
   - $N := N(Q)$
   - $R := R(Q)$
   - $Q_0 := R^2 I$
   - $x_0 := 0$
   - $k := 0$
2. **Iterative step:**
   while $k < N$
   call Strong Seperation (Q, x^k)
   if $x^k \in Q$
   halt
   else hyperplan $\{x \in R^d | g^T x = g_0\}$ separates x^k from Q
   update
   $\mathbf{b} := \frac{Q_k g}{\sqrt{g^T Q_k g}} x^{k+1} := x^k - \frac{b}{d+1} v$
   $Q_{k+1} := \frac{d^2}{d^2-1} \left(Q_k - \frac{2}{d+1} bb^T \right)$
   $k := k+1$ **endwhile**
3. **Empty polydron:**
   - **halt** and **declare** Q is empty
4. **End**

In order to solve a problem which has n variables and can be encoded in L input bits, this algorithm uses $O(n^4 L)$ pseudo-arithmetic operations on numbers with $O(L)$ digits. We can observe that in practice the simplex algorithm is faster than the ellipsoid method.

The next faster and more efficient for linear programming is **the interior-point method** (developed by N. Karmarkar in 1984) [18]. In contrast to the simplex algorithm, which finds the optimal solution by progressing along points on the boundary of a polyhedral set, interior point methods move through the interior of the feasible region. This method claimed to be up faster than the simplex method, in practice and for any instance $O(n^{3.5} L)$. It is based on the analytic center yc of a full dimensional polyhedron $D = \{c : A^T y + z = c\}$, the logarithmic barrier formulation of the dual D with positive barrier parameter μ is:

$$D_\mu = \max \left\{ b^T y + \mu \sum_{j=1}^n \ln(z_j) : A^T y + z = c \right\} \tag{7}$$

This method starts with μ at some positive value and approaches zero. We can define diagonal matrix that:

$$X = diag\{x_1^0, ..., x_n^0\}, Z_T = diag\{z_1^0, ..., z_n^0\} \text{ and } e^T = (1, ..., 1)$$

The optimality conditions for D_μ are given by:

$$\begin{cases} Ax - b = 0 \\ A^T y + z - c = 0 \\ XZe - \mu e = 0 \end{cases} \tag{8}$$

The Newton method gives the following system of equations:

$$\begin{bmatrix} A & 0 & 0 \\ 0 & A^T & 1 \\ Z & 0 & X \end{bmatrix} \begin{Bmatrix} \delta x \\ \delta y \\ \delta z \end{Bmatrix} = \begin{Bmatrix} d_p \\ d_D \\ \mu e - XZe \end{Bmatrix} \quad (9)$$

The strategy is to take one Newton step, toreduce μ and to iterate until the optimization is complete. The algorithm steps:

1. **Initialize:**
   - X, T
   - $k := 0$
   - $0 \leq \alpha_{(P,D)} \leq 1$
2. **Iterative step:**
   while $k < N$
   compute:
   $AZ^{-1}XA^T$
   $\delta z = d_D A^T \delta y$
   $\delta x = Z^{-1}[\mu e - XZe - X\delta z]$
   set:
   $\alpha_p = 0.994 min \left\{ \frac{-x_j}{\delta x_j} | \delta x_j < 0 \right\}$
   $\alpha_D = 0.994 min \left\{ \frac{-z_j}{\delta z_j} | \delta z_j < 0 \right\}$
   $x^{k+1} = x^k + \alpha_p \delta x$
   $y^{k+1} = y^k + \alpha_D \delta y$
   $z^{k+1} = z^k + \alpha_D \delta z$
   $k = k + 1$
   endwhile
3. **End**

Interior point methods have permanently changed mathematical programming theory. The importing characteristic of this method is that the number of iterations depends very little on the size of the problem. Many interior point methods have been proposed and analyzed. Of course methods depend on the problem but we can say that ipm method is currently most popular in LP.

16.2.3 *Nonlinear Programming*

It is a technique for the minimizing of a function subject, where some of the constraints or the objective function are nonlinear. There is no analytical formula for this solution. We can define four numbers of specific cases. These problems can be solved by using a variety of methods such as penalty and barrier function methods, gradient projection methods, and sequential quadratic programming (SQP) methods.

The first case is when the objective function f and some constraints g are convex functions. This kind of programming is called **convex programming**. This task has two interesting variations to simplify the solution. We have to find:

$min(f(x))$
with linear constraints:
$Ax \geq b$ and $x \geq 0$
where:
A is a matrix $m \times n$ and $\dim b = m$.

Secondly, the objective function is expressed as a sum of linear form and quadratic form. This kind of programming is called **Quadratic Programming Methods**. Restrictions are linear, convex programming problem is reduced to minimize:

$f(x) = \langle c, x \rangle + \langle x, Dx \rangle$
while constraints:
$Ax \geq b$ and $x \geq 0$
where:
D is a symetric matrix $n \times n$, $\dim x = n$ and $\dim b = m$.

The second case is when the objective function f and some constraints g_i, $i = 1, , m$ has the Distributive Property. It can be expressed as a sum of n components which is a function of only one variable x. This kind of programming is called **Separable Programming Methods**. This task can be written as:

$min f(x) = \sum_{j=1}^{n} f_j(x_j)$
with constraints:
$g_i(x) = \sum_{j=1}^{n} g_{ij}(x_j) \leq 0, i = 1,...,m$
$x_j \geq 0, j = 1,...,n$.

The next case is when the objective function f and some constraints g_i, $i = 1, , m$ are polynomials of positive. This kind of programming is called **Geometric Programming Methods** and the task can be written as:

$min(f(x))$
with constraints:
$g_i(x) \leq 1, i = 1,...,m,$
$x > 0.$

The last case is when the objective function f and some constraints g_i, $i = 1, , m$ are linear but variables x_j can get only integer values. This kind of programming is called **Integer Programming Methods**. This task can be written as:

$min(f(x)) = \langle c, x \rangle$
with constraints:
$Ax \geq b, x \geq 0$ and $x \in \zeta$
where:
ζ is the set integers an A is $m \times n$ matrix, $\dim c = n$ and $\dim b = m$.

Many powerful techniques developed for constrained optimization problems are based on unconstrained optimization methods. A more recent development in non-convex constrained optimization is the extension of the modern **interior-point approaches** [17]. In this method, we consider the general nonlinear programming problem:

$min\,(f(x))$,
with constraints:
$h(x)=0$,
$g(x)-s=0$
$s \geq 0$.
where:
$\nabla f(x) + \nabla h(x) y - \nabla g(x) w = 0$
$w - z = 0$
$h(x) = 0$
$g(x) - s = 0$
$ZSe = 0$
$(s,z) \geq 0$

At finite iterations on while k let vector $v_k = (x_k, y_k, s_k, w_k, z_k)$, we obtain correction $\Delta v_k = (\Delta x_k, \Delta y_k, \Delta s_k, \Delta w_k, \Delta z_k)$ corresponding to the μ_k, as the solution of the perturbed Newton linear system:

$$F'_\mu(v_k)\Delta v = -F_\mu(v_k) \tag{10}$$

For various components of v_k, we construct the expanded vector of steplengths:

$$\alpha_k = (\alpha_x, ..., \alpha_x, \alpha_y, ..., \alpha_y, \alpha_s, ..., \alpha_s, \alpha_w, ..., \alpha_w, \alpha_z, ..., \alpha_z)$$

Hence the subsequence iterate $v_k + 1$ can be written as:

$$v_{k+1} = v_k + \Lambda_k \Delta v_k \tag{11}$$

where: $\Lambda_k = diag(\alpha_k)$

The algorithm steps:

1. **Initialize:**
   - $v_0 = (x_0, y_0, s_0, w_0, z_0)$, where $(s_0, w_0, z_0) > 0$
   - $k := 0$
   - $\Lambda_k = diag(\alpha_k)$
2. **Iterative step:**
   while $k < N$
   test for convergence
   take $\mu_k > 0$
   solve linear system for $\Delta v = (\Delta x, \Delta y, \Delta s, \Delta w, \Delta z)$
   Compute:
   $\alpha_s = -\dfrac{1}{min\left((S_k)^{-1}\Delta s_k, -1\right)}$
   $\alpha_w = -\dfrac{1}{min\left((W_k)^{-1}\Delta w_k, -1\right)}$
   $\alpha_z = -\dfrac{1}{min\left((Z_k)^{-1}\Delta z_k, -1\right)}$
   take $\tau \in [0,1]$, $\alpha_p \in [0,1]$, $\beta \in [0,1]$:
   $\varphi(v_k + \Lambda_k \Delta v) \leq \varphi(v_k) + \beta \alpha_p \nabla \varphi(v_k)^T \Delta v_k$
   $\alpha_x = \alpha_p$

$$\alpha_y = \alpha_p$$
$$\alpha_s = min\,(1, \tau_k, \alpha_s)$$
$$\alpha_w = min\,(1, \tau_k, \alpha_w)$$
$$\alpha_z = min\,(1, \tau_k, \alpha_z)$$
set:
$$v_{k+1} = v_k + \Lambda_k \Delta v_k$$
$$k := k+1$$
endwhile

3. **End**

This algorithm solves many problems and, in most cases, the convergence. Many optimization algorithms need to start from a feasible point. One way to obtain such a point is to relax the feasibility conditions using a slack variable.

16.2.4 Unconstrained Optimization Problems

The usefulness of the unconstrained optimization algorithms applied to engineering problems considers an optimization method for finding local optima of smooth unconstrained optimization problems. We examine a problem of homogenization technique. In other words, the homogenization problem can be converted into an unconstrained minimization problem. Many powerful techniques are based on unconstrained optimization methods. If the constraints are simply given in terms of lower and/or upper limits on the parameters, the problem can at once be converted into an unconstrained problem. In its most general form, a constrained optimization problem is to find a vector x^* that solves the objective function $f : R_n \to R$, into $R\{+\infty\}$:

$$min_{x \in R^n} (f(x)) \tag{12}$$

Numerical methods for unconstrained optimization methods can be divided into two main groups:

- Stochastic methods
- Deterministic methods

These methods are different but on the basis of certain characteristics, can be divided into several types, assuming the following two criteria of division:

- Conjugate gradient methods,
- The partitioned quasi-Newton methods for large scale optimization,
- Newton's methods,
- Descent methods.

In all cases, iterations of the following form are considered:

$$x_{k+1} = x_k + \alpha_k d_k$$
$$d_k = -g_k + \beta_k d_{k-1} \tag{13}$$

where d_k is a search direction, α_k is a steplength by means of one-dimensional search, β_k is chosen.

The other broad class method that defines the search direction is:
$$d_k = -B_k^{-1} g_k$$
Where B_k is a nonsingular symmetric matrix. In special cases are given by:

$B_k = I$ (descent methods),
$B_k = \nabla^2 f(x_k)$ (Newton's methods).

The positive steplenght α_k determinates two conditions:
$$\begin{aligned} f(x_k, +\alpha_k d_k) &\leq f(x_k) + \sigma_k \alpha_k g_k^T d_k \\ g(x_k, +\alpha_k d_k)^T d_k &\geq \sigma_2 g_k^T d_k \end{aligned} \quad (14)$$

where: $0 < \sigma_1 < \sigma_2 < 1$

All the described methods are iterative methods. Most of the global convergence analyses use Zoutendijk's condition explicitly or follow similar approaches. Many variable metric methods are super-linearly convergent and this is proved by simply verifying that:
$$\alpha_k d_k = d_k^N + o\left(\|d_k^N\|\right) \quad (15)$$

A popular strategy called backtracking consists in successively decreasing the steplength starting from an initial guess until a sufficient function reduction is obtained.

The method of **Steepest Descent** is the simplest of the gradient methods. The choice of direction is where f decreases most quickly, which is in the direction opposite to $\nabla f(x_i)$. The search starts at an arbitrary point x_0 and slide down the gradient until close enough to the solution. The algorithm steps:

1. **Initialize:**
    - $k := 0$
    - $x := x_0$
    - $found = false$
2. **Iterative step:**
    while $(k < k_{max})$ or $found$
    Compute:
    $h_d := search_{direction(x)}$
    if no such h exist
    $found := true$
    else
    find step length α
    $x := x + \alpha h_d$
    $k := k + 1$
    $found := update\,(found)$
    endwhile
3. **End**

These methods produce series of steps leading from the starting position to the final result and the directions of the steps are determined by the properties of $f(x)$ at the current position. When the iteration steps are determined from the properties of a model of the objective function inside a given region we consider the **Descent method region methods** is considered. The steps of the algorithm are the following:

1. **Initialize:**
   - $k := 0$
   - $x := x_0$
   - $found = false$
   - $\Delta := \Delta_0$
   - $q(h) = f(x) + h^T f'(x)$
2. **Iterative step:**
   while $(k < k_{max})$ or $found$
   Compute:
   $k := k + 1$
   $h_{tr} := argmin_{h \in d} \{q(h)\}$
   $r := \frac{f(x) - f(x+h)}{q(0) - q(h)}$
   if $r > 0.75$
   $\Delta := 2\Delta$
   if $r < 0.25$
   $\Delta := \frac{\Delta}{3}$
   if $r > 0$
   $x := x + h_{tr}$
   Update $found$
   endwhile
3. **End**

Currently very popular optimization methods are **new line search methods** or modifications to known methods. An advantage of soft line search over exact line search is that it is the fastest of the two. The result of exact line search is normally a good approximation to the result, and this can make descent methods with exact line search find the local minimizer in less iteration than what is used by a descent method with soft line search.

The steps of the algorithm are the following:

1. **Initialize:**
   if $\left(\phi'(0) \geq 0\right)$
   $\alpha = 0$ **else** $k := 0$
   $\gamma := \beta \phi'(0)$
   $a := 0$
   $b := min\{1, \alpha_{max}\}$
2. **Iterative step:**
   while $(\Phi(b) \leq \lambda(b))$ **and** $\left(\Phi'(b) \leq \gamma\right)$ **and** $(b < \alpha_{max})$ **and** $(k < k_{max})$
   Compute:
   $k := k + 1$
   $a := b$
   $b := min\{2b, \alpha_{max}\}$

$\alpha := b$
endwhile while $(\phi(\alpha) > \lambda(\alpha))$ **and** $\left(\phi'(\alpha) \leq \gamma\right)$ **and** $(k < k_{max})$
Compute:
$k := k+1$
Refine α *and* $[a,b]$
if $(\phi(\alpha) \geq \lambda(0))$
$\alpha := 0$
endwhile
3. **End**

The Optimization line search is a very popular method and is used more and more often. Next methods of practical importance are **conjugate gradient methods**. The conjugate gradient methods are simple and easy to implement, and generally they are superior to the steepest descent method.

The steps of the algorithm are the following:

1. **Initialize:**
   - $k := 0$
   - $x := x_0$
   - $found = false$
   - $\gamma := 0$
   - $h_{cg} := 0$
2. **Iterative step:**
   while $found$ **or** $(k > k_{max})$
   Compute:
   $h_{prev} := h_{cg}$
   $h_{cg} := -f'(x) + \gamma h_{prev}$
   if $\left(f'(x)^T h_{cg} \geq 0\right)$
   $h_{cg} := -f'(x)$
   $\alpha := line\ search\ (x, h_{cg})$
   $x := x + \alpha h_{cg}$
   $\gamma := ...$
   $k := k+1$
   $found := ...$
   endwhile
3. **End**

These methods are simple and easy to implement, but **Newton's method** and its relatives are usually even better. If, however, the number of variables n is large, then the conjugate gradient methods may outperform Newton-type methods. Class of methods for unconstrained optimization which is based on Newton's method is called **Quasi-Newton methods**. The reason is that the latter rely on matrix operations, whereas conjugate gradient methods only use vectors. Ignoring sparsity, Newton's method needs $O(n3)$ operations per iteration step, Quasi-Newton methods need $O(n2)$, but the conjugate gradient methods only use $O(n)$ operations per iteration step.

The steps of the algorithm are the following:

1. **Initialize:**
   - $k := 0$
   - $x := x_0$
   - $found = false$
   - $\mu := \mu_0$
2. **Iterative step:**
   while $found$ **or** $(k > k_{max})$
   while $\left(f''(x) + \mu I < 0\right)$
   Compute:
   $\mu := 2\mu$
   $Solve \left(f''(x) + \mu I\right) h_{dn} = -f'(x)$
   $Compute Compute\ gain\ factor\ r$
   if $(r > \delta)$
   $x := x + h_{dn}$
   $\mu := \mu max\left\{\frac{1}{3}|1 - (2r-1)^3\right\}$
   else $\mu := 2\mu$
   $k := k + 1$
   $Updte\ found$
   endwhile
   endwhile
3. **End**

The more efficient modified Newton methods are constructed as either explicit or implicit hybrids between the original Newton method and the method of steepest descent. The idea is that the algorithm in some way should take advantage of the safe, global convergence properties of the steepest descent method whenever Newton's method gets into trouble.

There is no "Panacea" equation or method. Optimization provides a general framework in which a variety of design criteria and specifications can be readily imposed on the required solution. In a real life design problem, the design is carried out under certain physical limitations with limited resources. If these limitations can be quantified as equality or inequality constraints on the design variables, then a constrained optimization problem can be formulated whose solution leads to an optimal design that satisfies the limitations imposed.

16.2.5 Direct Search Method: Hook-Jeeves

The tasks of minimizing the energy of the magnetic field in anisotropic structures such as joints and air gaps in the transformer core correspond to Direct Search Method. In each iteration the value function is checked in several points contained in the vicinity of the point x_i. On this basis point x_{i+1} is found with the next iteration. For the solution of the problem the Hooke-Jeeves method is chosen. This method is not very effective, but it is reliable. Reliability follows from the insensitivity to interference, and most importantly from insensitivity to the choice of starting point. This is a very important feature in the calculation of the distribution of flux density. A useful result of this minimizing procedure is not the minimum value of energy, but the homogenized reluctivity of the structure. This value is found many times during the calculation of the magnetic field in a very large number of points of the structure. Frequent lack of possibilities of obtaining it requires a simplified method. This causes instability or inaccuracy of the calculation. This may lead to the possibility of not obtaining the final solution. Hooke-Jeeves method is not sensitive to the choice of starting point and, therefore, it is often slower than the method for finding the minimum of a non-differentiable function. However, it gives satisfactory results in almost every situation. The reduced speed of the method is not a factor to determine the characteristics of the group for the replacement structure. Indeed equivalent characteristics are determined only once for a given type of structure and specific type of materials. The structure type is characterized by the ratio of the number of sheets in each layer. This will be explained in detail later in this chapter.

For example, for splicing structure the characterized type of transformer is the ratio of yoke and column sheets, the replacement ratio for the number of yoke, column and also the number of layers of air in the virtual gap. The number of different structures encountered in the real core of transformers is not large. This will allow the creation of a database with the groups of characteristics.

Hook-Jeeves method is characterized by a very simple algorithm.

The pattern search method of Hook and Jeeves is a sequential technique in which each step consists in two kinds of moves, one called exploratory move and another called as pattern move. The first move is done to explore the local behavior of the objective function and the second move is made to take advantage of the pattern direction. The general procedure can be described by the following steps:

1. Start with an arbitrarily initial point $X_1 = [x_1\ x_2\ x_n]^T$, called the starting base point and prescribed step lengths Δx_i in each of the coordinate directions u_i, $i = 1, 2, ..., n$. set $k = 1$.
2. Compute $f_k = f(X_k)$. Set $i = 1$ and define new variable with initial value set as, $Y_{k,0} = X_k$ and start the exploratory move as stated in step 3.
3. The variable x_i is perturbed about the current temporary base point $Y_{k,i-1}$ to obtain the new temporary base point as follows:

$$Y_{k,i} = \begin{cases} Y_{k,i-1} + \Delta x_i U_i & \text{if } \begin{array}{l} f^+ = f(Y_{k,i-1} + \Delta x_i U_i) \\ < f = f(Y_{k,i-1}) \end{array} \\ Y_{k,i-1} - \Delta x_i U_i & \text{if } \begin{array}{l} f^- = f(Y_{k,i-1} + \Delta x_i U_i) \\ < f = f(Y_{k,i-1}) \\ < f^+ = f(Y_{k,i-1} + \Delta x_i U_i) \end{array} \\ Y_{k,i-1} & \text{if } f = f(Y_{k,i-1}) < \min(f^+, f^-) \end{cases} \quad (16)$$

This process of finding the new temporary base point is continued for $i = 1, 2, ...$ until x_n is perturbed to find $Y_{k,n}$.

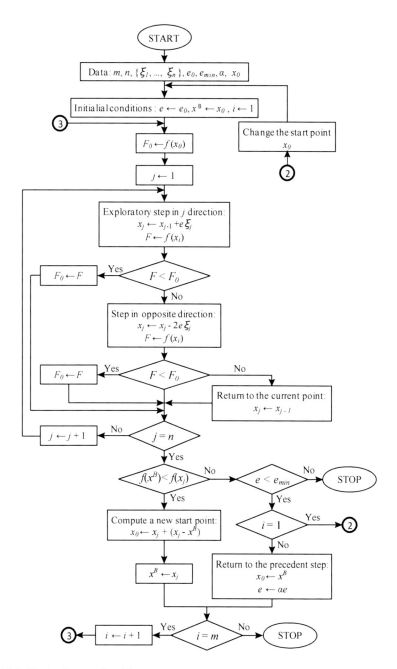

Fig. 16.1. Hook - Jeeves algorithme

4. If the point $Y_{k,n}$ remains same as the X_k, reduce the step lengths Δx_i (say by a factor of two), set $i = 1$ and go to step 3.
   If $Y_{k,n}$ is different from X_k, obtain the new base point as $X_{k+1} = Y_{k,n}$ and go to step 5.
5. With the help of the base points X_k and X_{k+1} establish a pattern direction S as
   $S = X_{k+1} - X_k$
   and find a point $Y_{k+1,0}$ as $Y_{k+1,0} = X_{k+1} + lS$
   The point $Y_{k,j}$ indicates the temporary base point obtained from the base point X_k by perturbing the jth component of X_k.
   where l is the step length which can be taken as 1 for simplicity.
6. set $k = k + 1$, $f_k = f(Y_{k,0})$, $i = 1$ and repeat step 3, if at the end of step 3, $f(Y_{k,n}) < f(X_k)$, we take the new base point as $X_{K+1} = Y_{k,n}$, and go to step 5. On the other hand if $f(Y_{k,n}) \geq f(X_k)$, set $X_{k+1} = X_k$, reduce the step length Δx_i, set $k = k + 1$ and go to step 2.
7. The process is assumed to be converged whenever the step lengths fall below a small quantity e, thus the process is terminated if $max(\Delta x_i) < e$

The algorithm of Hook-Jeeves method is shown in Fig. 16.1, where:

- n is the number of unknowns.
- m is the limit of iterations number.
- e is the initial step of change of unknowns value.
- $e = 0.1T$ for the structures of joints of different RD, 0.01T for structures containing air gaps.
- e_{min} is the permissible length of the step being used in the criterion $e < e_{min}$ end of an iterative process. The energy minimization assumed $e_{min} = 0.001T$.
- $0 < \alpha < 1$ is the length correction factor change step. In our calculations $\alpha = 0.618$, with base mutually orthogonal directions, e.g. for $n = 2$: $\xi_1 = [1,0]$ and $\xi_2 = [0,1]$.
- $X = [x_1, x_2, ...x_n]$ vector of independent variables.
- x_0 and x^B starting point and base point.

The objective function values are calculated in our case according to algorithms presented later in this chapter.

16.3 Homogenization Method Principle

The developed homogenization method is based on the discretization of an anisotropic steel sheets stack. A volume made of different materials - air, sheets with different angles of magnetization relative to their rolling direction - is replaced by a homogeneous material which is magnetically equivalent. The objective is to determine the equivalent magnetic characteristics of the original structure. Thus, a complex 3D structure can be replaced by a 2D simplified structure called quasi-3D model, providing an obvious reduction of simulation time. For example, Fig. 16.2 shows the equivalence of a step-lap 90° joint. The equivalent magnetic characteristics can be used to model in 2D the magnetic behavior of real 3D structures. The

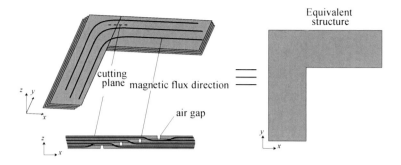

Fig. 16.2. Scheme of equivalence between a real 3D structure (left) and a quasi-3D structure (right)

time savings for the geometry capture and calculations are considerable. That allows repeating the simulation many times to optimize a structure with one or more specific criteria.

The discretization of real 3D structures requires to distinguish between the regions that are not made with the same material. The example in Fig. 16.2 allows to explain the method. In this case, the macrostructures are defined by a cut in the plane (z,x). The Fig. 16.3 shows the step lap arrangement. To apprehend the homogenization method, two basic concepts have to be considered: the macrostructure and microstructure. In the considered problem the macrostructure is a complete assembly of the layers in the overlap, while the microstructure is a repeatable structure of layers: the column sheet layer, the yoke sheet layer and the air gap. In Fig. 16.3, step lap arrangements, can be constituted of one or more sheets.

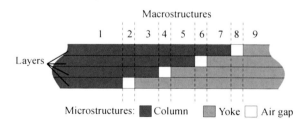

Fig. 16.3. Macrostructure and microstructure in a step lap 90°

The homogenization technique enables the replacement of heterogeneous structure by an equivalent homogeneous one. The homogenization can be performed in three stages. The first one is the discretization, where we assume that the division of magnetic flux density vector through the microstructures results from the tendency to reach the minimum of the magnetic field energy in the macrostructure. Then, the magnetic reluctivity of the real material is measured in different directions regarding to the rolling direction and the result is extrapolated to attain 2T. In the final stage, the equivalent reluctivity of homogeneous 2-D structure (which replaces the real 3-D structure of the overlap) is calculated.

To determine the equivalent magnetic characteristics of the macrostructres, the method uses the principle of the energy minimizing W_r stored in a macrostructure which has an equivalent reluctivity v_r and volume V_r. W_r depends on the magnetic energy of the n materials which constitute the macrostructure. Each material (cell) is characterized by:

- an index i,
- a volume V_i,
- a number of layers n_i,
- a reluctivity v_i,
- and a magnetic energy W_i.

$\vec{H_i}$ and $\vec{B_i}$ are respectively the magnetic field and the magnetic flux density of a cell i. $\vec{H_r}$ and $\vec{B_r}$ are the similar quantities of the macrostructure r. α_i determines the angle of the vector $\vec{B_i}$ regarding to the axis x. Similarly, α_r is the angle between $\vec{B_r}$ and the axis x.

Under these conditions, W_r can be written:

$$W_r = \sum_{i=0}^{n} W_i\left(\vec{B_i}, v_i\right) = \frac{1}{2} V_r \vec{B_r}\vec{H_r} = \frac{1}{2} V_r B_r^2 v_r(B_r, \alpha_r) \qquad (17)$$

The energy stored in any step of the overlap (the cell) is given by:

$$W_i = \frac{1}{2} V_i \vec{B_i}\vec{H_i} = \frac{1}{2} V_i B_i^2 v_i(B_i, \alpha_i) \qquad (18)$$

Equations (17) and (18) take into account the anisotropy of the material: in the plan (x,y). The values of α_r and B_r are imposed and they define a point of the equivalent magnetic characteristic $v_r(B_r, \alpha_r)$.

Denoting N the number of elements in a macrostructure, the expression (19) gives $\vec{B_r}$ with $\vec{B_i}$. The indications x and y identify the components of the different vectors in the plane (x,y).

$$\vec{B_r} = \frac{1}{N} \sum_{i=0}^{n} n_i \vec{B_i} = \begin{pmatrix} \frac{1}{N} \sum_{i=0}^{n} n_i B_{ix} \vec{x} \\ \frac{1}{N} \sum_{i=0}^{n} n_i B_{iy} \vec{y} \end{pmatrix} \qquad (19)$$

To define the magnetic flux density vectors in the microstructures, the relation (20) is the objective function to minimize and the equality (19) is considered as the constraint. The equivalent reluctivity of the macrostructure is ginen by equation (21).

$$f_{min} = \sum_{i=0}^{n} n_i B_i^2 v_i(B_i, \alpha_i) \qquad (20)$$

$$v_r(B_r, \alpha_r) = f_{min}/B_r^2 \qquad (21)$$

The proposed model requires to know the real magnetic characteristics of the material $v_i(B_i, \alpha_i)$. Before their use in the calculations, a preliminary stage is required.

16.3.1 Magnetic Characteristics $v_i(B_i, \alpha_i)$

The magnetic characteristics $B(H, \theta)$, of the column and the yoke elements are measured with the Epstein frame for different angles θ regarding to the rolling direction. Fig. 16.4 shows the measurements results for $\theta = 0, 5, 10, 15, 20, 30, 35, 45, 55, 60, 75$ and $90°$.

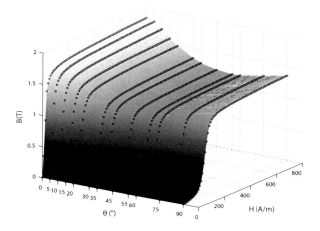

Fig. 16.4. Magnetic characteristics of a GO sheet $B(H, \theta)$. The measurement points are marked with 'o'

The characteristic $v(B, \theta)$ is calculated from the measurements of $B(H, \theta)$ is shown in Fig. 16.5. On the plane $v(B, \theta)$, there is a discontinuity of magnetic flux density B. To avoid this discontinuity and divergences in the calculation of the equivalent characteristics, an extrapolation and a smoothing of the measured characteristics are required. To extrapolate the magnetic characteristic, the judicious way is the extrapolation of the dynamic permeability in function of magnetic flux density $\mu_d(B)$. The dynamic permeability is calculated in any point A of the magnetization curve $B(H)$ as defined by the relation (22). The extrapolation is done by matching an analytical model given at equation (23) for the descending part of the curve $\mu_d(B)$ and assuming that at 2T, μ_d is equal to the magnetic constant μ_0. Extrapolation is performed for all measured values $v(B, \theta)$. Fig. (16.5) shows $\mu_d(B)$ extrapolated with this methodology for $\theta = 90°$. The value of magnetic field H for the extrapolated part is determined by a numerically integration of $\mu_d(B)$ with equation (24). The results are presented in Fig. 16.7 for $\theta = 90°$. The operation is performed for all characteristics $B(H, \theta)$ to reach a bounded area between 0 and 2T.

$$\mu_d = \left.\frac{dB}{dH}\right|_A \quad (22)$$

$$\mu_d(B) = ae^{bB} + ce^{db} \quad (23)$$

Methods of Homogenization of Laminated Structures 297

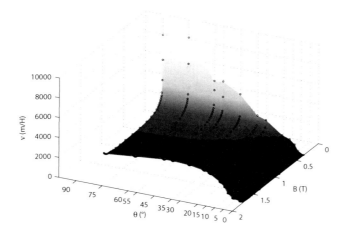

Fig. 16.5. Reluctivity versus magnetic flux density for different angles, $\nu(B,\theta)$

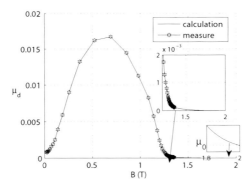

Fig. 16.6. Extrapolation of $\mu_d(B)$ for $\theta = 90°$

where a,b,c and d are real numbers. They are constant for a given θ value and calculated to respect the limit condition: $\mu_d = \mu_0$ when $B \mapsto 2T$.

$$H = \int \mu_d dB \qquad (24)$$

When the equivalent magnetic characteristics are defined for different angles, $0°$ is supposed to be the direction of the x axis. This assumption is invalid in the case of laminations with different RD orientations. Regarding the example of the step lap and $90°$, the characteristic $\nu_1(B_1,\alpha_1)$ of the yoke is the same as the measured characteristic $\nu(B,\theta)$ because the x axis coincides with the RD of the sheet. But for the column stacking, the RD is perpendicular to the x axis: $\nu_1(B,\alpha_1) = \nu_2(B,\alpha_1 + \pi/2)$.

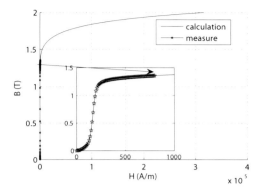

Fig. 16.7. $B(H)$ curve, bounded [0,2T]

Thus, the characteristic $v_i(B, \alpha_i)$ requires an arrangement in terms of α_i compared to those measured $v(B, \theta)$.

16.3.2 Homogenization Method Application

The joint of the transformer corner is studied. The characteristic feature of the step-lap joint is a 'comb' displacement ('parallel' both vertically and horizontally) of the laminations in subsequent cycles of the core assembly. For the purpose of analysis, we have assumed four cycles resulting in four 'steps', with each step comprising two sheets. Five cycles is standard for unskewed corners and joints. Such designs lead to a smaller area taken by the 'overlap' and allow the gaps to be spread over bigger space, resulting in better conditions for flux distribution. The neighboring layers of laminations are displaced by 5mm, which is typical for power transformers. In the region of an overlap, air-gaps are created due to imperfections of the yoke and limb assembly (note that a term 'column' is also used in literature, instead of a 'limb', to describe the "vertical" section of the transformer core). Hence in the core cross section sequential presence of magnetic sheets (laminations) and gaps may be observed. Two regions need to be considered: the virtual air-gap and the overlapping area. The flux density $\vec{B}$ of the equivalent structure of the air gap may then be defined as:

$$(n_1 + n_2 + n_0)\vec{B_r} = n_0 \vec{B_0} + n_1 \vec{B_1} + n_2 \vec{B_2} \qquad (25)$$

where $i = 0$ for the air gap, $i = 1$ for the yoke and $i = 2$ for the limb. The energy stored in region is:

$$W_r = \frac{1}{2}V\left(n_0 v_0(B_0, \alpha_0)B_0^2 + n_1 v_1(B_1, \alpha_1)B_1^2 + n_2 v_2(B_2, \alpha_2)B_2^2\right) \qquad (26)$$

where $V = V_0 = V_1 = V_2$ is elements volume, which is the same for all the elements of the different region. That is assuming because the sheets insulation is neglicted. Then, the objective function in the minimization process is defined as:

Methods of Homogenization of Laminated Structures

$$f_{min} = \left(n_0 v_0(B_0, \alpha_0) B_0^2 + n_1 v_1(B_1, \alpha_1) B_1^2 + n_2 v_2(B_2, \alpha_2) B_2^2\right) \tag{27}$$

During the minimization process, the unknowns are the components of $\vec{B_1}$ and $\vec{B_2}$. In the otherwise, $\vec{B_0}$ components are calculated from 28.

$$\begin{cases} B_{0x} = \frac{1}{n_0}\left(nB_r x - n_1 B_{1x} - n_2 B_{2x}\right) \\ B_{0y} = \frac{1}{n_0}\left(nB_r y - n_1 B_{1y} - n_2 B_{2y}\right) \end{cases} \tag{28}$$

The equivalent characteristics for different regions are computed by varying the equivalent flux density B from 0.01T up to saturation. For each equivalent flux density the equivalent reluctivity v_r was found through optimization. From the assumption, the energy of the heterogeneous microstructure must be equal to the energy of the equivalent homogeneous structure. The equivalent homogenized reluctivity is obtained by (21).

A single point of $B_r(H)$ was determined according to the following algorithm:

1. calculate B_1, B_2, α_1, α_2 from the minimization task (where x_j represents $B_i x$ and $B_i y$, and f is f_{min}),
2. $v_1(B_1, \alpha_1)$ and $v_2(B_2, \alpha_2)$ are determined by interpolation,
3. the homogenized value v_r is calculated,
4. the field intensity $H = v_r B_r$ is calculated.

Equivalent static $B_r(H)$ curves have been computed for every region. The curves are drawn for various inclination angles of B_r with respect to the x axis. The calculations are performed for the step lap joint made of grain oriented electrical steel sheets (M 140-35S), of 0.35mm thickness. The region (6), indicated in Fig. 16.2 contains laminations with different direction of rolling and at least one air-gap are called "virtual gaps". The equivalent characteristics shown in Fig. 16.8 refer to the virtual gap region consisting of two yoke sheets ($n_2 = 2$), one limb sheet ($n_1 = 1$) and one air gap ($n_0 = 1$), at four anisotropic angles of 0°, 30°, 60° and 90°. 0° has been assumed to coincide with the yoke (x axis) while 90o with the limb (y axis). Selected representative experimental and computed characteristics are summarised in Fig. 16.9. The measured static curves refer to the laminations used in the assembly and are shown for both the rolling and transverse directions (TD), while the computed values are given for the region 2, 5 and 8 refered in Fig. 16.3.Fig. 16.9 should not be interpreted as a comparison between experiment and computation; instead the curves are merely superimposed to demonstrate that the characteristics of the different structures lie between the two extremes as given by the two directions of rolling. This result was to be expected as the actual structures are built from laminations of various combinations of anisotropic angles. The computed equivalent static characteristics are then used to represent the 3D effects using 2D simulations.

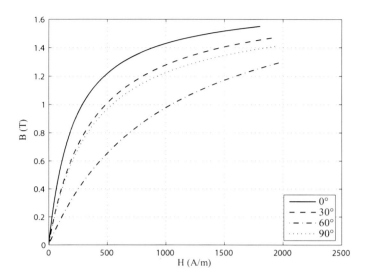

Fig. 16.8. Equivalent static $B_r(H)$ of region (6) at four different anisotropic angles

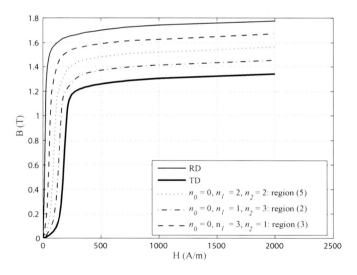

Fig. 16.9. Measured static characteristics for the lamination in the RD and TD, with computed characteristics for various structures (2, 5 and 8) for the case of an angle of 0°

16.4 Conclusion

The proposed homogenization method serves to simplify the characterization of anisotropic laminated magnetic circuits. It is based on the real magnetic characteristics of the electrical steel sheets and the energy minimization. The various steps required to achieve the equivalent magnetic characteristics are the extrapolation of measured magnetic characteristics of the sheets and the discretization of the original structure. To be more accurate, the extrapolation of the measured magnetization curves is done for the dynamic permeability. The discretization of the original structure is based on the differentiation of the constitutional material in the overlapping regions. The equivalent properties are calculated for different regions. These equivalent characteristics make it possible to replace the original 3D structure by an equivalent 2D model. For calculations, a 2D finite element model is faster than a 3D one. The parameters governing the effectiveness of a step lap joint are:

- the steps length,
- the number of sheets per layer,
- and the thickness of the column-yoke gap.

These parameters can be optimized by repeating the calculations of the homogenized finite elements model in order to save time.

References

1. Bastos, J.P.A., Quinchaud, G.: 3D modeling of a nonlinear anisotropic lamination. IEEE Trans. Magn. 21, 2366–2369 (1985)
2. De Rochebrune, A., Dedulle, J.M., Sabonnadiere, J.C.: A technique of homogenization applied to the modeling of transformers. IEEE Trans. Magn. 26, 520–523 (1990)
3. Brauer, J.R., Nakamoto, E.: Finite element analysis of nonlinear anysotropic B-H curves. Int. J. Appl. Electromagn. Mater. 3 (1992)
4. Brauer, J.R., Cendes, Z.J., Beihoff, B.C., Phillips, K.P.: Laminated steel eddy current losses versus frequency computed with finite elements. IEEE Int. Elec. Mach. Drives, 543–545 (1999)
5. Silva, V.C., Meunier, G., Foggia, A.: A 3-D finite-element computation of eddy currents and losses in laminated iron cores allowing for electric and magnetic anisotropy. IEEE Trans. Magn. 31, 2139–2141 (1995)
6. Chiampi, M., Chiarabaglio, D.: Investigation on the asymptotic expansion technique applied to electromagnetic problems. IEEE Trans. Magn. 28, 1917–1923 (1992)
7. Bertotti, G., Boglietti, A., Chiampi, M., Chiarabaglio, D., Fiorillo, F., Lazzari, M.: An improved estimation of iron losses in rotating electrical machines. IEEE Trans. Magn. 6, 5007–5009 (1991)
8. Chevalier, T., Kedous-Lebouc, A., Cornut, B., Cester, C.: A new dynamic hysteresis model for electrical steel sheet. Physica B 275, 197–201 (2000)
9. Sadowski, N., Batistela, N.J., Bastos, J.P.A., Lajoie-Mazenc, M.: An inverse Jiles-Atherton model to take into account hysteresis in time stepping finite-element calculations. IEEE Trans. Magn. 38, 797–800 (2002)
10. Hollaus, K., Bíró, O.: A FEM formulation to treat 3D eddy currents in laminations. IEEE Trans. Magn. 36, 1289–1292 (2000)

11. Hihat, N., Napieralska Juszczak, E., Lecointe, J.P., Sykulski, J.K.: Computational and experimental verification of the equivalent permeability of the step-lap joints of transformer cores. In: IET Conference CEM 2008, Brighton, U.K., pp. 38–39 (2008)
12. Gyselinck, J., Vandevelde, L., Melkebeek, J., Dular, P., Henrotte, F., Legros, W.: Calculation of Eddy Currents and Associated Losses in Electrical Steel Laminations. IEEE Trans. Magn. 35, 1191–1194 (1999)
13. Antoniou, A., Lu, W.S.: Practical Optimization: Algorithms and Engineering Applications. Springer, Heidelberg (2007)
14. Chandru, V., Rao, R.M.: Linear Programming, Department of Computer Science and Automation, Indian Institute of Science (1998), `http://www.eecs.harvard.edu/~parkes/cs286r/spring02/papers/lp.eps`
15. Lawden, D.F.: Analytical Methods of Optimization. Dover Books on Mathematics (2006)
16. El-Bakry, A.S., Tapia, R.A., Tsuchiya, T., Zhang, Y.: On the formulation and theory of the Newton interior-point method for nonlinear programming. Journal of Optimization Theory and Applications 89, 507–540 (1996)
17. El-Bakry, A.S., Tapia, R.A., Tsuchiya, T., Zhang, Y.: On the formulation and theory of the Newton interior-point method for nonlinear programming. Journal of Optimization Theory and Applications 89, 507–540 (1996)
18. Astfalk, G., Lustig, I., Marsten, R., Shanno, D.: The Interior-Point method for linear programming. IEEE Software 9, 61–68 (1992)

Chapter 17
Applications Examples

In this Chapter, several examples of applications of optimisation methods in the design of electrical devices are presented.

Section 1 is focused on the study of the torque ripple in a switched reluctance motor. The maximisation of the torque and minimisation of the torque ripple are the major objectives of the proposed procedure.

In Section 2 a lumped-parameter model is presented that computes the electromechanical parameters of a linear electrostatic induction micromotor, and simulates its behaviour. A Genetic Algorithm (GA) approach for tuning the parameters of the proposed model is used. The quality of the circuit model is evaluated in terms of how well the computed parameters (like force density or voltage at the interface) match the actual performance.

In Section 3 a reluctance network technique is applied to model the behaviour of a linear stepping motor. The choice of this particular method is due to the speed of calculation, a benefit especially when a large number of configurations need to be tried in the search of an optimal solution. The parameter optimisation of the motor is presented using a sequential quadratic programming (SQP) method with the aim of maximising the magnetic forces induced in the movable part of the motor.

In Section 4, optimisation and computer-aided design of special transformers are considered, namely resistance-welding transformers (RWT) and combined current-voltage instrument transformers (CCVIT). The electrical parameters of both kinds of transformers are calculated using a finite element method (FEM-3D). The FEM results are the input data for a Genetic Algorithm (GA) based optimal design procedure. The CCVIT is optimised for metrological performance, while RWT is optimised for energy efficiency.

In Section 5 the effects of geometric parameters on the performance of a switched reluctance machine (SRM) are studied. The influence of the shape of the tooth shape on the performance of the SRM is investigated based on a finite-element method coupled with the magnetic circuit approach (FEM-EMC). The subsequent example deals with optimisation approach to the design of electrical machines. The proposed method couples computational fluid-dynamics and heat-transfer equations. The properties of turbulent flow are obtained from a 2D model, and implemented as the boundary conditions in a 3D model for the thermal analysis.

Finally, Section 7 investigates particle swarm optimisation for the reconstruction of groove profiles in two dimensions, in the area of non destructive evaluation. The proposed method is based on the use of signal processing techniques; it presents an inversion approach of the eddy-current testing signals in order to recover the profile of an axi-symmetric groove. The method uses a finite-element model to solve the forward problem, and a particle-swarm optimization algorithm to solve the inverse problem.

17.1 Comparative Finite-Elements and Permeance-Network Analysis for Design Optimization of Switched Reluctance Motors

Dan Ilea[1], Frédéric Gillon[2], Pascal Brochet[2], and Mircea M. Radulescu[1]

[1] SEMLET Group, Technical University of Cluj-Napoca
 P.O. Box 345, RO – 400110 Cluj-Napoca, Romania
[2] L2EP, Ecole Centrale de Lille
 BP 48, F- 59651 Villeneuve d'Ascq CEDEX, France

17.1.1 Introduction

The design of a switched reluctance motor depends on the computation of the permeances in the airgap, poles and back-iron. In order to obtain these permeances, two methods are being comparatively used.

The first, the 'Permeance Network Analysis' uses an equivalent electric circuit that replicates the magnetic geometry of the motor. The advantage of this method is that it requires very little computation time, and it can be, in some cases even computed on-line. However, the method lacks in precision and the magnetic circuit models can become cumbersome.

The second method, the 'Finite Element Analysis' has the advantage of an accurate outcome, but is more time-consuming. In order to obtain even more accurate results, the analysis is made on the dynamic model instead of the static one.

17.1.2 Permeance Network Analysis

17.1.2.1 Airgap Permeance

A static permeance network analysis was done in order to express the permeances in the machine as a function of position θ and current i [1]. The airgap permeances are divided into three categories: aligned, half-aligned and unaligned, and are computed [2] from the basic expression:

$$P = \mu \cdot \frac{S}{l} \qquad [H] \qquad (1)$$

where S is the area of section crossed by the respective flux, l is the length of the flux path and μ is the magnetic permeability. The MMFs created in the excited phase are expressed as voltage sources.

These permeances are directly linked to the production of torque in the SRM and are function of the position θ of the rotor relative to the stator. The equation of torque in the SRM [4] is presented in (2), where L is the phase inductance and i is the phase current:

$$T = \frac{1}{2} i^2 \frac{\partial L(\theta, i)}{\partial \theta} \quad [\text{Nm}] \qquad (2)$$

The phase inductance and permeance are linked by equation (3), where n is the number of coils.

$$L = n * P \quad [\text{Wb*A-1}] \qquad (3)$$

17.1.2.2 Poles and Back-Iron Permeances

The permeances in the poles and back-iron are computed taking into consideration the nonlinearity of the machine. Because of the frequent use in the saturated region of the switched reluctance motor, these permeances are corrected through the B-H characteristics, expressed by:

$$H = \alpha_1 B + \alpha_n B^n \quad [\tfrac{A}{m}] \qquad (4)$$

In order to obtain the analytical form of the B-H characteristic, only its first magnetization curve is taken into consideration, neglecting the hysteresis effect. Its slope μ_i is approximated using the least square method [5] which gives the possibility of obtaining the magnetic permeability μ_i as a function of the magnetic induction B.

$$\frac{1}{\mu_i} = \frac{1}{\mu_0} [\varepsilon + (c - \varepsilon) \frac{B^{2\alpha}}{B^{2\alpha} + \tau}] \qquad (5)$$

where c, ε, τ and α are coefficients determined so that the analytical B(H) curve is as close as possible to the real characteristic.

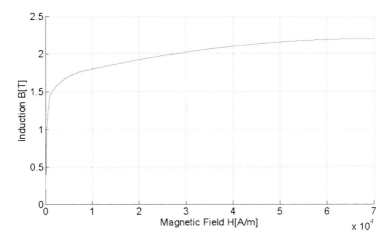

Fig. 17.1 B-H characteristic for the used material

An iterative method is used in order to obtain the new permeances of the stator poles and back-iron from the values of the magnetic induction. The iterative process ends when the error between two consequent values of μ_i is below a set limit.

17.1.2.3 Complete Permeance Network

The model of the machine is used to obtain the magnetic circuit configuration. The permeance network is further solved using the classical circuit methods.

Computation of the permeance network is related to the linear development of the 6/8 SRM (Fig. 17.2).

The computed values of the permeances from the stator poles, rotor poles and back-iron as well as those of the airgap are considered in order to obtain the total flux and ultimately the produced torque.

The rotor position-dependent electronic commutation and drive control have to be integrated into the permeance network analysis [3]. Hence, a dynamic model of the machine is obtained.

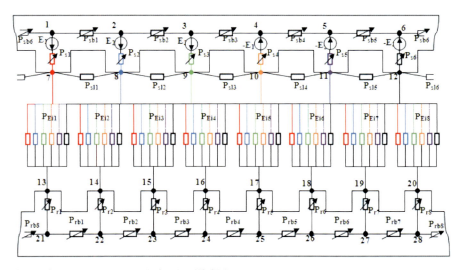

Fig. 17.2 Permeance Network for the 6/8 SRM

17.1.3 Finite-Element Analysis

A two dimensional model of the 6/8 SRM is being used in order to determine the magnetic field distribution and the produced torque and its ripple. The analysis has been made using the commercial software JMag-Studio.

The effects of saturation and mutual coupling between adjacent poles can be better observed in the FEA model. In order to validate the Permeance Network Analysis, a number of rotor positions are chosen and the values of magnetic induction in the poles from the two methods are compared.

Only one phase of the motor is energized at a time in order to better assess the magnetic effects in the SRM.

17.1.4 Motor Specifications and Simulation Results Analysis

The simulation model of the motor is a three phase 6/8 switched reluctance motor with the parameters presented in Tab. 17.1. The value of the phase current is 1 A and the material characteristic is presented in Fig. 17.3.

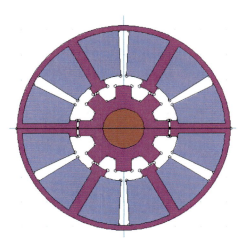

Fig. 17.3 Geometric structure of the three phase 6/8 SRM

Table 17.1 Dimensions of the three phase 6/8 SRM

Parameter	Value	Unit
Stator core outer diameter	41.6	[mm]
Rotor core outer diameter	17.6	[mm]
Airgap length	0.4	[mm]
Shaft diameter	8.8	[mm]
Stator pole arc	19.6	[°]
Rotor pole arc	20.18	[°]
Stator length	50	[mm]

As it can be seen from Fig. 17.4, the flux lines form a closed loop from one charged stator pole to the other, trough the airgap, rotor poles, rotor back iron and stator back iron. It can also be observed that there is virtually no flux passing through the other stator poles, which means that phase coupling is minimum in the case of one phase-on strategy.

The magnetic inductance in the stator pole is at peak value at the aligned position, when the airgap is at its minimum, as it can be seen from Fig. 17.5. The FEA results confirm the Permeance Network Analysis, the values from both methods for the computation of the magnetic inductance in the energized pole being compared in Fig. 17.6.

The computation of torque and torque ripple produced in the three phase 6/8 SRM in dynamic mode are the final purpose of this paper. In Fig. 17.7 we can observe the correlation, given by eq. (2) and (3), between the airgap permeances (computed using the Permeance Network Analysis model of the SRM) and the dynamic torque.

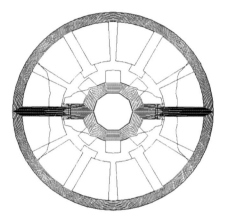

Fig. 17.4 Flux paths in the 6/8 SRM

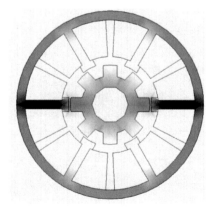

Fig. 17.5 Magnetic inductance at the aligned position

Comparative Finite-Elements and Permeance-Network Analysis

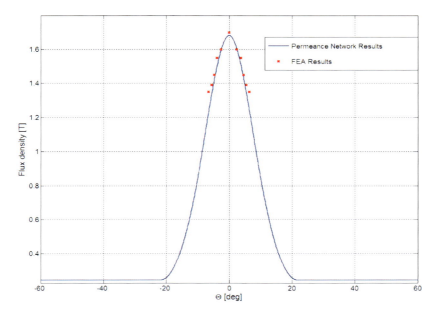

Fig. 17.6 Comparative FEA- Permeance Network computation of the flux density for one stator and one rotor tooth

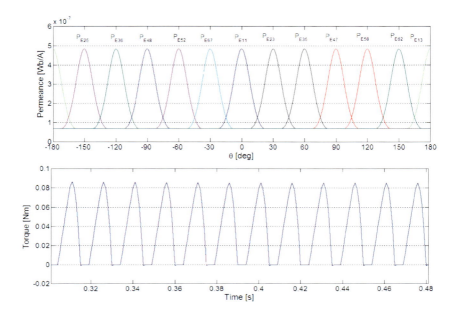

Fig. 17.7 Dynamic torque and torque ripple of the used 6/8 SRM

17.1.5 Conclusions

In this paper a comparison between two electromagnetic analyses (Permeance Network Analysis and Finite-Element Analysis) used in the electromagnetic modeling of a three phase 6/8 SRM is presented. In order to obtain the dynamic torque and its ripple, it is first necessary to compute the dynamic evolution of the flux and inductances for the whole machine, taking into consideration the characteristics of the used materials. The analytical results from the Permeance Network Analysis, although not as exact as those from the FEA are nevertheless precise enough for the modeling of the SRM.

The results from the two analyses are necessary in the further development of the design optimization of the motor, where torque maximization and torque ripple minimization are the major objectives.

17.2 Lumped Parametric Model of an Electrostatic Induction Micromotor

Francisco Jorge Santana-Matin[1], José Miguel Monzon-Verona[1], Santiago Garcia-Alonso[2], and Juan Antonio Montiel-Nelson[2,*]

[1] Electrical Eng. Dept.,
 University of Las Palmas de Gran Canaria, Institute for Applied Microelectronics
[2] Electronic and Automatic Eng. Dept.
 University of Las Palmas de Gran Canaria, Institute for Applied Microelectronics,
 Campus Univ. de Tafira, E35017 Las Palmas de Gran Canaria, Spain
 montiel@iuma.ulpgc.es

17.2.1 Introduction

Research advances on electromagnetic induction machines have been established in fundamental models, as the classic equivalent circuit model in stationary state [6]. Once the duality of the electrostatic induction micromachine in the microscale with the electromagnetic induction machine at the macroscale has been demonstrated [7], the authors propose a new lumped parametric equivalent circuit model for the micromotor. Figure 17.8 illustrates the physical model of the electrostatic induction micromotor and the lumped parametric equivalent model.

The quality of this circuit model is based on how its magnitudes of interest match the main phenomena under study in the micromotor (force density, voltage at the interface, etc.). The proposed lumped parametric model defines a micromotor in terms of circuit linear passive elements (C_1, C_2, G'_2, etc.) and a network or topological structure. Each element has its own constitutive equations that can be expressed in a mathematical form. The network or topological structure defines the system configuration. The composition of these mathematical relations for each element gives us a system of algebraic equations.

Figure 17.8(a) represents the micromotor physical model. Figure 17.8(b) shows the proposed lumped parametric equivalent circuit model. As is shown in Figure 17.8(a) and (b), the equivalence between the physical model and the proposed lumped parametric equivalent circuit model is straightforward.

The proposed model explains and predicts the behavior of the electrostatic induction micromotor, without expensive temporal simulations, saving computational resources. To our knowledge, no lumped parametric equivalent circuit model per phase has been found in the literature for the electrostatic induction micromotor.

---

[*] Corresponding author.

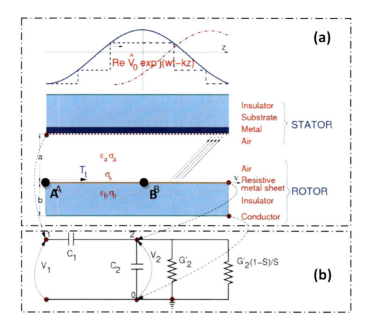

Fig. 17.8 a) Physical model and b) lumped parametric equivalent models for the electrostatic induction micromotor per face.

17.2.2 Lumped Parametric Equivalent Circuit Model

In this parametric model for an electrostatic induction micromotor, there are not inductive impedances because they are neglected for second order effects. Instead, we introduce capacitive impedances that are characteristic of the physical nature of the micromotor. The correspondence between the circuit and the physical model is based on next premises:

a) The ground of the electric circuit (see node 0 in Figure 17.8(b)) coincides with the ground terminal of the electric supply applied to the micromotor.

b) The input applied voltage in the active electric circuit terminal (see node 1 in Figure 17.8(b)) coincides with the terminal of one of the six phases of the electric supply applied to the micromotor.

c) The output terminal voltage of the electrical circuit (see node 2 in Figure 17.8(b)) coincides with the voltage at the interface of the micromotor, which is the potential of the rotor. The interface is the surface of the resistive metal sheet of the rotor that is in contact with the air.

In the micromotor, the electrical resistance of the stator does not exist, and the inductive impedance is replaced by its dual capacitive impedance, which is C_1 in Figure 17.8(b). The resistance and the inductive impedance of the rotor, referred to the stator, are replaced by their dual conductance G'_r and capacitive impedance C_2, respectively. Conductance G'_r is expressed as:

$$G'_r = G'_2 + G'_2\left(\frac{1}{S} - 1\right) \qquad (1)$$

where G'_2 is the conductance –resistance loss due to metallic resistive sheet of the rotor – and S the slip of the micromotor. The slip S is expressed as follows

$$S = \frac{\omega/k - v}{\omega/k} \qquad (2)$$

where ω/k is the wave speed and v the rotor velocity.

The relationship between the lumped parametric equivalent circuit model and the physical nature of components of the micromotor is summarized as follows:

a) The capacitor C_1, placed in Figure 17.8(b) between nodes 1 and 2 of the equivalent circuit, is the capacitance between two parallel plates separated by a distance, and air as dielectric material. These two planes contain the fixed and mobile elements of the electrostatic induction micromotor.

b) The capacitor C_2, placed in Figure 17.8(b) between nodes 0 and 2 of the circuit, is the capacitance between the superior and inferior sheets of the mobile element of the electrostatic induction micromotor. Between these sheets there is a dielectric material, with a permittivity ε_b.

c) The conductances G'_2 and $G'_2(1-S)/S$ are, respectively, the resistance losses due to the metallic resistive sheets of the rotor, and the mechanic power generated by the micromotor which is a function of S, respectively.

17.2.3 Fitting of Lumped Parameters Using GA in Steady State

As an optimization method to obtain the value of the lumped parameters of the equivalent circuit, we use Genetic Algorithm techniques [8]. Genetic Algorithm (GA) is an optimization technique inspired in the Darwin's evolution theory that has raised a great interest in the scientific community all over the world in recent years. This technique imitates the selection mechanics of the nature where only the more capable individuals of a population survive. John Holland was the pioneer in this topic and his main contribution was to develop the foundations that allowed the incorporation of these techniques to a computer [9]. Reference [10] provides a formal definition of the GA optimization method.

Both the general convergence constraints and the functional parameters —population size, crossover and mutation probability, number of generations and number of pairs for generation— of the used GA method are defined as follows.

17.2.3.1 Objective Function

The objective function Ψ to be minimized consists of the quadratic error between V_2 and $\Phi^b(\omega)$, where V_2 and $\Phi^b(\omega)$ are the potential at the interface in the proposed lumped parametric circuit model, and in the physical model of the induction electrostatic micromotor, respectively. That is to say:

$$\Psi = \sum \left(V_2(\omega, C_1, C_2, G'_2) - \Phi^b(\omega)\right)^2 \qquad (3)$$

We calculate the voltage at the interface in the equivalent circuit model, V_2, by a conventional method of Circuit Theory, obtaining the following equations:

$$Z_{1e} = -\dfrac{\dfrac{S}{C_2 G'_2 \omega} j}{\dfrac{S}{G'_2} - \dfrac{1}{C_2 \omega} j} \qquad (4)$$

$$Z_{2e} = Z_{1e} - \dfrac{1}{C_1 \omega} j \qquad (5)$$

$$Z_{ed} = Z_{1e} Z \qquad (6)$$

$$V_2 = |(V_0 Z_{ed})| \qquad (7)$$

as is expressed in Eqs. (4) – (7), V_2 is a function of C_1, C_2, G'_2 and ω.

The voltage at the interface of the physical model, Φ^b, is obtained by the following expression [7]:

$$\Phi^b = \dfrac{V_0}{\sinh(ka)} \dfrac{\dfrac{\sigma_a}{\sigma_{eff}} + \dfrac{\varepsilon_a}{\varepsilon_{eff}} \omega S j}{\left(1 + \dfrac{\varepsilon_{eff}}{\sigma_{eff}} \omega S j\right)} \qquad (8)$$

where,

$$\sigma_{eff} = \sigma_a \coth(ka) + \sigma_b \coth(kb) + \sigma_s k \qquad (9)$$

$$\varepsilon_{eff} = \varepsilon_a k \coth(ka) + \varepsilon_b k \coth(kb) \qquad 10)$$

Please, note that parameters and symbols are introduced in Table 17.2.

In this work, we obtained the parameter values for the proposed model using $S=1$, therefore, the rotor is stopped. Because, inductive impedances are neglected at the microscale [7], another set of parameter values is obtained following the same procedure when slip conditions change — $S \in [0,1]$. This implies that in the proposed equivalent circuit model, the conductance given by $G'_2(1-S)/S$ remains in open circuit (the mechanic power is equal to zero because the mobile part is stopped).

For a given surface conductivity σ_S, the objective function Ψ is minimized with GA when the frequency ω is swept. Based on these equations (see Eqs. (3) – (10)), we obtain the parameters of the equivalent circuit model, C_1, C_2 and G'_2, that models the electric induction micromotor.

Table 17.2 Definitions and Symbols

Symbol	Name	Unit
a	Height of dielectric a, air	(m)
b	Height of dielectric b, rotor	(m)
C_1	Capacitance 1	(F/m)
C_2	Capacitance 2	(F/m)
G'_2	Conductance 2	(S/m)
G'_r	Variable Conductance	(S/m)
k	Number of waves per metre	—
j	Imaginary unity	—
S	Slip	—
T_t	Force density tangential component	(N/m^2)
v	Linear speed of mobile part	(m/s)
V	Interelectrodic voltage	(V)
V_0	Supply voltage	(V)
V_1	Voltage in node 1	(V)
V_2	Voltage in node 2	(V)
ε_a	Permittivity of the dielectric a	(F/m)
ε_b	Permittivity of the dielectric b	(F/m)
ε_{eff}	Effective permittivity	(F/m)
ω	Angular frequency of the signal	(Hz)
σ_a	Conductivity of the dielectric a	(S/m)
σ_b	Conductivity of the dielectric b	(S/m)
σ_{eff}	Effective Conductivity	(S/m)
σ_S	Surface Conductivity	(S)
Φ^b	Voltage at the interface	(V)
Ψ	GA's objective function	—

17.2.3.2 GA Parameters

We have applied Genetic Algorithms for tuning the parameters of the equivalent circuit model to the linear electrostatic induction micromotor. Both physic and geometric parameters are shown in Tables 17.2 and 17.3, respectively.

A toolbox of Scilab [11] has been used as a platform of numerical calculation. The problem has been solved for different values of the surface conductivity σ_S.

We have fitted the lumped parameters, C_1, C_2 and G'_2, for four different conductivities, and we have observed that the capacitances C_1, C_2 do not vary with the conductivity, they remain constant. However, the value of the conductances depend on conductivity.

Table 17.4 shows the values and parameters that have been used for the GA optimization. The GA converges in 50 iterations. Parameter values in Table 17.3 and 17.4 have been used in all simulations and for each surface conductivity of the material.

Table 17.3 Physical and geometrical parameters of the micromachine

Symbol	Name	Value	Unit
L	Length of the structure	$44 \cdot 10^{-6}$	(m)
hm	Height of the metallic plates	$0.01 \cdot 10^{-6}$	(m)
a	Height of dielectric 2	$3 \cdot 10^{-6}$	(m)
b	Height of dielectric 1	$10 \cdot 10^{-6}$	(m)
k	Number of waves per metre	$2\pi/L$	(m^{-1})
v	Linear speed of mobile part	0	(m/s)
f	Temporal frequency of excitation	$2.6 \cdot 10^6$	(Hz)
V_0	Maximum value of excitation	200	(V)

Table 17.4 G A parameters

Name	Value
Population size	400
Crossover probability	0.7
Mutation probability	0.05
Number of generations	50
Number of pairs	150

17.2.3.3 Lumped Parameters in Stationary State

Table 17.5 contains the optimized values for the parameters of the lumped model per phase and σ_S unit of width (m). Note that when the material conductivity of the metallic sheet of the rotor increases, the values of the capacities C_1 and C_2 remain, approximately, invariable. However, the conductivity G'_2 increases. From the results, we demonstrate that the relationship between the surface conductivity) of the physic model and the conductance G'_2 of the lumped parametric equivalent circuit model is linear.

Figure 17.9 shows the force density at the interface versus slip for the physical model. This was calculated through analytical equations using Maxwell's field equations [7], and for the equivalent lumped circuit model, calculated through GA. Both curves are coincident, so the model has been validated. The mean square error for 50 generations is lower than $15 \cdot 10^{-7}$ %.

Table 17.5 Calculated values for lumped parameters per phase

σ_S	C_1	C_2	G'_2
$(1/\Omega)$	(F/m)	(F/m)	(Ω^{-1}/m)
$1/(1800 \cdot 10^6)$	$238.71857 \cdot 10^{-9}$	$483.966 \cdot 10^{-9}$	$9.4506 \cdot 10^{-1}$
$1/(600 \cdot 10^6)$	$238.58074 \cdot 10^{-9}$	$483.283 \cdot 10^{-9}$	$28.3215 \cdot 10^{-1}$
$1/(200 \cdot 10^6)$	$238.58724 \cdot 10^{-9}$	$483.306 \cdot 10^{-9}$	$84.9662 \cdot 10^{-1}$
$1/(144 \cdot 10^6)$	$238.58507 \cdot 10^{-9}$	$483.294 \cdot 10^{-9}$	$118.007 \cdot 10^{-1}$

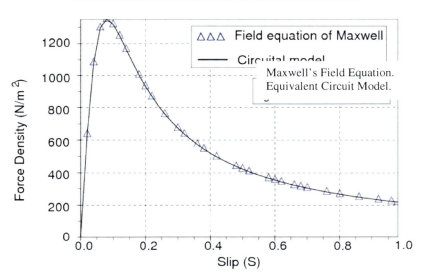

Fig. 17.9 Force Density vs. slip value for both the lumped parametric model and analytical Maxwell´s field equation

17.2.4 Time Domain Analysis

In this section, we analyze the transient state of the physical and equivalent circuit models. We have used the lumped parameters obtained in previous section 2.3.

17.2.4.1 Transient State of the Physical Model

In order to analyze the transient state of the electrostatic induction micromotor, we start from the following equation:

$$\nabla \cdot \sigma \nabla \Phi + \frac{\partial}{\partial t} \nabla \cdot \varepsilon \nabla \Phi = 0 \quad (11)$$

Equation (12) is expressed as follows

$$\frac{\partial}{\partial t} \nabla \cdot \varepsilon \nabla \Phi + \nabla \cdot \sigma \nabla \Phi = 0 \quad (12)$$

Equation (12) is a Differential–Algebraic system of Equations (DAE) that is represented as

$$M \cdot \frac{\partial \Phi}{\partial t} + N \cdot \Phi = f \quad (13)$$

where M and N are matrix of coefficients, Φ is the potential at the interface, and $f(t)$ is the excitation signal on each node when Dirichlet conditions are used to solve the FEM equation system. This time–dependent equation has to be discretizated in both space and time domain. We used Gmsh software [12]. The discretization in time domain has been realized applying the θ–method [13]. The Equation (14) is transformed as follows:

$$M \frac{\Phi_{n+1} - \Phi_n}{\Delta t} + N(\theta \Phi_{n+1} + (1-\theta)\Phi_n) = \theta b_{n-1} + (1-\theta)b_n \quad (14)$$

where index n and $n+1$ refer to Φ quantities at time t and $t+\Delta t$, respectively. Depending on the values for parameter θ we obtain a particular equation that is solved by a classical method. For example, if $\theta=1$ we use implicit Euler method, $\theta=0$ explicit Euler method and $\theta=0.5$ is Crank–Nicholson [8]. In this work we use $\theta=1$.

At $t=0$, the initial conditions for the electric potential is $V=0$ for the entire domain. The total time for the transient analysis is 14 cycles of the applied sinusoidal voltage with a maximum value of 200 (V). The time step used in θ–method is equal to $T/40$ s, where T is the signal period (see Table 17.3). The potential at the interface has been calculated with the Finite Element Method (FEM), using the GetDP [14] software solution. At $t=0$ the initial condition for the electric potential is $\Phi=0$ in the entire domain.

Figures 17.10, 17.11 and 17.12 show the transient state of the voltage in the interface region between geometrical points $z=0$ y $z=L/2$ (see points A and B in Figure 17.8(a) at the time instants $1.92 \cdot 10^{-7}$ (s), $4.81 \cdot 10^{-7}$ (s) and $3.85 \cdot 10^{-6}$ (s), respectively. For a given time t, we have observed that the voltage at points A and B are identical in magnitude with opposite sign. Note that voltage magnitude tends to 65.9 (V) —steady state voltage magnitude.

Figure 17.10 illustrates the voltage distribution at the time instant $1.92 \cdot 10^{-7}$ (s). The maximum voltage value (117 (V)) is greater than the maximum voltage magnitude in steady state (65.9 (V)). As the transient analysis evolves, this maximum voltage value tends to the maximum of the steady state voltage value, see Figures 17.11 and 17.12. In addition, the traveling wave changes its wave form until it reaches the definitive sinusoidal slope (see Figure 17.12).

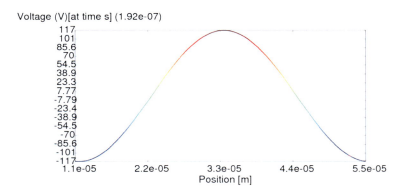

Fig. 17.10 Voltage distribution at the interface for time instant $1.92 \cdot 10^{-7}$ s

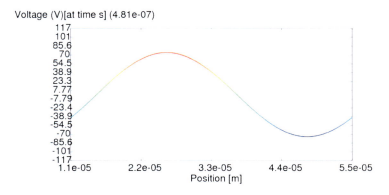

Fig. 17.11 Voltage distribution at the interface for time instant $4.81 \cdot 10^{-7}$ s

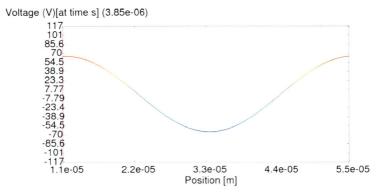

Fig. 17.12 Voltage distribution at the interface for time instant $3.85 \cdot 10^{-6}$ s

Transient State of the Equivalent Circuit Model

To calculate the transient state of the equivalent circuit model shown in Figure 17.8(b), we have applied the state variable method [15]. In this circuit, we have used the values of the parameters calculated previously for $S=1$ using GA. The resolution of this problem leads to the following equation

$$\frac{dV_{C2}(t)}{dt} = -\frac{G_2}{C_1+C_2}V_{C2}(t) + \frac{C_1}{C_1+C_2}\frac{dV(t)}{dt} \quad (15)$$

This is an Ordinary Differential Equation (ODE), type $y'=-Ay+Bu$, with initial conditions $V_{C2}(t)=0$. This is a classic problem of Initial Value Problem (IVP) that has been resolved applying the explicit method of Runge-Kutta-Fehlberg (RKF45) or adaptative step method [13]. The obtained results have been represented in Figure 17.13. The curve shows the output potential V_2 of the equivalent circuit and how this potential tends to the steady state value. Figure 17.14 illustrates the temporal variation of the integration step of the RKF45 method.

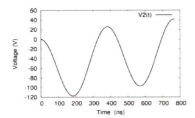
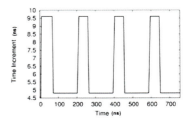

Fig. 17.13 Transient state of the output potential for the lumped equivalent model

Fig. 17.14 Temporal variation of the integration step for the lumped equivalent model

Comparisons

Figure 17.15 shows the obtained voltage values in transient state for both models. The dot points represent the FEM solution of the physic model and lines represent the solution for the equivalent circuit model obtained through RKF45 algorithm. We observe that both results are coincident, because the error is neglected. The mean quadratic error is neglected. This result validates the proposed lumped parametric model.

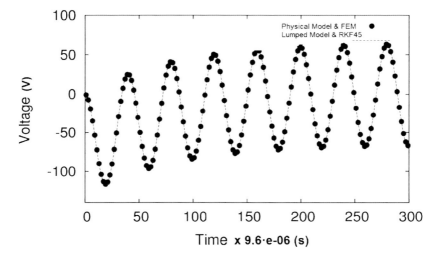

Fig. 17.15 Transient state for circuital and physical models

17.2.5 Conclusions

We have introduced a novel lumped parametric equivalent circuit model per phase in stationary state that couples electromechanical parameters, describing the behavior of a linear electrostatic induction micromotor. We have used Genetic Algorithms for tuning the parameters, of the proposed model. Our model has been validated by comparison against analytical solution. The comparison results demonstrate that the fitting error between our proposed equivalent circuit and the analytical solution —calculated applying the field analytical Maxwell's equations— for the interface potential is neglected. Based in the lumped model parameters obtained in steady state, we have also calculated the potential in transient state. We calculated the potential in the transient state in the physical model with FEM and we obtained the same results.

17.3 Contribution to Optimization and Modeling by Reluctances Network

Abdelghani Kimouche, Mohamed Rachid Mekideche, Ammar Boulassel, Tarik Hacib, and Hicham Allag

LAMEL laboratory, University of Jijel, BP 98,Ouled Aissa,18000 Jijel, Algeria

17.3.1 Introduction

Well that the numerical approaches, as the finite elements method, recorded a lot of advantage concerning the precision and the study of complex geometry, they present again the inconvenience of important time calculation. Generally, when it is about a research of optimizations problems, the analytic approaches are often solicited because they are fast in time of calculation. It is for this reason that we chose the modelling by reluctance network. The modelling by magnetic equivalent circuit (MEC) consists in defining a topology of the reluctances network. The knowledge of magnetic flux distribution is necessary and considering the hypothesis on this flux target is interesting. However when the magnetic material is saturated, the flux is canalized rightly in iron and therefore the reluctance network becomes easy to construct (the tubes of flux follow the magnetic circuit). Nevertheless, an important saturation implicates an increase of the leakage fluxes, what makes difficult to construct the reluctance network. Also, the number of reluctance influence on the topology of the network [16].

Our objective is focused on study of linear stepping motor and application the method of SQP in order to maximum magnetic force. We use the finite elements method to obtain the flux target in stepping motor, then the reluctance network technique is applied for calculation the magnetic forces.

17.3.2 Linear Stepping Motor Study

17.3.2.1 Finite Element Method Analysis

Considering the symmetrical nature of the structure of studied motor (fig. 17.16), it is interesting to analyze the magnetic field in 2D axisymmetric case [17].

The magnetic field is calculated in axisymmetric case using the formulation in magnetic vector potential [16] [17] as following equation in cylindrical coordinate (r,z):

$$\frac{\partial}{\partial r}\left(\frac{v}{r}\frac{\partial \vec{A}}{\partial r}\right) + \frac{\partial}{\partial z}\left(\frac{v}{r}\frac{\partial \vec{A}}{\partial z}\right) = \vec{J} \qquad (1)$$

Where v is the magnetic reluctivity, $\vec{A}$ is magnetic vector potential and $\vec{J}$ is the current density.

Contribution to Optimization and Modeling by Reluctances Network

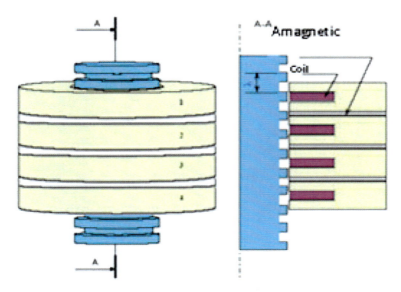

Fig. 17.16 Linear stepping motor structure

The resolution of the electromagnetic problem by FEM, bring us, after assembly of the elementary complete shapes to solve an algebraic system of equation as following:

$$[S][A] = [J] \qquad (2)$$

[S] Represent the magnetic stiffness matrix and *[J]* the source vector. The fig. 17.17 shows the flux distribution in the model proposed.

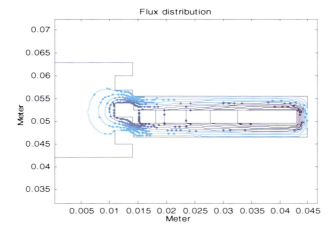

Fig. 17.17 Flux distribution

17.3.2.2 Magnetic Equivalent Circuit Modeling

The essential component of the MEC model is a flux tube. It is considered as a volumetric space between two planes of equal magnetic scalar potential. A diagram of flux tube is shown in fig. 17.18.

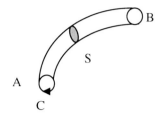

Fig. 17.18 Tube Flux

A flux tube has elements that are similar conceptually to electric circuit element. In particular, the ends of the tube are similar to node potentials in electric circuit, and the flux through a tube can be treated as current in an electric conductor.

The ration of difference of the magnetic potentials to the total flux in the volume is defined as the tube reluctance.

$$R = \frac{E_A - E_B}{\phi} \tag{3}$$

Where E_A and E_B are the magnetic scalar potentials of the tube's equipotential planes. ϕ is the magnetic flux.

The reluctance is similar conceptually to a resistance in electric circuit. Moreover, reluctance is proportional to tube length and inversely proportional to tube area [16]. Further, the reluctance is a function of the magnetic permeability inside the tube.

$$R = \int_A^B \frac{dl}{\mu.s} \tag{4}$$

where l is the tube length, μ the permeability and s the tube area. The Magneto motive force sources (MMF) source is used to represent the effects of electrical conductors in a magnetic system. The MMF source is similar to a voltage source in an electrical circuit. The value of an MMF source is determined using Ampere's current law [16].

In particular, Ampere's law can be expressed in integral form as:

$$\oint \vec{H}.\vec{dl} = i_{enclosed} \tag{5}$$

Where $\vec{H}$ is the magnetic vector field and $i_{enclosed}$ is the current enclosed by the line of integration. For a multiple conductor winding, (5) is often written in the form:

$$\vec{H}.\vec{l} = Hl = Ni \qquad (6)$$

Where N is the number of turns and Ni is defined as the MMF, i.e.

$$F = Ni \qquad (7)$$

The fig.17.19 shows the two dimensional magnetic circuit equivalent of model studied.

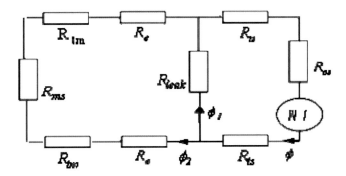

Fig. 17.19 MEC model reluctance network

$R_{cs}, R_{ts}, R_e, R_{tm}, R_{cm}$ And R_{leak} are respectively the stator magnetic core reluctance, stator tooth reluctance, air gap reluctance, mobile part tooth reluctance, mobile part shaft reluctance and coil leakage reluctance, which have the expressions as following.

The radial reluctance:

$$R_{ts} = \frac{r_5 - r_3}{\mu_0 \mu_r . \pi (r_5 + r_3).d} \qquad (8)$$

$$R_e = \frac{r_3 - r_2}{\mu_0 . \pi (r_3 + r_2).d} \qquad (9)$$

$$R_{tm} = \frac{r_2 - r_1}{\mu_0 \mu_r . \pi (r_2 + r_1).d} \qquad (10)$$

The axial reluctance:

$$R_{cs} = \frac{3.d}{\mu_0 \mu_r . \pi (r_6^2 + r_5^2)} \qquad (11)$$

$$R_{cm} = \frac{3.d}{\mu_0 \mu_r . \pi (r_1^2)} \qquad (12)$$

The coil leakage reluctance [19]:

$$R_{leak} = \frac{6.b}{\mu_0 \pi . r_b (3.r_b + 4.r_4)} \tag{13}$$

With: $r_b = r_5 - r_4$

The fig. 17.20 shows the different geometrical parameters used to calculate the reluctances.

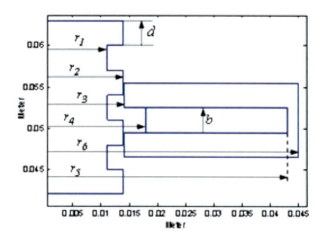

Fig. 17.20 Model geometrical parameters

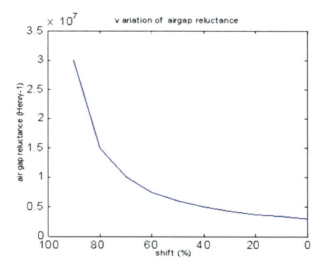

Fig. 17.21 Variation of air gap reluctance

The fig. 17.21 shows the variation of the air gap reluctance as function the displacement of mobile part. This variation of reluctance is at origin of the useful axial force able to displace the mobile part. The variation of reluctance is at origin of the useful axial force able to displace the mobile part.

The Kirshoff laws are used to calculate the magnetic flux circulating in the reluctance network. The evolution of magnetic flux as function the displacement of mobile part is shows in fig. 17.22.

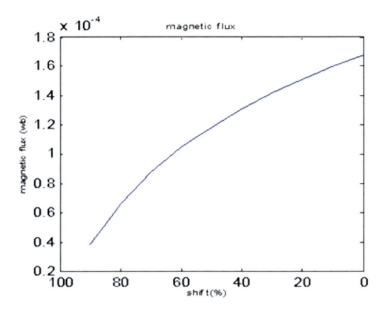

Fig. 17.22 Evolution of total magnetic flux

17.3.3 *Magnetic Force Computation*

The magnetic force in the mobile part, for one position of the mobile part, is determinate by the resolution of the electromagnetic problem. We use the principal of virtual works [16]. The expression of the magnetic force is given by (14):

$$F_z = \left.\frac{\partial W'}{\partial z}\right|_{NI=cst} \quad (14)$$

With W' is the magnetic co-energy given by the (15).

$$W' = \int_\Omega \left(\int_0^H B^T . dH \right) d\Omega \quad (15)$$

Where $\vec{B}$ and $\vec{H}$ are respectively the magnetic induction and magnetic field.

The fig. 17.23 shows the magnetic force as function the displacement of mobile part using the principal of virtual works [16].

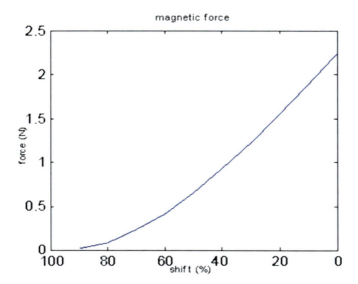

Fig. 17.23 Magnetic force as function shift

17.3.4 Optimization

The method of sequential quadratic programming (SQP) has been applied to obtain the optimal configuration of slot (geometrical parameters) for maximum the magnetic force. The process of optimization (fig. 17.24) uses intensively the calculation of objectif function by reluctance network for each iterations.

The geometry for optimization of the linear stepping motor is described by two parameters $X(r_b, b)$. ($r_b = r_5 - r_4$) and b are the parameter of the stator slot indicated in the fig. 17.20.

The limits of variation of these parameters have been chosen in order to avoid a configuration for which the geometry of the stator is not realizable. These limits as well as the initial values used for optimization are shown in the Table 17.6.

The optimization problem is defined as following

$$\begin{cases} Max\ F_z(x) \\ b^{k\ min} \leq b^k \leq b^{k\ max} & k = 1,.....,n \\ r_b^{k\ min} \leq r_b^k \leq r_b^{k\ max} \end{cases} \quad (16)$$

The Table 17.7 shows the values gotten of the parameters and of the function at the end of optimization.

Contribution to Optimization and Modeling by Reluctances Network

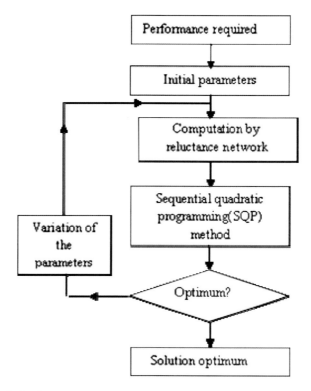

Fig. 17.24 Optimization process

Table 17.6 Limit of the parameters variations

Parameters	b	r_b
Minimal Values (m)	0.0030	0.014
Maximal Values (m)	0.0040	0.025
Initial values (m)	0.0032	0.016

Table 17.7 Values of the parameters optimized and the objective function

Parameters and objective function	Values
b	0.0034m
r_b	0.020m
F(x)	0.6466N

The geometry initial and the geometry optimized are represented respectively on fig. 17.25 and fig. 17.26.

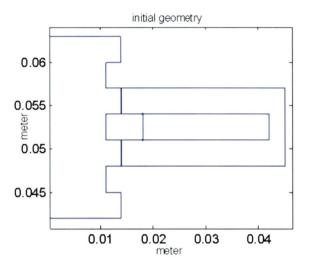

Fig. 17.25 Initial geometry

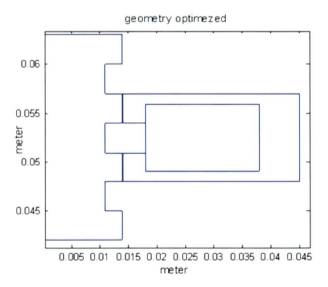

Fig. 17.26 Geometry optimized

17.3.5 Conclusion

In this paper the SPQ method has been applied to obtain the optimal slot dimension of a stator structure of a linear stepping motor to maximum the magnetic forces. These forces are calculated by the method of the virtual works. The magnetic equivalent circuit is used to modelling the motor in the aim to benefice of the time calculation. We remark the strong influence of the geometrical parameters used in this optimization for maximizing the magnetic forces.

This study can be extended as a multi-objective problem considering both maximizing forces and minimizing the losses.

17.4 Optimization and Computer Aided Design of Special Transformers

Marija Cundeva-Blajer, Snezana Cundeva[*], and Ljupco Arsov

Ss. Cyril & Methodius University, Faculty of Electrical Engineering and
Information Technologies, Skopje, R. Macedonia
scundeva@feit.ukim.edu.mk

17.4.1 Introduction

The process of optimal design of special transformers like the resistance welding transformer (RWT) or combined current-voltage instrument transformer (CCVIT) is a complex procedure. The both design objects are complex non-linear electromagnetic systems, as described in [20] and [21]. The procedure of optimal design begins with thorough study of the initial design of the both transformers. For the purpose of the optimization the mathematical models of the transformers are defined. A thorough finite element (FEM-3D) analysis of the electromagnetic parameters is performed, [20, 21] and the derived results are input data in the genetic algorithm (GA) optimal design procedure. The CCVIT is a complex electromagnetic device with two measurement cores: voltage measurement core (VMC) and current measurement core (CMC) with four windings and two magnetic cores with mutual non-linear electromagnetic influence in one housing (VMC ratio $20000V/\sqrt{3} : 100V/\sqrt{3}$ and CMC ratio 100 A: 5 A). The resistance welding transformer has the following rated data: primary voltage 380 V; secondary no-load voltage (1.41 – 4.63) V; conventional power 24 kVA; rated frequency 50 Hz; thyristor controlled switching; number of primary tap positions 9. The transformer is a single phase with shell type core.

17.4.2 Optimization Design Procedure of CCVIT

The optimization design procedure of the CCVIT starts with the definition of the initial design according to the classical analytical transformer theory. The geometrical CCVIT construction is given in details in [20]. The optimal design goal of the CCVIT is the minimum of the metrological parameters (VMC voltage error p_u, VMC phase displacement error δ_u, CMC current error p_i, CMC phase displacement error δ_i). The main contributors to the measurement uncertainty budget of the CCVIT metrological parameters are the leakage reactances of the four CCVIT windings: VMC high voltage primary winding (indexed $1u$); VMC low voltage secondary winding (indexed $2u$); CMC high current primary winding (indexed $1i$); CMC low current secondary winding (indexed $2i$). The detailed finite element

---
[*] Corresponding author.

analysis of the magnetic field distribution in the CCVIT by using the original program package FEM-3D is given in [20] and [21]. The initial metrological parameters of the CCVIT are: accuracy class 3 of the VMC and accuracy class 1 of the CMC.

The flow chart of the design procedure is presented in Fig. 17.27.

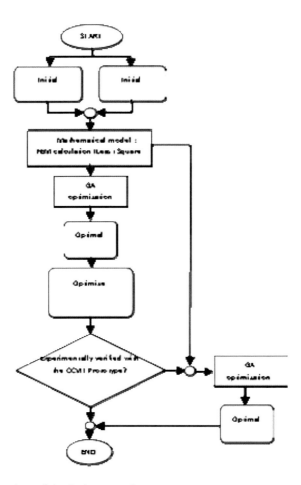

Fig. 17.27 Flow chart of the design procedure

17.4.2.1 CCVIT Mathematical Model: FEM-3D Coupled with Least Squares Method

The previously derived results (the dependences of the four leakage reactances via the magnetic flux density) are discreet. The FEM-3D calculation has been accomplished for 196 combinations of the input VMC voltage and CMC input current (from plug-out regime to 120 % of the rated values on the input VMC voltage and CMC current, with step of 10 % of the rated values). For the purposes of the genetic algorithm optimal design the continual mathematical models of the dependences of the leakage reactances via the magnetic flux density must be defined. The mathematical interpolation of the discreet values has been done by using the *least squares method* (LSM). For the GA input mathematical model of the CCVIT the best suitable form of the leakage reactances functions are the polynomial functions. The VMC leakage reactances functions are modeled at rated regime of the CMC and the CMC leakage reactances functions are modeled at rated regime of the VMC. The mathematical model of the VMC high and low voltage winding leakage reactance in dependence of the magnetic flux density in the middle of the VMC magnetic core derived by the FEM-3D coupled with LSM interpolation method are given in (1) and (2), respectively,

$$x_{\sigma 1u} = 0{,}00003 \cdot B_u^3 - 0{,}00009 \cdot B_u^2 + 0{,}0001 \cdot B_u + 0{,}00004 \tag{1}$$

$$x_{\sigma 2u} = 0{,}00002 \cdot B_u^3 - 0{,}00006 \cdot B_u^2 + 0{,}00007 \cdot B_u + 0{,}00003 \tag{2}$$

The mathematical model of the CMC high and low current winding leakage reactance in dependence of the magnetic flux density in the middle of the CMC magnetic core derived by the FEM-3D coupled with LSM interpolation method are given in (3) and (4), respectively,

$$x_{\sigma 1i} = 0{,}0019 \cdot B_i^5 - 0{,}0027 \cdot B_i^4 + 0{,}0015 \cdot B_i^3 - 0{,}0004 \cdot B_i^2 + 5 \cdot 10^{-5} \cdot B_i + 2 \cdot 10^{-5} \tag{3}$$

$$x_{\sigma 2i} = 0{,}0006 \cdot B_i^5 - 0{,}0008 \cdot B_i^4 + 0{,}0005 \cdot B_i^3 - 0{,}0001 \cdot B_i^2 + 2 \cdot 10^{-5} \cdot B_i + 8 \cdot 10^{-5} \tag{4}$$

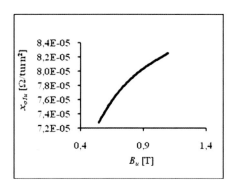

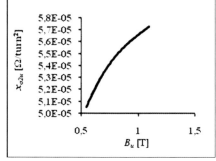

Fig. 17.28 Leakage reactance per turn of the VMC high voltage winding via the magnetic flux density derived by FEM-3D coupled with LSM

Fig. 17.29 Leakage reactance per turn of the VMC low voltage winding via the magnetic flux density derived by FEM-3D coupled with LSM

The leakage reactances functions derived by FEM-3D coupled with LSM are incorporated into the global mathematical model of the CCVIT originally developed for the GA optimization procedure. The graphical representation of leakage reactances dependences rated per turn derived by FEM-3D coupled with LSM are given in Figures 17.28-17.31.

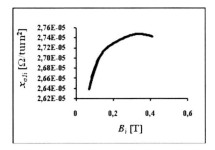

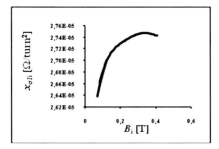

Fig. 17.30 Leakage reactance per turn of the CMC high current winding via the magnetic flux density derived by FEM-3D coupled with LSM

Fig. 17.31 Leakage reactance per turn of the CMC low current winding via the magnetic flux density derived by FEM-3D coupled with LSM

17.4.2.2 CCVIT GA Optimization

The GA tool applied for the optimization purposes is described in [20] and [22]. To set up the design problem in a formal way the geometrical parameters of the CCVIT have been selected as design variables. GA with 11 chromosomes with a population size of 16 was run 30000 generations. The crossover and mutation probabilities were set to 0.65 and 0.03 respectively. Data representation was real-valued. In the GA optimal design process the objective function was the minimum of the VMC voltage error (at rated load and at 25% of the rated load, according to the IEC standard [23]), and the CMC current error (at rated load, according to the IEC standard [23]). The VMC and CMC phase displacement errors were constraints (according to the IEC standard [23]). Table 17.8 shows the design variables, the mapping range and the results from the GA process.

Table 17.8 CCVIT 11 optimization variables

Design variable	Mapping range		GA input	GA output
VMC primary winding number of turns	23584	24000	24000	23655
VMC primary winding current density [A/mm^2]	1,5	3,0	2,04	1,509
VMC secondary winding current density [A/mm^2]	2,0	3,0	2,61	2,00
VMC magnetic core outside length [mm]	183	193	185	191
VMC magnetic core depth [mm]	49	54	50	54
CMC secondary winding number of turns [turns]	115	125	120	119
CMC primary winding current density [A/mm^2]	1,0	1,6	1,36	1,0198
CMC secondary winding current density [A/mm^2]	2,0	3,0	2,55	2,495
CMC primary winding copper width [mm]	9	11	10,5	9,48
CMC magnetic core outside length [mm]	136	162	142	156,31
CMC magnetic core depth [mm]	25	60	25	39,91

17.4.2.3 Results CCVIT

The *optimal design* of the CCVIT is derived from the GA procedure. The solution which matches best the assigned criteria is: VMC accuracy class 1 and CMC accuracy class 0,1. The ideal design is adapted into *optimized design* for a realization of a prototype. For practical realization of the CCVIT prototype the GA derived ideal parameters have been corrected according to the real constructive restrictions of the production process. The prototype of the optimally designed CCVIT is realized in the Instrument Transformer Production Company EMO A.D.- Ohrid, R. Macedonia. The prototype is given in Figure 17.32. The CCVIT prototype is experimentally tested in a metrological laboratory. The prototype is one accuracy class higher than the initial design: VMC accuracy class 1 and CMC accuracy class 0,5 (Figure 17.33 and 17.34). The experiments have verified the CAD procedure.

Optimization and Computer Aided Design of Special Transformers

Fig. 17.32 Combined instrument transformer prototype

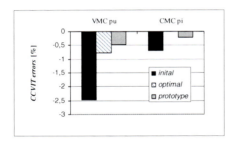

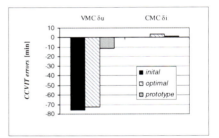

Fig. 17.33 Comparison of the VMC voltage and CMC current errors of the two CCVIT designs and the prototype (derived by experiment)

Fig. 17.34 Comparison of the VMC voltage and CMC current errors of the two CCVIT designs and the prototype (derived by experiment)

17.4.3 *Optimization Design Procedure of RWT*

The verified optimization procedure is applied on the second design object the RWT. In this case the procedure consisted of huge analysis of the initial design, identification of the undesirable features, setting of a suitable mathematical model for GA optimization, setting an optimization goal and GA search of a new optimized design.

17.4.3.1 Initial Study of RWT

Figure 17.35 shows the initial RWT design. The tap changer, part of the primary winding and the transformer type can be clearly seen. At the beginning huge analytical and numerical analyses of the initial design were performed [21]. These analyses have identified the main characteristics of the initial design as well as its undesirable features. It was found that the primary winding is composed of two different cross sections, and that there is irrationally large number of sections. The winding space was found to be not rationally filled, that consequently increases the total length of the transformer core and thus the total weight. The last is in contrary with the main design requirement for this kind of a transformer – minimal weight. On the basis of the FEM field analysis it was found that the flux density is lower than the allowed value for this kind of transformers that operate with low duty factor. These considerations were taken into account when the mathematical model suitable for GA optimization was set. The mathematical model suitable for GA optimization assumes three tap positions (instead of 9 in the existing design) with unique cross section (instead of 2 in the existing). Position 3 is selected to be the nominal. The assigned goal of the designing process was to derive optimized design variables that will result in a RWT with optimal dimension of the magnetic circuit, decreased number of tap positions and decreased total weight compared to the initial design.

Fig. 17.35 RWT initial design

17.4.3.2 RWT GA Optimization

The GA tool applied for the optimization is described in [22, 24]. To set up the design problem in a formal way the number of turns of the semisections that form the nominal position, core height and length have been selected as design variables. GA with 4 chromosomes with a population size of 50 was run 700 generations. The crossover and mutation probabilities were set to 0.8 and 0.15 respectively. Data representation was real-valued. In the GA optimal design process the efficiency was the objective function. Attention was paid to those optimized results that fulfill the previously assigned optimization goal. Table 17.9 shows the design variables, the mapping range and the results from the GA process. The main parameters of the new design are given in Table 17.10 The RWT efficiency, the numbers of turns at nominal position in the GA searching process are shown in Fig. 17.36 and Fig. 17.37, respectively.

Table 17.9 RWT design variables

Design variables	Mapping range		GA output	Corrected values
Core height	70	90	79.578	80
Core length	150	154	150.008	150
Turns semisec 1	58	62	61.993	62
Turns semisec 2	22	26	25.976	26

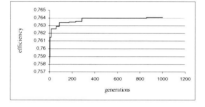
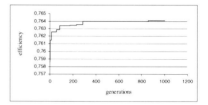

Fig. 17.36 Efficiency in the optimization process

Fig. 17.37 Number of turns in the optimization process

17.4.3.3 Results CAD RWT

Table 17.11, compares the main parameters of the initial and the optimally computer aided design RWT. The main improvement of the new design is the simplified structure with 3 tap positions, decreased dimension and weight of the transformer for 16.8 %, considerably more rational winding system, good working parameters and improved cooling conditions.

Nominal position 3 of the novel design efficiently replaces the nominal position 8 of the initial design. In the new design, the rms of the welding current between the neighboring positions can be achieved by the phase control. The simplified design can be easily manufactured.

Table 17.10 Main parameters of the CAD RWT

Transformer tap position	3	2	1
Number of turns	88	124	155
Secondary no-load voltage [V]	4.32	3.06	2.45
Secondary rms current [A]	11668	8625	6988
Flux density [T]	1.67	1.18	0.948
Iron losses [W]	298	149	95
Copper losses [W]	9414	4160	2496
Efficiency	0.759	0.788	0.802
Duty factor [%]	11.66	47	permanent

Table 17.11 Initial versus CAD RWT

	Initial RWT	CAD RWT
Primary voltage [V]	380	380
Rated frequency [Hz]	50	50
Secondary no load voltage [V]	1.41 – 4.63	2.45 - 4.32
Rated power [kVA]- duty factor 50%	27.237	20.152
Rated power [kVA] - duty factor 100%	19.260	14.250
Thermal current [A]	4160	3300
Tap positions	9	3
Nominal position	Eight	Third
Nominal current [A]	11325	11668
Maximal duty factor	12.30	11.66
Type	Shell	Shell
Thermal class	F	H
Cooling (secondary winding)	Water	Water
Transformer dimension [mm]	210/200/152	207/171/150
Iron weight [kg]	41.4	35.3
Copper weight [kg]	9.374	6.684

17.4.4 Conclusions

The GA optimization coupled with FEM results given in the paper derives positive design results. The instrument transformer has been metrologically optimized and the resistance welding transformer is optimized from energy efficiency aspect. The procedure presented is universal and has been experimentally verified.

17.5 Effects of Geometric Parameters on Performance of a SRM by Numerical-Analytical Approach

A. Bentounsi, R. Rebbah, F. Rebahi, H. Djeghloud, H. Benalla, S. Belakehal, and B. Batoun

Lab Electrotechnique Constantine, LEC, Mentouri University, Algeria

17.5.1 Introduction

The robustness, simplicity, reduced cost and high mass torque ratio of the SRM allow them to have various applications at high speed or low speed and high torque [25], [26]. The drawback of these machines is the presence of a pulsatory torque. Several research works try to minimize this latter by optimizing the geometric and control parameters [27-33]. The prediction of the behavior of an electromagnetic system from the knowledge of its non-linear parameters, also connected to external sources, has always been a difficult problem to solve. The numerical solution is *a priori* appealing for solving complex problems with better accuracy but the computing time is penalizing [34]. While easy to implement, the analytical models are relatively inaccurate due to the simplifying assumptions [35].

Consequently, we choose a numerical-analytical method using FEM-EMC modeling to properly design a prototype of a doubly salient 6/4 SRM [36] and to study the influence of many geometrical parameters on its characteristics. The initial results are validated and are encouraging for an optimization procedure of geometric parameters being developed in this paper. The originality of this work is mainly the study of the influence of the teeth shape, particularly their slope.

17.5.2 Preliminary Design Process

The choice of the number of stator Ns and rotor Nr teeth is important since they have significant implications on the torque. Usually, Ns is selected to be greater than Nr with some conditions [37]. Since the speed is related to the frequency of the supply ($f=Nr*\Omega$), to minimize the iron losses without using material of high quality, we try to reduce the number of rotor poles. Among the most frequently used structures, we finally choose to study a three phase 6/4 SRM represented in Figure 17.38. The parameters of the 110 Nm, 3000 rpm, 6/4 SRM studied machine are given in Table 17.12 [36].

The choice of the stator and rotor angles teeth (βs ; βr) has significant effects on the torque ripple, duration of output torque, winding space and is an important factor in motor design optimization. Initially, they can be selected in the middle of

Effects of Geometric Parameters on Performance of a SRM

the lower half of the feasible triangle where βs ≤ βr and with the three following conditions [31], [32], [37].

$$(2\pi/3Nr) = 30° \leq \beta s \leq 45° = (\pi/Nr) \quad (1)$$

$$(2\pi/3Nr) = 30° \leq \beta r \leq 60° = (4\pi/3Nr) \quad (2)$$

$$(\beta s + \beta r) < 90° = (2\pi/Nr) \quad (3)$$

17.5.3 Results by FEM Simulation

By combining all the electromagnetic equations, the vector potential equation governing the problem is given by

$$\nabla \times \left(\frac{1}{\mu} \nabla \times A\right) = J \quad (4)$$

where A is the vector potential, J is the current density and μ is the permeability.

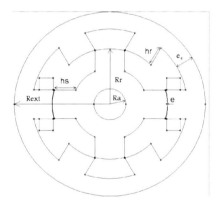

Fig. 17.38 Diagram of the studied 6/4 SRM

For the resolution of (4), we use the finite element method (FEM). The implementation was carried out under the software package FEMM [38], a very user-friendly software which allowed us to draw the equal flux lines from which we have plotted the magnetic flux versus stator magneto motive force characteristics at different rotor positions (Figure 17.39a).

The expression of linkage flux is given by

$$\lambda = N \oint_C A dl \quad (5)$$

The expression of average torque is given by

$$<T> = \left(\frac{qNr}{2\pi}\right) * (Wa - Wu) \tag{6}$$

where the aligned co-energy

$$Wa = (\lambda_1 + \lambda_2 + \ldots + \frac{1}{2}\lambda_n) * \Delta I \tag{7}$$

is calculated using n points on the λ vs. Ni curve (Figure 17.39a) with the trapezoidal integration algorithm and $\Delta I = I/n$

and the unaligned co-energy is

$$Wu = \frac{1}{2}\lambda u * I \tag{8}$$

The inductance being the ratio of linkage flux and excitation current, we have deduced from the previous Figure 17.39a their characteristics vs. rotor position at different excitations (Figure 17.39b).

Figure 17.40a shows the effect of the air gap on flux from which we deduce that the average torque is inversely proportional to the air-gap e (Figure 17.40b) as confirmed by the following expression [39]

$$Te = \left[\frac{1}{8}k_1\mu_0\right]\left[\frac{N^2 DL}{e}\right]I^2 \tag{9}$$

The manufacturing tolerances and minimum air gap that could be supported in the production environment are two important factors driving the minimum air gap length determination.

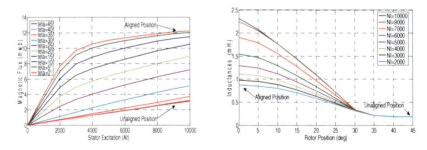

Fig. 17.39a Magnetic flux vs. stator excitation   **Fig. 17.39b** Inductances vs. rotor position

From Figures 17.41a, 17.42a, 17.43a and 17.44a of magnetic flux vs. excitation curves we deduce the influence of the rotor pole arc βr, the stator pole arc βs, the inner/outer radius ratio Kr and the stator yoke thickness ratio Kc, respectively on the average torque (Figure 17.41b to Figure 17.44b) from which we notice the existence of optimum values given by: $\beta r^* \approx 35°$, $\beta s^* \approx 36°$, $Kr^* \approx 0.6$ and $Kc^* \approx 1$. The various results obtained by FEM approach are validated in [27-32]. Now, we present our contribution which consists of studying the influence of the rotor teeth

Effects of Geometric Parameters on Performance of a SRM 345

slope (Figure 17.45) on the performance of the machine. In the case of the slope inclined towards the outside ($\alpha s>0$), Figure 17.46a shows an increase in the area included between the magnetic characteristics of aligned and unaligned positions which induces an increase in the average torque (Figure 17.46b).

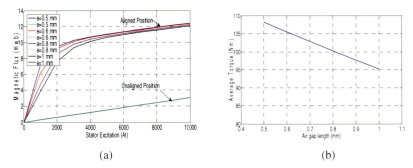

Fig. 17.40 Effect of air gap length on (a) magnetic flux and (b) average torque

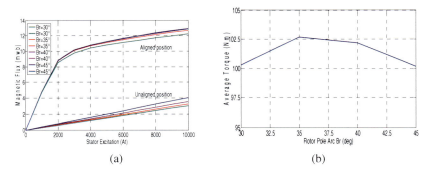

Fig. 17.41 Effect of rotor pole arc on (a) magnetic flux and (b) average torque

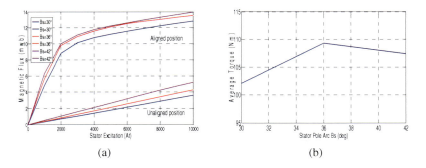

Fig. 17.42 Effect of stator pole arc on (a) magnetic flux and (b) average torque

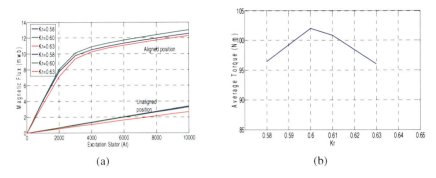

Fig. 17.43 Effect of radius ratio Kr on (a) magnetic flux curves and (b) average torque

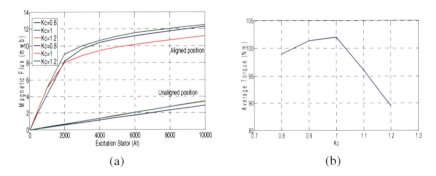

Fig. 17.44 Effect of stator yoke thickness ratio Kc on (a) magnetic flux curves and (b) average torque

Fig. 17.45 The slope of the rotor teeth

The same results are obtained for the slope of the stator teeth [27].

Effects of Geometric Parameters on Performance of a SRM

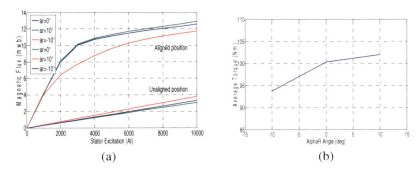

Fig. 17.46 Effect of the rotor teeth slope α_r on (a) the magnetic flux and (b) the average torque

17.5.4 Results by EMC Simulation

For the two extreme positions of the rotor (Figure 17.47), we had to work with seven equal flux tubes quite representative of the field lines [39]. This is a classic analytical method based on the calculation of reluctance portions of the equivalent magnetic circuit (Figure 17.48). It is possible to calculate the flux linkages for the aligned position analytically due to the fact that the leakage flux is negligible in this position but the same is not true for the unaligned position because the leakage paths are not known *a priori*. Finite element analysis techniques are used to estimate the flux linkages. We have exploited the previous results of the FEM to carry out our calculations under *MATLAB* environment with very user-friendly software (Figure 17.49).

The main difficulty with this program lies in the subroutine that computes the unaligned inductance which is based on an iterative process related to each flux tube. The process of optimization is based on the constancy of the following expression of the copper losses.

$$Pcu = K\rho Lm(NI)J \tag{10}$$

where K represents the current waveform coefficient, ρ the copper resistivity, Lm the mean length of turn, N the number of winding turns per stator phase, I the current level and J the current density.

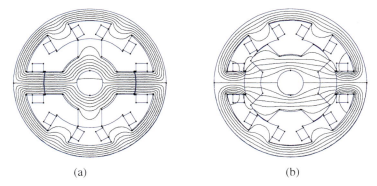

Fig. 17.47 Flux plot at fully (a) aligned and (b) unaligned positions of the studied 6/4 SRM

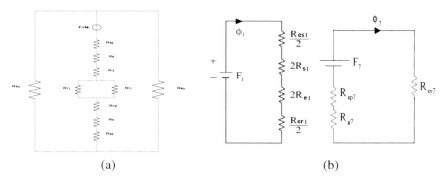

Fig. 17.48 Equivalent magnetic circuits for (a) aligned and (b) unaligned positions (paths 1-7)

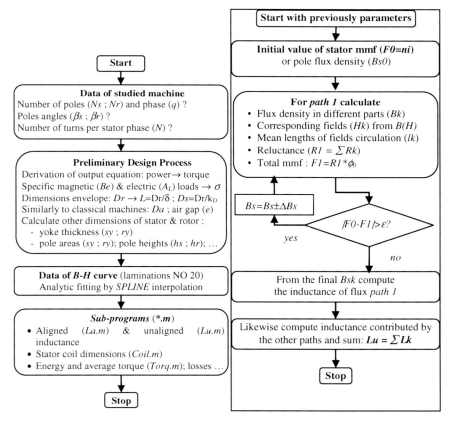

Fig. 17.49 Flowcharts of main program and sub-program for the analytic calculation of inductance

Effects of Geometric Parameters on Performance of a SRM

Since the excitation *NI* is a constant for the design process, copper losses are thus proportional to the current density. Among the various results obtained by EMC approach, we show the optimal values: $\beta r^* \approx 0.45(2\pi/Nr)$ for the rotor teeth (Figure 17.50) and $Kr^* \approx 0.6$ for the inner/outer radius ratio (Figure 17.51).

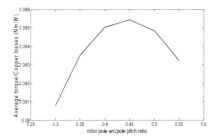 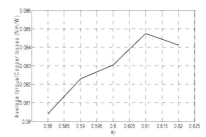

Fig. 17.50 Effect of rotor pole arc/pole pitch ratio on average torque/copper losses ratio

Fig. 17.51 Effect of inner/outer radius *Kr* on average torque/copper losses ratio

Table 17.12 Parameters of the studied 6/4 SRM

Quantity	Symbol	Value
Stack length	L	150 mm
Outer radius	Rext	125 mm
Rotor radius	Rr	75 mm
Shaft radius	Ra	21 mm
Air-gap length	e	0.8 mm
Height of stator teeth	hs	26.2 mm
Height of rotor teeth	hr	28 mm
Stator yoke thickness	e_c	24.8 mm
Number of turns per phase	N	23

17.5.5 Conclusion

In this paper, we presented a sufficiently simple numerical-analytical procedure by FEM-EMC modeling under *FEMM* to *MATLAB* software to properly design a prototype of a 6/4 SRM and to study the effects of the geometrical parameters on its characteristics. Although complicated, the program of analytic calculations under *MATLAB* environment is fast and less time consuming allowing us to predict the optimization performances of the machine. But, in order to study the effect of the teeth slope, we had to use a numerical approach based on finite elements method.

17.6 Multiphysics Method for Determination of the Stator Winding Temperature in an Electrical Machine

Zlatko Kolondzovski

Aalto University, School of Science and Technology, Finland

17.6.1 Introduction

The main benefits of using high-speed applications in the industry are their increased efficiency and their significantly reduced dimensions in comparison to the conventional applications. In fact, the volumes of the stator and rotor in a high-speed electrical machine are much smaller than in a conventional one of the same power. This leads to a very high loss density in the high-speed machine and the thermal design of this type of machine is a very demanding task. A well designed high-speed electrical machine operates at high temperatures that are close to the critical temperatures of its components. That is why an extensive and reliable thermal analysis for a high-speed electrical machine should be performed. This kind of analysis must be performed during the design process of the machine since one should be sure that the critical temperatures in all machine parts will never be exceeded in order to achieve a long lasting life of the machine.

Many references consider the thermal analyses of electrical machines but they are mainly based on traditional analytical and empirical methods. A thermal-network model that is intended for the analysis of an induction motors is given in [40]. Thermal network models intended for high-speed electrical machines are elaborated in [41, 42, 43]. In many cases the thermal-network methods are quite reliable and give good agreement with measured results, but they have some disadvantages mainly arising from their coarse nature. The coarse nature of the thermal-network method usually leads to rough thermal view of the machine. This method gives only the mean temperatures of the machine parts but does not give a detailed temperature distribution in the whole machine domain. The empirical calculation of the coefficients of thermal convection over a cooling surface usually results only in mean values of the coefficients but not in their exact local values.

The numerical techniques give the opportunity for an advanced thermal multiphysics analysis of the electrical machines. Using the computational fluid dynamics (CFD) coupled with the numerical heat-transfer method, a more realistic thermal modelling of the machine can be done. The complex machine geometry strongly influences the turbulence of the flow and more detailed turbulent analysis could be done in comparison with the empirical methods that are mostly referred to simple geometries. There are some references in which a numerical thermal analysis is performed but the numerical estimation of temperature distribution in an electrical machine is not coupled with the numerical estimation of the turbulent flow of the cooling fluid. For example in [44], a numerical thermal analysis of a conventional induction machine is done but the coefficients of thermal convection are calculated by empirical equations. In [45], a numerical CFD approach for estimation of the turbulent fluid properties in the air gap in a high-speed induction machine is reported without estimation of the stator and rotor temperatures. In order to perform a complete numerical thermal analysis of the electrical machine the equations of the heat transfer must be coupled with the equations of the turbulent flow i.e. a multiphysics method is required [46]. The reason for this coupling is that the temperature in the turbulent cooling fluid and the temperature in the solid machine parts are inherently dependent on each other. For example the temperature rise of the cooling fluid in a high-speed machine is significant since the coolant extracts a large amount of heat due to the very high loss density. That temperature rise in the cooling fluid is a reason for an unequal temperature distribution in the machine so that the outlet side of the machine has a higher temperature than the inlet one. The turbulent properties of the coolant are also influenced by its temperature change and this is taken into account in the CFD models. A multiphysics thermal analysis for different high-speed rotors is performed in [47] and is validated with measurements.

In the present paper, a multiphysics thermal analysis of the stator of a high-speed permanent-magnet electrical machine with nominal speed and power $n_{nom} = 31{,}500$ rpm, $P_{nom} = 130$ kW is performed. The analysis is performed using the COMSOL Multiphysics® commercial software. The multiphysics method that is used here is validated with direct measurements of the temperatures of the stator winding using thermocouples. The analysis is particularly dedicated to the stator winding that is the most sensitive part of the stator from thermal point of view because its class of insulation F limits the maximum allowed temperature of the winding to a value $t_{Fcrit} = 155^0 C$.

17.6.2 Method

17.6.2.1 2D Axi-symmetric Numerical Multiphysics Model

The temperatures in the fluid and solid domains of the machine are simultaneously estimated using a multiphysics coupled CFD heat-transfer model. This model is a 2D axi-symmetric one and it is primary intended for estimation of the turbulent properties of the flow. The idea of creating a 2D turbulent model was only to estimate the parameters of the air flow such as the temperatures of the fluid and the coefficients of convection. An approximation that the flow has a 2D axi-symmetric geometry is reliable since we can assume that the turbulent properties of the flow are changing only along the axial direction but they are not changing in the tangential direction. The finite-element mesh of the 2D geometry of the whole machine domain is presented in Figure 17.52. The air flow in the high-speed electrical machine should be always turbulent since in the case of turbulent flow the heat extraction due to convection is much more effective than in laminar flow. The turbulent flow is modelled using COMSOL Multiphysics® and the complete CFD analysis of this method is elaborated in [48]. The flow of an incompressible fluid can be described using the Reynolds Averaged Navier-Stokes Equations.

$$\rho \frac{\partial U}{\partial t} - \eta \nabla \cdot \nabla U + \rho U \cdot \nabla U + \nabla P + \nabla \overline{(\rho u' \otimes u')} = F; \quad \nabla \cdot U = 0 \quad (1)$$

Here η denotes the dynamic viscosity, U is the averaged velocity field, u is the velocity vector, ρ is the density of the fluid, P is the pressure, and F is the volumetric force vector. The last term on the left-hand side in the first equation is called the Reynolds stress tensor and it represents fluctuations around a mean flow. The solution for the turbulent flow variables is feasible by involving assumptions about the flow that is often called the closure of a turbulence model. One of the most used turbulence models is the κ-ε model and it is used for the CFD analysis in this paper. The κ-ε turbulence model is implemented in [45, 46, 47, 49, 50], in which an extensive numerical CFD approach for estimation of the turbulent fluid properties in the fluid domain in different types of electrical machines is reported. This model gives a closure to the system and introduces the following equations

$$\rho \frac{\partial U}{\partial t} - \nabla \cdot \left[\left(\eta + \rho C_\mu \frac{\kappa^2}{\varepsilon} \right) \cdot \left(\nabla U + (\nabla U)^T \right) \right] + \rho U \cdot \nabla U + \nabla P = F; \quad \nabla \cdot U = 0 \quad (2)$$

Here κ is the turbulent kinetic energy and ε is the dissipation rate of the turbulent energy. These new variables introduce two extra equations

$$\rho \frac{\partial \kappa}{\partial t} - \nabla \cdot \left[\left(\eta + \rho \frac{C_\mu \kappa^2}{\sigma_\kappa \varepsilon} \right) \nabla \kappa \right] + \rho U \cdot \nabla \kappa = \rho C_\mu \frac{\kappa^2}{2\varepsilon} \left(\nabla U + (\nabla U)^T \right)^2 - \rho \varepsilon$$

$$\rho \frac{\partial \varepsilon}{\partial t} - \nabla \cdot \left[\left(\eta + \rho \frac{C_\mu \kappa^2}{\sigma_\varepsilon \varepsilon} \right) \nabla \varepsilon \right] + \rho U \cdot \nabla \varepsilon = \rho C_{\varepsilon 1} \frac{\kappa}{2} \left(\nabla U + (\nabla U)^T \right)^2 - \rho C_{\varepsilon 2} \frac{\varepsilon^2}{\kappa} \quad (3)$$

The values of the model constants are: $C_\mu = 0.09$, $C_{\varepsilon 1} = 1.44$, $C_{\varepsilon 2} = 1.92$, $\sigma_\kappa = 0.9$ and $\sigma_\varepsilon = 1.3$ and they are determined from experimental data.

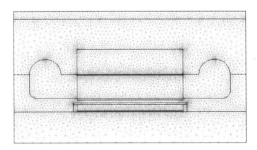

Fig. 17.52 Finite-element mesh of the 2D geometry of the high-speed electrical machine

The κ-ε turbulence model is an isotropic model i.e. it considers the turbulence of the fluid to be constant in all directions. The problem that arises here is that close to solid walls the fluctuations in the turbulence vary greatly in magnitude and direction so in these places the turbulence cannot be considered as an isotropic one. The best approach for modelling the thin boundary layer of the flow close to the solid wall is by using empirical equations and the following boundary conditions for κ and ε are obtained.

$$\kappa = \frac{u_\tau^2}{\sqrt{C_\mu}}; \quad \varepsilon = \frac{u_\tau^3}{k_a y} \tag{4}$$

where u_τ is the friction velocity and $k_a \approx 0.42$ is the Kármán's constant. The presented multiphysics model performs a simultaneous estimation of the velocity, pressure and temperature fields.

17.6.2.2 3D Numerical Heat-Transfer Model

The 2D turbulent axi-symmetric model is quite rough for the thermal modelling of the solid domain of the stator. For example, the domain of the slots and teeth was modelled as a domain of equivalent average thermal conductivity between the two materials. It means that the 2-D model would give only an average value for the temperature related to the equivalent domain but it would not give the exact temperatures for the copper winding in the slots and the laminated iron of the teeth region, separately. For all axi-symmetric parts as the stator yoke this 2D model gives reliable results. The temperatures in the solid domain of the stator can be determined by introducing a 3-D thermal model. The 3D finite element thermal analysis is also performed using the COMSOL Multiphysics® software. The application of the FEM for solving heat-transfer problems is elaborated in [51]. The finite-element mesh of the 3D geometry of the stator is presented in Figure 17.53 The turbulent fluid parameters like the fluid temperature next to the solid wall and the heat transfer coefficients of convection are taken from the 2D multiphysics model. In fact they represent input parameters for the boundary conditions of the 3D model. The conditions on the stator surface that is in a contact with the cooling fluid are modelled using the boundary condition for the heat flux q

$$-\boldsymbol{n} \cdot \boldsymbol{q} = h(T_f - T) + q_0 \tag{5}$$

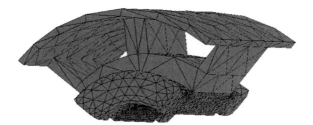

Fig. 17.53 Finite-element mesh of the 3D geometry of the stator

Here h is the heat transfer coefficient of convection that is calculated using the 2D multiphysics model and T_f is the temperature of the fluid close to the solid wall. The term q_0 models a general heat source on the solid surface such as the friction losses on the inner stator surface in the air gap. The boundary condition that models a thermal insulation or symmetry of the model is given by the equation.

$$-\mathbf{n} \cdot \mathbf{q} = 0 \qquad (6)$$

This boundary condition is used for reduction of the model size by taking advantage of the symmetry. The electrical machine that is under consideration is a 4-pole machine and only a quarter of the stator i.e. one pole pitch is modelled.

17.6.3 Results and Discussion

17.6.3.1 Results for the Temperatures in a Steady State Performance

The results that are shown here are related to the steady state performance of the high-speed electrical machine when its rotational speed is n_{nom}=31,500 rpm and the mechanical power on the shaft is P_{nom}=130 kW. The machine operates as an electric motor intended to drive a high-speed compressor. The air that supplies the compressor previously passes as a coolant in the electrical machine. When it flows through the machine it is distributed into two parts. The main part of the coolant flows between the stator yoke and the frame and a small part of the coolant flows in the air gap.

For a successful determination of the temperature distribution an accurate determination of the losses is needed. The electromagnetic losses were calculated using FEM [52] and the mechanical losses were calculated with empirical equations [53]. The results for all types of losses are presented in Table 17.13.

Table 17.13 Results for all types of losses in the stator

Types of losses		Power [W]
Electromagnetic losses	Resistive losses in the active part of the stator winding	397
	Resistive losses in the end-winding	538
	Core losses in the stator teeth	355
	Core losses in the stator yoke	223
Mechanical losses	Air friction losses in the air gap	166
	Cooling losses in air gap	23

The 2D multiphysics method couples the equations from CFD and heat transfer so it gives a solution for the turbulent and thermal properties of the fluid. A result of the temperature distribution of the fluid is presented in Figure 17.54 We can notice that the temperature of the outlet side is much higher than the temperature of the inlet side. Besides the temperature distribution of the coolant, another important parameter that is obtained with this method is the coefficient of thermal convection on each surface between the solid and fluid domains. The local value of the coefficient of convection is calculated in the postprocessor using the equation.

$$h_{local} = \frac{q_{local}}{T_w - T_f} \tag{7}$$

In this equation q_{local} is the local heat flux, T_w is the local wall temperature of the solid domain and T_f is the local average temperature of the fluid next to the wall. In this way, all local heat-transfer coefficients of convection on each outer surface of the stator are calculated and the values are exported to the 3-D model.

The 3D heat transfer method serves for estimation of the temperature distribution only in the solid domain of the machine. It gives a precise final 3D view of the temperature distribution in all parts of the stator including those parts that cannot be successfully modelled with the 2D axi-symmetric model. Using boundary conditions for symmetry, only a quarter of the machine, i.e. one pole pitch, has been modelled. At this point should be stated that there is some small asymmetry of the stator geometry. The place at which the line cables of the stator winding are conducted out of the stator in order to reach the stator terminals causes asymmetry of the stator cooling. In fact, these cables pass through the region between the stator yoke and frame and therefore block the normal flow of the cooling fluid and this causes a significant reduction of the convection in that place of the machine. Only the inlet side of the end winding is effectively cooled, but the stator yoke and the outlet side of the end winding are exposed to significantly reduced cooling. This part of the stator which is affected by the terminals in the 3D model is modelled with zero convection on the outer surfaces of the stator yoke and the outlet side of the end-winding. The calculated temperature distributions of all stator parts at nominal operation of the machine are presented in Figure 17.55 and Figure 17.56. Figure 17.55 shows the temperature distribution of the part of the stator affected by the terminals and Figure 17.56 shows the temperature distribution in the rest of the stator which has normal cooling. These two parts of the stator are separately modelled. We can notice that the critical thermal part is the outlet of the end-winding close to the terminal side. Its highest temperature is almost $t_{max}=130^0C$ which is about 15 degrees hotter than the same part of the end winding which is normally cooled. A good circumstance is that the maximum temperature of the hottest spot of the end winding is lower than the allowed temperature of the class F of insulation of the winding $t_{Fcrit}=155^0C$ which means that the stator winding is in a safe steady-state operation.

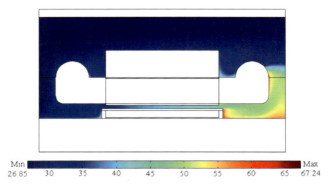

Fig. 17.54 Distribution of the temperature in the fluid domain of the machine. The inlet is on the left side and the outlet is on the right side of the figure.

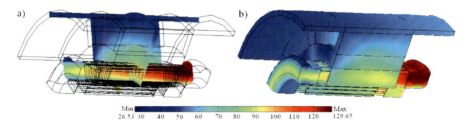

Fig. 17.55 Calculated temperature distribution t [0C] of the stator part with reduced cooling due to the terminal cables: a) inner stator parts; b) outer stator surface.

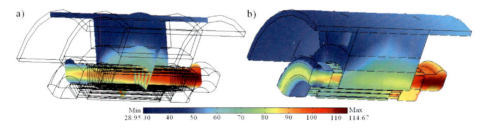

Fig. 17.56 Calculated temperature distribution t [0C] of the stator part with normal cooling: a) inner stator parts; b) outer stator surface

17.6.3.2 Experimental Validation of the Multiphysics Method

Since the multiphysics method is based on a theoretical nature, its accuracy should be validated with measurements. This is especially very important for accurate determination of the temperature of the stator winding because it is the most thermally sensitive part in the stator. The direct measurement of the winding temperature is performed using ten thermocouples distributed in different positions of the winding. The distribution of the thermocouples is shown in Fig. 17.57 The thermocouples from 1 to 5 are placed along the terminals' side of the stator in order to measure the highest temperatures of the winding that are expected on that side due to the worse cooling. The thermocouples from 6 to 10 are placed on the opposite side of the stator exposed to normal cooling. Two measurements are performed for validation of the multiphysics method at high-speed operation of the electrical machine, at speeds 18,000 rpm and 22,000 rpm which is the maximum speed that can be achieved in our case in laboratory conditions. The measured results are compared with the simulated ones in Fig. 17.58 and Fig. 17.59 Fig. 17.58 shows the results for operation point $n_1 = 18{,}000$ rpm, $P_1 = 45$ kW and Fig. 17.59 shows the results for operation point $n_2 = 22{,}000$ rpm, $P_2 = 65$ kW. The points that actually represent the results from the thermocouples in the figures are numerated according to the number of the thermocouple. The simulated and the measured results both show that there is a higher temperature in the stator winding on the terminal side of the stator. The comparison of the results proves that there is a very good agreement between the simulated and the measured results.

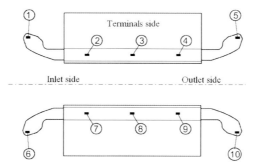

Fig. 17.57 Position of the thermocouples in the stator winding

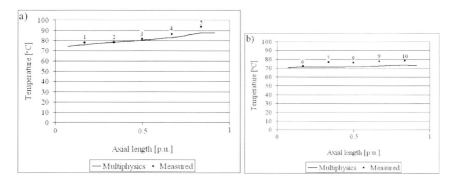

Fig. 17.58 Simulated and measured temperatures [0C] for the operation point n1 = 18,000 rpm, P1 = 45 kW at different positions along the stator winding: a) terminals side; b) side opposite of the terminals.

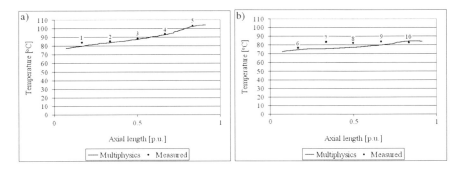

Fig. 17.59 Simulated and measured temperatures [0C] for the operation point n2 = 22,000 rpm, P2 = 65 kW at different positions along the stator winding: a) terminals side; b) side opposite of the terminals

17.6.4 Conclusion

The thermal design of a high-speed electrical machine is a very demanding task because due to the decreased size of the machine the density of the losses is very high and some parts can be heated very close to their critical temperatures. That is why, reliable theoretical methods for accurate prediction of the temperature distribution in the whole machine domain should be developed. In this paper, a combined 2D-3D finite-element thermal analysis of the stator of a high-speed electrical machine is presented. First, using a 2D multiphysics method, the thermal properties of the flow such as the coefficients of thermal convection and temperature rises of the flow were estimated. The temperature distribution in the stator was determined by the 3D numerical heat-transfer method. The method shows that the hottest spot in the stator is in the outlet side of the end-winding and its temperature is about $t_{max}=130^0C$ which is still lower than the allowed temperature of the class F of insulation of the winding $t_{Fcrit}=155^0C$. The asymmetry of the cooling that occurs from some practical reasons is taken into account in the model. The numerical multiphysics method is successfully validated with measured values of the temperatures along the stator winding using thermocouples.

17.7 Particle Swarm Optimization for Reconstitution of Two-Dimentional Groove Profiles in Non Destructive Evaluation

Ammar Hamel[1], Hassane Mohellebi[2], Mouloud Feliachi[3], and Farid Hocini[4]

[1] Electrical Engineering Laboratory, Departement of Electrical Engineering,
University of Béjaïa, Algeria
`hamelkane@yahoo.fr`
[2,4] Electrical Engineering Laboratory, Departement of Electrical Engineering,
University of Tizi-Ouzou, Algeria
`b_mohellebi@yahoo.fr, d_hocini.farid@yahoo.fr`
[3] IREENA Laboratory, IUT of Saint-Nazaire, University of Nantes, France
`mouloud.feliachi@univ-nantes.fr`

17.7.1 Introduction

The inversion problem in eddy currents non-destructive evaluation has been intensively studied during last years. The majority of the inversion methods can be classified according to whether they are phenomenological or non-phenomenological. These methods are based typically on the use of the signal processing techniques [54] - [55]. Such methods are advantageous whenever a fast inversion is required as in many industrial configurations. The main disadvantage of these methods is that they require a large database of signals for the various forms, with an accurate knowledge of the geometries. In addition, it is often difficult to obtain good results with these methods. Therefore, there is a growing interest for methods that do not require a database. In contrast to non-phenomenological approaches, these methods include a phenomenological model in the inversion process and do not require any database. Such approaches are usually based on a physical model to simulate accurately the fundamental physical phenomenon to predict the response of the sensor [56]-[58]. Thus, the problem of forms reconstruction from the signals provided by a sensor is formulated as an optimization problem to search for all geometrical parameters, and iteratively minimizing an objective function that represents the difference between the calculated signal and the measured one. A considerable amount of works using these approaches has been reported. Many of these approaches use deterministic optimization methods with a direct model suitable for solving the inverse problem [59]-[60]. These approaches use the methods of the gradient to take advantage of their rapid convergence. However, such methods often do not converge to a global optimum when the objective function has many local optima. Therefore, increasing interest has been granted in recent years, with the use of stochastic optimization methods that guarantee the convergence to the global optimum of the function to optimize. Several algorithms based on these methods are used in eddy currents non-destructive evaluation, for the treatment of the inverse problem namely simulated annealing, tabu search and genetic algorithms...

The present work involves the use of Particle Swarm Optimization (PSO) method for the treatment of the inverse problem applied in eddy currents non-destructive evaluation. This optimization method is known for its robustness in terms of global exploration of the parameters space, thus avoiding any local minimum, and especially in terms of local exploitation of the research area, which allows a better refinement of the result around the global optimum. This explains the interest to apply this method for the recovery of forms in the case of eddy currents non-destructive evaluation where the correctness of the optimized parameters is of considerable importance. The aim is to reconstruct, from an impedance measurement, the profile of the grooves inside an aluminum tube, examined by a differential probe. The solution is obtained using a direct model based on the finite element method and a PSO algorithm to solve the optimization problem.

17.7.2 Particle Swarm Optimization

Particle swarm optimization (PSO) is a population based on stochastic optimization technique developed by Dr. Eberhart and Dr. Kennedy in 1995, inspired by social behaviour of bird flocking or fish schooling. PSO shares many similarities with evolutionary computation techniques such as Genetic Algorithms (GA). The system is initialized with a population of random solutions and searches for optima by updating generations. However, unlike GA, PSO has no evolution operators such as crossover and mutation. In PSO, the potential solutions, called particles, fly through the problem space in the direction of the current optimum particles.

The PSO algorithm takes as a starting point the social behaviour of particles designated as individuals moving in a multidimensional and common space of research. These individuals tend to imitate the successful behaviours which they observe in their neighbourhood, while bringing their personal experiments. Each particle adjusts its position being useful itself of the best position produced by itself (pbest) and its neighbours (gbest) figure 17.60, according to the following equations [59]-[61]:

$$v_{i+1} = w v_i + c_1 r_1 (pbest_i - x_i) + c_2 r_2 (gbest_i - x_i) \qquad (1)$$

$$x_{i+1} = x_i + \Delta t\, v_{i+1} \qquad (2)$$

Where v is the particles speed. r_1 with r_2 are two random numbers generated in the interval [0, 1], and represent the intensities of attraction towards pbest and gbest respectively. The time parameter Δt symbolizes the advance step of the particles and w is a factor of inertia which controls the velocity influence.

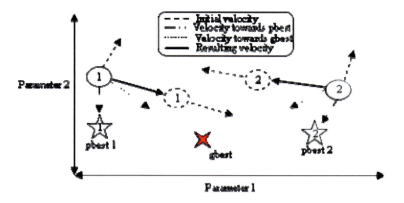

Fig. 17.60 Movement of particles in the space of parameters

At iteration $i+1$, the velocity of a particle is changed from its current value, assigned by a coefficient of weight inertia, and two forces that attract the particle to its own past best position and best position of the whole swarm. The attraction intensity is given by the coefficients c_1 and c_2. The position of the particle is changed from the current position and the new calculated velocity. So, the process of research is based on two rules:

- Each particle is equipped with a memory which enables him to memorize the best point by which it already passed and it tends to go back there.
- Each particle is informed of the best known point of all the swarm and it tends to go towards this point.

Experience shows that good exploration of the research area is obtained by introducing the random numbers r_1 and r_2, in general with a uniform distribution between 0 and 1. A linearly decreasing inertia weight is used, starts at 0.9 and ends at 0.4. The values of the used parameters were determined during preliminary studies on a set of classical test functions mentioned in literature. The number of particles is fixed at 30. Indeed, this number has proved well suited for many problems.

17.7.2.1 Validation with Test Functions

The particle swarm optimization algorithm (PSO) was first implemented and applied to test functions which are known analytically. These functions, with different characteristics, are used to test the performance of the method. The parameters of the algorithm are the same for all the examined functions. The function of De Jong was used to test the performance of the algorithm in terms of exploitation and refinement around the optimum value. The obtained result, with an almost null error, shows that the algorithm converges with a satisfactory precision (see Table.17.14). All other functions present a global minimum and several local minimums. They have been used to show the performance of the algorithm in terms of global exploration of the research area: and in this case, the results are also satisfactory.

Table 17.14 PSO algorithm performances

Function	Stopping criterion	Success rates for 100 executions
De Jong (2 variables)	10^{-20}	100%
Griewank (2 variables)	10^{-10}	99%
Goldstein Price (2 variables)	10^{-10}	99%
Zhu-Fu (3 variables)	10^{-10}	100%
Rastrigin (3 variables)	10^{-6}	100%
Hartmann (3 variables)	10^{-10}	98%
Hartmann (6 variables)	10^{-6}	97%

The function of which we give the results of simulation hereafter is the Hartmann function with six variables. It is written as:

$$H_{n,m}(x) = -\sum_{i=1}^{m} c_i \cdot \exp\left(-\sum_{j=1}^{n} a_{i,j} \cdot (x_j - p_{i,j})^2\right) \quad (3)$$

with $n = 6$:

Table 17.15 PSO algorithm performances

i	$a_{i,j}$	c_i
1	10.0 3.00 17.0 3.50 1.70 8.00	1.0
2	0.05 10.0 17.0 0.10 8.00 14.0	1.2
3	3.00 3.50 1.70 10.0 17.0 8.00	3.0
4	17.0 8.00 0.05 10.0 0.01 14.0	3.2

This function has 5 minima including one global which is -3.3223 and it is given for a solution vector $x^* = $ (0.2017, 0.1500, 0.4768, 0.2753, 0.3116, 0.6573). The research area is: $0 < x_i < 1$. Figures 17.61 and figure 17.62 show the evolution of the Hartmann function value with 6 variables and the variations of its parameters according to the iteration count of the PSO algorithm respectively.

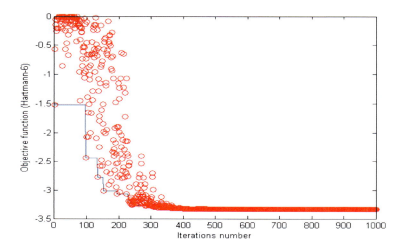

Fig. 17.61 Objective function evolution (Hartmann_6) with iterations number

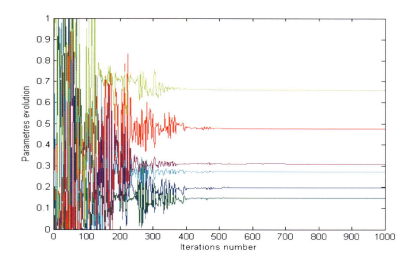

Fig. 17.62 Parameters variation versus number of iterations

17.7.3 Problem Description

The problem geometry is described in figure 17.63 A differential probe is used to scan a conducting tube (aluminum). The supposed groove, having the shape of an internal throat, is characterized by its height and its depth. The impedance change of the probe reflects the material in-homogeneities of the inspected tube. We define a cost function.

$$J = (Z_c - Z_m)^2 \tag{4}$$

where is represents the computed coil impedance at scanning position, and Z_m is the corresponding probe impedance from actual measurement.

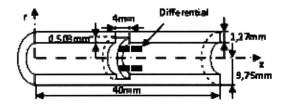

Fig. 17.63 Problem geometry

The inversion is based on an iterative approach that employs a direct finite element model to simulate the fundamental physical process, as shown in figure 17.64. The inversion algorithm starts with an initial estimate of the groove profile and then determines the signal by solving a finite element direct problem. The squared error between the measured and the calculated signals is minimized iteratively by updating the groove parameters by keeping the best profile of the previous iteration. When the error is below a threshold, the profile determined is the desired solution.

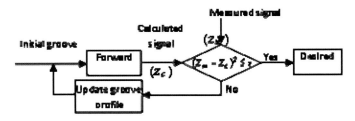

Fig. 17.64 Inversion principle method

17.7.4 Results

As one can notice on figure 17.63, a groove was practiced inside an aluminum tube having an electrical conductivity $\sigma_c = 1M\,[s/m]$. This one is examined from the inside with an eddy currents differential probe for which the geometrical and electrical data are given as follows:

- Height of a coil according to z: 0.75e-3 m.
- Inner radius of a coil: 7.75e-3 m.
- Outer radius of a coil: 8.5e-3 m.
- Vertical distance between the coils: 0.5e-3 m.
- Number of coils turns: 70

The probe is supplied by a current having an intensity of $5mA$ and a frequency of $100kHz$. The measured impedance is $Z_m = (0.55 - 1.45j)\,\Omega$. The coil impedance is obtained when the medium of this one is opposite the lower edge of the groove. Figures 17.65 and 17.66 illustrate the evolution of the objective function and of the two parameters corresponding to groove profiles respectively, according to the iteration count.

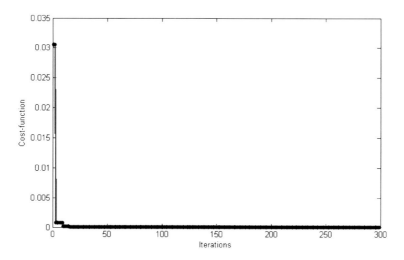

Fig. 17.65 Cost function

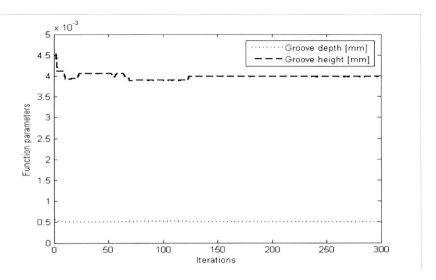

Fig. 17.66 Groove parameters evolution

Convergence is practically obtained with the iteration 124 for which the objective function is about $7e-9$. We extended the calculation up to 300 iterations in order to show the stability of the results. This also shows that the optimum found is global. With iteration 300 the value of the function cost is about $1,2e-10$, and compared to the value found for iteration 124 this shows that the algorithm is able to refine calculation around the best optimum. We notice that the convergence of the cost function towards its minimum is then obtained with a satisfactory precision. Indeed, the value of the height and the depth of the groove are $h=3.98mm$ and $p=0.508mm$ respectively. So, the error made on these two parameters vanishes.

17.7.5 Conclusion

This paper presents an inversion approach of the eddy currents testing signals in order to reconstitute the profile of an axi-symmetrical groove. The method uses a finite-element forward model to simulate the physical process and particle swarm optimization algorithm to solve the inverse problem. The results obtained are of a good precision and show that the application of PSO algorithm for more complex profiles reconstitution is promising.

References

[1] Ilea, D.: Modeling and optimization of an axial-flux switched reluctance motor, M.Sc. Thesis, Ecole Centrale de Lille, France (2008) (in French)
[2] Krishnan, R.: Switched Reluctance Motor Drives - Modeling, Simulation, Analysis, Design and Applications, pp. 27–28. CRC Press, Boca Raton (2001)

[3] Nakamura, K., Kimura, K., Ichinokura, O.: Electromagnetic and motion-coupled analysis for switched reluctance motor based on reluctance network analysis. Journal of Magnetism and Magnetic Materials 290, 1309–1312 (2005)
[4] Bae, H.-K.: Control of switched reluctance motors considering mutual inductance, PhD. Thesis, VirginiaTech, Blacksburg, pp.22-24 (2000)
[5] Marroco, A.: Analyse numérique de problemés d'électrotechniques. Ann. Sc. Math., Quebec 1, 271–296 (1977)
[6] Ayasun, S., Nwankpa, C.O.: Induction Motor Test Using Matlab/Simulink and Their Integration Into Undergraduate Electric Machinery Courses. IEEE Trans. On Education 48(1) (Feburary 2005)
[7] Santana, F.J., Monzón, J.M., García–Alonso, S., Montiel–Nelson, J.A.: Analysis and Modeling of an Electrostatic Induction Micromotor. In: Proc. of the IEEE ICEM 2008, Villamoura, Portugal (September 2008)
[8] Abdelhadi, B., Benoudjit, A., Nait–Said, N.: Application of Genetic Algorithm with a Novel Adaptive Scheme for the Identification of Induction Machine Parameters. IEEE Trans. On Energy Conversion 20(2) (June 2005)
[9] Holland, J.H.: Adaptation in Natural and Artificial Systems. MIT Press, Cambridge (1975)
[10] Koza, J.R.: Genetic Programming: On the Programming of Computers by Means of Natural Selection. MIT Press, Cambridge (1992) ISBN 0 262 11170 5
[11] Scilab, Optimization Tools: Genetic Algorithms, scilab@inria.fr, http://www.scilab.org
[12] Geuzaine, C., Remacle, J.F.: Gmsh: a Three–Dimensional Finite Element Mesh Generator with Built–in Pre–and Post–Processing Facilities. Int. J. for Numerical Methods in Engineering (May 2009)
[13] Nicolet, A., Delincé, F.: Implicit Runge–Kutta Methods for Transient Magnetic Field Computation. IEEE Transactions on Magnetics 32(3) (1996)
[14] Dular, P., Geuzaine, C., Henrotte, F., Legros, W.: A General Environment for the Treatment of Discrete Problems and its Application to the Finite Element Method. IEEE Transactions on Magnetics 34, 3395–3398 (1998)
[15] Herdem, S., Koksal, M.: A Fast Algorithm to Compute the Steady-State Solution of Nonlinear Circuits by Piecewise Linearization. Computers & Electrical Engineering 28(2), 91–101 (2002)
[16] Elamraoui, L.: Conception électromécanique d'une gamme d'actionneurs linéaire tubulaire à reluctance variable, doctoral thèses of university Lille (December 2002)
[17] Bastos, J.P.A.: Electromagnetic Modeling By Finite Element Methods. Marcel Dekker Inc., New York (2003) ISBN:0-8247-4269-9
[18] Chong, E.K.P.: An introduction to optimization. Wiley-Interscience, Hoboken (2001) Sec Edit ISBN 0-471-39126-3
[19] Dlforge, C.: Modélisation d'une machine asynchrone par réseaux de perméances en vue de sa commande. J. Phys. III (December 1996)
[20] Cundeva, M., Arsov, L., Cvetkovski, G.: Genetic Algorithm Coupled with FEM-3D for Metrological Optimal Design of Combined Current-Voltage Instrument Transformer. COMPEL: The International Journal for Computation and Mathematics in Electrical and Electronic Engineering 23(3), 670–676 (2004)
[21] Cundeva-Blajer, M., Cundeva, S., Arsov, L.: Nonlinear Electromagnetic Transient Analysis of Special

[22] Cvetkovski, G., Petkovska, L., Cundev, M.: Hybrid Stochastic-Deterministic Optimal Design of Electrical Machines. In: Proc. of 11th Int. Symp. on Electromagnetic Fields in El. Eng. ISEF 2003, pp. 725–730 (2003)
[23] IEC (International Electrotechnical Commission) 60044-2: Instrument transformers, Part 3: Combined transformers, Geneva (1980)
[24] Holland, J.H.: Adaptation in natural and artificial systems. University of Michigan Press, Ann Arbor (1995)
[25] Ku, H., et al.: Design of a high SRM for spindle drive. In: 5th Int. Conf. Workshop CPE (2007)
[26] Angle, T.L., et al.: A New Unique HP Pump System. In: Proc. 22nd Int. Pump Users Symposium (2005)
[27] Geoffroy, M.: Etude de l'influence des paramètres géométriques du circuit magnétique sur les formes d'onde de perméances et de couple des machines cylindriques à reluctance variable à double saillance, Thèse Doctorat, Janvier 27 , Univ. Paris-Sud, France (1993)
[28] Wichert, T., Ku, H.: Influence of modified rotor geometry on torque ripple and average torque of a 6/4 SRM. In: Proc. of XLII SME 2006, Cracow, Poland, July 3-6 (2006)
[29] Deihimi, A., Farhangi, S.: A conceptual optimal design of switched reluctance motors under similar asynchronous motors constraints. In: Proc. of EPE 1999, Lausanne, Switzerland (1999)
[30] Hoang, E., et al.: Influence of stator yoke thickness and stator teeth shape upon ripple and average torque of SRM. In: SPEEDAM, Taormina, Italy, June 8-10, pp. 145–149 (1994)
[31] Le Chenadec, J.Y., et al.: Torque ripple minimisation in SRM by optimization of current wave-forms and of tooth shape with copper losses and V.A. silicon constraints. In: ICEM 1994, Paris, September 5-7, vol. 2, pp. 559–564 (1994)
[32] Multon, B., et al.: Pole Arcs Optimization of Vernier Reluctance Motors Supplied with Square Wave Current. Electric Machines and Power Systems 21(6) (1993)
[33] Bizkevelci, E., Ertan, H.B., Leblebicioglu, K.: Effects of control parameters on SRM. In: Proc. of ICEM 2007, Turkey, September 10-12 (2007)
[34] Parreira, B., Rafael, S., Pires, A.J., Costa Branco, P.J.: Obtaining the magnetic characteristics of an 8/6 SRM: from FEM analysis to the experimental tests. IEEE Trans. on Ind. Electronics 52(6), 1635–1643 (2005)
[35] Lovatt, H.C.: Analytical model of a classical SRM. In: IEE Proc. Electr. Power Appl. (2004)
[36] Bentounsi, A., et al.: Design and modeling of a doubly salient variable reluctance machine. In: ICEM 2008, Vilamoura, Portugal, September 6-9 (2008)
[37] Lawrenson, P.J., et al.: Variable-Speed SRM. Proc. IEE 127(4), 253–265 (1980)
[38] FEMM, finite elements software by D. Meeker free downloaded on Web
[39] Krishnan, R.: SRM Drives : Modeling, Simulation, Analysis, Design & Applications, Industrial Electronics Series
[40] Melor, P.H., Roberts, D., Turner, D.R.: Lumped parameter thermal model for electrical machines of TEFC design. IEE Proceedings-B 138, 205–218 (1991)
[41] Saari, J.: Thermal Analysis of High-Speed Induction Machines, Acta Polytechnica Scandinavica, Espoo (1998), web address for guidance:
http://lib.tkk.fi/Diss/199X/isbn9512255766/isbn9512255766.pdf

[42] Kolondzovski, Z.: Determination of a critical thermal operation for high-speed permanent magnet electrical machines. COMPEL-The International Journal for Computation and Mathematics in Electrical and Electronic Engineering 27, 720–727 (2008)

[43] Aglén, O., Anderson, Å.: Thermal analysis of a high-speed generator. In: Conference Record of the 38th IAS Annual Meeting, pp. 547–554 (2003)

[44] Huai, Y., Melnik, R.V.N., Thogersen, P.B.: Computational analysis of temperature rise phenomena in electric induction motors. Applied Thermal Engineering 23, 779–795 (2003)

[45] Kuosa, M., Salinen, P., Larjola, J.: Numerical and experimental modeling of gas flow and heat transfer in the air gap of an electric machine. Journal of Thermal Science 13, 264–278 (2004)

[46] Kolondzovski, Z., Belahcen, A., Arkkio, A.: Multiphysics thermal design of a high-speed permanent-magnet machine. Applied Thermal Engineering 29, 2693–2700 (2009)

[47] Kolondzovski, Z., Belahcen, A., Arkkio, A.: Comparative thermal analysis of different rotor types for a high-speed permanent-magnet electrical machine. IET- Electric Power Applications 3, 279–288 (2009)

[48] Comsol Multiphysics, Heat Transfer Module, User's Guide, Version 3.3 (2006)

[49] Kolondzovski, Z.: Numerical modelling of the coolant flow in a high-speed electrical machine. In: XVIII International Conference on Electrical Machines ICEM 2008, p. 5 (2008)

[50] Trigeol, J.F., Bertin, Y., Lagonotte, P.: Thermal modeling of an induction machine through the association of two numerical approaches. IEEE Transactions on Energy Conversion 21, 314–323 (2006)

[51] Minkowycz, W.J., Sparrow, E.M., Murthy, J.Y.: Handbook of Numerical Heat Transfer, 2nd edn. Wiley, New York (2006)

[52] Arkkio, A.: Analysis of induction motors based on the numerical solution of the magnetic field and circuit equations, Acta Polytechnica Scandinavica, Helsinki (1987), web address for guidance:
http://lib.tkk.fi/Diss/198X/isbn951226076X/
isbn951226076X.pdf

[53] Saari, J.: Thermal modelling of high-speed induction machines. Acta Polytechnica Scandinavica, Helsinki (1995)

[54] Hoole, S., Subramaniam, S., Saldanha, R., Coulomb, J.: Inverse Problem Methodology and Finite Elements in the Identification of Cracks, Sources, Materials, and their Geometry in Inaccessible Locations. IEEE Trans. Magn. 27, 3433–3443 (1991)

[55] Li, Y., Udpa, L., Udpa, S.S.: Three-Dimensional Defect Reconstruction From Eddy-Current NDE Signals Using a Genetic Local Search Algorithm. IEEE Trans. Magn. 40(2), 410–417 (2004)

[56] Rebican, M., Chen, Z., Yusa, N., Janousek, L., Miya, K.: Shape Reconstruction of Multiple Cracks From ECT Signals by Means of a Stochastic Method. IEEE Trans. Magn. 42(4), 1079–1082 (2006)

[57] Norton, S., Bowler, J.: Theory of ECT Inversion. J. Appl. Phys. 73, 501–512 (1993)

[58] Boeringer, D.W., Werner, D.H.: Particle Swarm Optimization Versus Genetic Algorithms for Phased Array Synthesis. IEEE Transactions on Antennas Propagation 52, 771–779 (2004)

[59] Eberhart, R.C., Kennedy, J.: A New Optimizer Using Particle Swarm Theory. In: Proc. Sixth International Symposium on Micro Machine and Human Science, Nagoya, Japon, pp. 39–43. IEEE Service Center, Piscataway (1995)
[60] Eberhart, R.C., Kennedy, J.: Particle Swarm Optimization. In: Proc. IEEE International Conference on Neural Networks, Perth, Australia, pp. 1942–1948. IEEE Service Center, Piscataway (1995)
[61] Shi, Y., Eberhart, R.C.: A Modified Particle Swarm Optimizer. In: Proc. IEEE Congress on Evolutionary Computation, CEC 1998, Piscataway, NI, pp. 69–73 (1998)